Frontiers in Mathematics

M. Elena Luna-Elizarrarás • Michael Shapiro
Daniele C. Struppa • Adrian Vajiac

Bicomplex Holomorphic Functions

The Algebra, Geometry and Analysis
of Bicomplex Numbers

 Birkhäuser

M. Elena Luna-Elizarrarás
Escuela Sup. de Física y Matemáticas
Instituto Politécnico Nacional
Mexico City, Mexico

Michael Shapiro
Escuela Sup. de Física y Matemáticas
Instituto Politécnico Nacional
Mexico City, Mexico

Daniele C. Struppa
Schmid College of Science and Technology
Chapman University
Orange, CA, USA

Adrian Vajiac
Schmid College of Science and Technology
Chapman University
Orange, CA, USA

ISSN 1660-8046   ISSN 1660-8054   (electronic)
Frontiers in Mathematics
ISBN 978-3-319-24866-0   ISBN 978-3-319-24868-4   (eBook)
DOI 10.1007/978-3-319-24868-4

Library of Congress Control Number: 2015954663

Mathematics Subject Classification (2010): 30G35, 32A30, 32A10

Springer Cham Heidelberg New York Dordrecht London
© Springer International Publishing Switzerland 2015

Printed on acid-free paper

Springer International Publishing AG Switzerland is part of Springer Science+Business Media
(www.birkhauser-science.com)

Contents

Introduction

The best known extension of the field of complex numbers to the four-dimensional setting is the skew field of quaternions, introduced by W.R. Hamilton in 1844, [36], [37]. Quaternions arise by considering three imaginary units, $\mathbf{i}, \mathbf{j}, \mathbf{k}$ that anticommute and such that $\mathbf{ij} = \mathbf{k}$. The beauty of the theory of quaternions is that they form a field, where all the customary operations can be accomplished. Their blemish, if one can use this word, is the loss of commutativity. While from a purely algebraic point of view, the lack of commutativity is not such a terrible problem, it does create many difficulties when one tries to extend to quaternions the fecund theory of holomorphic functions of one complex variable. Within this context, one should at least point out that several successful theories exist for holomorphicity in the quaternionic setting. Among those the notion of Fueter regularity (see for example Fueter's own work [27], or [97] for a modern treatment), and the theory of slice regular functions, originally introduced in [30], and fully developed in [31]. References [97] and [31] contain various quaternionic analogues of the bicomplex results presented in this book.

It is for this reason that it is not unreasonable to consider whether a four-dimensional algebra, containing $\mathbb{C}$ as a subalgebra, can be introduced in a way that preserves commutativity. Not surprisingly, this can be done by simply considering two imaginary units $\mathbf{i}, \mathbf{j}$, introducing $\mathbf{k} = \mathbf{ij}$ (as in the quaternionic case) but now imposing that $\mathbf{ij} = \mathbf{ji}$. This turns $\mathbf{k}$ into what is known as a *hyperbolic imaginary* unit, i.e., an element such that $\mathbf{k}^2 = 1$. As far as we know, the first time that these objects were introduced was almost contemporary with Hamilton's construction, and in fact J.Cockle wrote, in 1848, a series of papers in which he introduced a new algebra that he called the algebra of *tessarines*, [15, 16, 17, 18]. Cockle's work was certainly stimulated by Hamilton's and he was the first to use tessarines to isolate the hyperbolic trigonometric series as components of the exponential series (we will show how this is done later on in Chapter 6). Not surprisingly, Cockle immediately realized that there was a price to be paid for commutativity in four dimensions, and the price was the existence of zero-divisors. This discovery led him to call such numbers *impossibles*, and the theory had no further significant development for a while.

It was only in 1892 that the mathematician Corrado Segre, inspired by the work of Hamilton and Clifford, introduced what he called *bicomplex numbers* in

[82], their algebra being equivalent to the algebra of tessarines. It was in his original papers that Segre noticed that the elements $\dfrac{1+\mathbf{ij}}{2}$ and $\dfrac{1-\mathbf{ij}}{2}$ are idempotents and play a central role in the theory of bicomplex numbers. Following Segre, a few other mathematicians, in particular Spampinato [88, 89] and Scorza Dragoni [83], developed the first rudiments of a function theory on bicomplex numbers.

The next major push in the study of bicomplex analysis was the work of J.D.Riley, who in 1953 published his doctoral dissertation [57] in which he further developed the theory of functions of bicomplex variables. But the most important contribution was undoubtedly the work of G. B. Price, [56], where the theory of holomorphic functions of a bicomplex variable (as well as multicomplex variables) is widely developed. Until this monograph, the work of G. B. Price had to be regarded as the foundational work in this theory.

In recent years, however, there has been a resurgence of interest in the study of holomorphic functions on one and several bicomplex variables, as well as a significant interest in developing functional analysis on spaces that have a structure of modules over the ring of bicomplex numbers. Without any pretense of completeness, we refer in this book to [2, 12, 13, 14, 19, 20, 29, 32, 34, 45, 59, 61, 62, 63, 65, 96]. Most of this new work indicates a need for the development of the foundations of the theory of holomorphy on the ring of bicomplex numbers, that better expresses the similarities, and differences, with the classical theory of one complex variable.

This is the explicit and intentional purpose of this book, which we have written as an elementary, yet comprehensive, introduction to the algebra, geometry, and analysis of bicomplex numbers.

We describe now the structure of this work. Chapter 1 introduces the fundamental properties of bicomplex numbers, their definitions, and the different ways in which they can be written. In particular, we show how hyperbolic numbers can be recognized inside the set of bicomplex numbers. The algebraic structure of this set is described in detail in the next chapter, where we define linear spaces and modules on $\mathbb{BC}$ and we introduce a partial order on the set of hyperbolic numbers. Maybe the most important contribution in this chapter is the definition of a hyperbolic-valued norm on the ring of bicomplex numbers. This norm will have great importance in all future applications of bicomplex numbers. In Chapter 3 we move into geometry, and we spend considerable time in discussing how to visualize the 4-dimensional geometry of bicomplex numbers. We also discuss the way in which the trigonometric representation of complex numbers can be extended to the ring of bicomplex numbers. In Chapter 4 we remain in the geometric realm and discuss lines in $\mathbb{BC}$; in particular we study real, complex, and hyperbolic lines in $\mathbb{BC}$. We then extend this analysis to the study of hyperbolic and complex curves in $\mathbb{BC}$. With Chapter 5 we abandon geometry and begin the study of analysis of bicomplex functions. We discuss here the notion of limit in the bicomplex context, which will be necessary when we study holomorphy in the bicomplex setting.

Chapter 6 is devoted to a careful and detailed study of the elementary bicomplex functions such as polynomials, exponentials, trigonometric (and inverse trigonometric) functions, radicals, and logarithms. This chapter is particularly interesting because, while it follows rather closely the exposition one would expect for complex functions, it also shows the significant, and interesting, differences that arise in this setting. Chapter 7 is, in some sense, the core of the book, as it explores the notions of bicomplex derivability and differentiability. It is in this chapter that the different ways in which bicomplex numbers can be written play a fundamental role. The fundamental properties of bicomplex holomorphic functions are studied in detail in Chapter 8. As one will see throughout the book, bicomplex holomorphic functions play an interesting role in understanding constant coefficients second order differential operators (both complex and hyperbolic). This role is explored in detail in Chapter 9. In Chapter 10 we discuss the theory of bicomplex Taylor series. Finally, this book ends with a chapter in which we show the way in which the Stokes' formula can be used to obtain new and intrinsically interesting integral formulas in the bicomplex setting.

Acknowledgments. This work has been made possible by frequent exchanges between the Instituto Politécnico Nacional in Mexico, D.F., and Chapman University in Orange, California. The authors express their gratitude to these institutions for facilitating their collaboration. A very special thank you goes to M. J. C. Robles–Casimiro, who skillfully prepared all the drawings that are included in this volume.

Chapter 1

The Bicomplex Numbers

1.1 Definition of bicomplex numbers

We start directly by defining the set $\mathbb{BC}$ of *bicomplex* numbers by

$$\mathbb{BC} := \{z_1 + \mathbf{j}z_2 \,|\, z_1, z_2 \in \mathbb{C}\},$$

where $\mathbb{C}$ is the set of complex numbers with the imaginary unit $\mathbf{i}$, and where $\mathbf{i}$ and $\mathbf{j} \neq \mathbf{i}$ are commuting imaginary units, i.e., $\mathbf{ij} = \mathbf{ji}$, $\mathbf{i}^2 = \mathbf{j}^2 = -1$. Thus bicomplex numbers are "complex numbers with complex coefficients", which explains the name of bicomplex, and in what follows we will try to emphasize the similarities between the properties of complex and bicomplex numbers. As one might expect, although the bicomplex numbers share some structures and properties of the complex numbers, there are many deep and even striking differences between these two types of numbers.

Bicomplex numbers can be added and multiplied. If $Z = z_1 + \mathbf{j}z_2$ and $W = w_1 + \mathbf{j}w_2$ are two bicomplex numbers, the formulas for the sum and the product of two bicomplex numbers are:

$$Z + W := (z_1 + w_1) + \mathbf{j}(z_2 + w_2) \tag{1.1}$$

and

$$Z \cdot W := (z_1 + \mathbf{j}z_2)(w_1 + \mathbf{j}w_2) = (z_1 w_1 - z_2 w_2) + \mathbf{j}(z_1 w_2 + z_2 w_1). \tag{1.2}$$

Of course there is no need to memorize these formulas; we have just to multiply term-by-term and take into account that $\mathbf{j}^2 = -1$.

The commutativity of the product of the two imaginary units together with definitions (1.1) and (1.2) readily imply that both operations possess the usual properties:

$$Z + W = W + Z, \qquad Z + (W + Y) = (Z + W) + Y,$$

that is, the addition is commutative and associative;

$$Z \cdot W = W \cdot Z, \qquad Z \cdot (W \cdot Y) = (Z \cdot W) \cdot Y,$$

which means that the multiplication is commutative and associative;

$$Z \cdot (W + Y) = Z \cdot W + Z \cdot Y,$$

that is, the multiplication distributes over addition.

The bicomplex numbers $0 = 0 + 0 \cdot \mathbf{j}$ and $1 = 1 + 0 \cdot \mathbf{j}$ play the roles of the usual zero and one:

$$0 + Z = Z + 0 = Z,$$

$$1 \cdot Z = Z \cdot 1 = Z.$$

Until now, we have used the denotation $\mathbb{C}$ for the field of complex numbers. Working with bicomplex numbers, the situation becomes more subtle since inside the set $\mathbb{BC}$ there are more than one subset which has the "legitimate" right to bear the name of the field of complex numbers; more exactly, there are two such subsets. One of them is the set of those bicomplex numbers with $z_2 = 0 :\ Z = z_1 + \mathbf{j}0 = z_1$; we will use the notation $\mathbb{C}(\mathbf{i})$ for it. Since $\mathbf{j}$ has the same characteristic property $\mathbf{j}^2 = -1$, then another set of complex numbers inside $\mathbb{BC}$ is $\mathbb{C}(\mathbf{j}) := \{z_1 + \mathbf{j}z_2 \mid z_1, z_2 \in \mathbb{R}\}$. Of course, $\mathbb{C}(\mathbf{i})$ and $\mathbb{C}(\mathbf{j})$ are isomorphic fields but coexisting inside $\mathbb{BC}$ they are different. We will see many times in what follows that there is a certain asymmetry in their behavior.

The set of *hyperbolic* numbers $\mathbb{D}$ can be defined intrinsically (independently of $\mathbb{BC}$) as the set

$$\mathbb{D} := \{x + \mathbf{k}y \mid x, y \in \mathbb{R}\},$$

where $\mathbf{k}$ is a hyperbolic imaginary unit, i.e., $\mathbf{k}^2 = 1$, commuting with both real numbers x and y. In some of the existing literature, hyperbolic numbers are also called *duplex*, *double* or *bireal* numbers.

Addition and multiplication operations of the hyperbolic numbers have the obvious definitions, we just have to replace $\mathbf{k}^2$ by 1 whenever it occurs. For example, for two hyperbolic numbers $\mathfrak{z}_1 = x_1 + \mathbf{k}y_1$ and $\mathfrak{z}_2 = x_2 + \mathbf{k}y_2$ their product is

$$\mathfrak{z}_1 \cdot \mathfrak{z}_2 = (x_1 x_2 + y_1 y_2) + \mathbf{k}(x_1 y_2 + x_2 y_1).$$

Working with $\mathbb{BC}$, a hyperbolic unit $\mathbf{k}$ arises from the multiplication of the two imaginary units $\mathbf{i}$ and $\mathbf{j}$: $\mathbf{k} := \mathbf{ij}$. Thus, there is a subset in $\mathbb{BC}$ which is isomorphic as a ring to the set of hyperbolic numbers: the set

$$\mathbb{D} = \{x + \mathbf{ij}y \mid x, y \in \mathbb{R}\}$$

inherits all the algebraic definitions, operations and properties from $\mathbb{BC}$.

The following subset of $\mathbb{D}$:

$$\mathbb{D}^+ := \{x + \mathbf{k}y \,|\, x^2 - y^2 \geq 0, \ x \geq 0 \}$$

will be especially useful later. We will call its elements "non-negative hyperbolic numbers"; the set

$$\mathbb{D}^+ \setminus \{0\} = \{x + \mathbf{k}y \,|\, x^2 - y^2 \geq 0, \ x > 0 \}$$

will be called the set of "positive hyperbolic numbers". Such definitions of "non-negativeness" and of "positiveness" for hyperbolic numbers do not look intuitively clear but later on we will give them other descriptions clarifying the reason for such names. It turns out that the non-negative hyperbolic numbers play with respect to all hyperbolic numbers a role deeply similar to that of real non-negative numbers with respect to all real numbers.

The set

$$\mathbb{D}^- := \{x + \mathbf{k}y \,|\, x^2 - y^2 \geq 0, \ x \leq 0 \} = \{\mathfrak{z} \,|\, -\mathfrak{z} \in \mathbb{D}^+ \}$$

will bear the name of non-positive hyperbolic numbers; and of course the set

$$\mathbb{D}^- \setminus \{0\} = \{x + \mathbf{k}y \,|\, x^2 - y^2 \geq 0, \ x < 0 \}$$

is the set of negative hyperbolic numbers. Clearly, there are hyperbolic numbers which are neither non-negative nor non-positive.

1.2 Versatility of different writings of bicomplex numbers

A bicomplex number defined as $Z = z_1 + \mathbf{j} z_2$ admits several other forms of writing, or representations, which show different aspects of this number and which will help us to understand better the structure of the set $\mathbb{BC}$. First of all, if we write $z_1 = x_1 + \mathbf{i}y_1$, $z_2 = x_2 + \mathbf{i}y_2$ with real numbers x_1, y_1, x_2, y_2, then any bicomplex number can be written in the following different ways:

$$\begin{align}
Z &= (x_1 + \mathbf{i}y_1) + \mathbf{j}\,(x_2 + \mathbf{i}y_2) =: z_1 + \mathbf{j}\,z_2 \tag{1.3} \\
&= (x_1 + \mathbf{j}x_2) + \mathbf{i}\,(y_1 + \mathbf{j}y_2) =: \zeta_1 + \mathbf{i}\,\zeta_2 \tag{1.4} \\
&= (x_1 + \mathbf{k}y_2) + \mathbf{i}\,(y_1 - \mathbf{k}x_2) =: \mathfrak{z}_1 + \mathbf{i}\,\mathfrak{z}_2 \tag{1.5} \\
&= (x_1 + \mathbf{k}y_2) + \mathbf{j}\,(x_2 - \mathbf{k}y_1) =: \mathfrak{w}_1 + \mathbf{j}\,\mathfrak{w}_2 \tag{1.6} \\
&= (x_1 + \mathbf{i}y_1) + \mathbf{k}(y_2 - \mathbf{i}x_2) =: w_1 + \mathbf{k}\,w_2 \tag{1.7} \\
&= (x_1 + \mathbf{j}x_2) + \mathbf{k}(y_2 - \mathbf{j}y_1) =: \omega_1 + \mathbf{k}\,\omega_2 \tag{1.8} \\
&= x_1 + \mathbf{i}y_1 + \mathbf{j}x_2 + \mathbf{k}y_2 \,, \tag{1.9}
\end{align}$$

where z_1, z_2, w_1, $w_2 \in \mathbb{C}(\mathbf{i})$, ζ_1, ζ_2, ω_1, $\omega_2 \in \mathbb{C}(\mathbf{j})$, and $\mathfrak{z}_1$, $\mathfrak{z}_2$, $\mathfrak{w}_1$, $\mathfrak{w}_2 \in \mathbb{D}$. Equation (1.9) says that any bicomplex number can be seen as an element of $\mathbb{R}^4$; meanwhile formulas (1.3) and (1.7) allow us to identify Z with elements in $\mathbb{C}^2(\mathbf{i})$ and formulas (1.4) and (1.8) with elements in $\mathbb{C}^2(\mathbf{j})$; similarly formulas (1.5) and (1.6) identify Z with elements in $\mathbb{D}^2 := \mathbb{D} \times \mathbb{D}$.

1.3   Conjugations of bicomplex numbers

The structure of $\mathbb{BC}$ (there are two imaginary units of complex type and one hyperbolic unit in it) suggests three possible conjugations on $\mathbb{BC}$:

(I) $\overline{Z} := \overline{z}_1 + \mathbf{j}\,\overline{z}_2$ (the bar-conjugation);

(II) $Z^\dagger := z_1 - \mathbf{j}\,z_2$ (the $\dagger$-conjugation);

(III) $Z^* := \left(\overline{Z}\right)^\dagger = \overline{(Z^\dagger)} = \overline{z}_1 - \mathbf{j}\,\overline{z}_2$ (the $*$-conjugation),

where $\overline{z}_1$, $\overline{z}_2$ are usual complex conjugates to z_1, $z_2 \in \mathbb{C}(\mathbf{i})$.

Let us see how these conjugations act on the complex numbers in $\mathbb{C}(\mathbf{i})$ and in $\mathbb{C}(\mathbf{j})$ and on the hyperbolic numbers in $\mathbb{D}$. If $Z = z_1 \in \mathbb{C}(\mathbf{i})$, i.e., $z_2 = 0$, then $Z = z_1 = x_1 + \mathbf{i}y_1$ and one has:

$$\overline{Z} = \overline{z}_1 = x_1 - \mathbf{i}y_1 = z_1^* = Z^*, \qquad Z^\dagger = z_1^\dagger = z_1 = Z,$$

that is, both the bar-conjugation and the $*$-conjugation, restricted to $\mathbb{C}(\mathbf{i})$, coincide with the usual complex conjugation there, and the $\dagger$-conjugation fixes all elements of $\mathbb{C}(\mathbf{i})$.

If $Z = \zeta_1$ belongs to $\mathbb{C}(\mathbf{j})$, that is, $\zeta_1 = x_1 + \mathbf{j}x_2$, then one has:

$$\overline{\zeta}_1 = \zeta_1, \qquad \zeta_1^* = x_1 - \mathbf{j}x_2 = \zeta^\dagger,$$

that is, both the $*$-conjugation and the $\dagger$-conjugation, restricted to $\mathbb{C}(\mathbf{j})$ coincide with the usual conjugation there. In order to avoid any confusion with the notation, from now on we will identify the conjugation on $\mathbb{C}(\mathbf{j})$ with the $\dagger$-conjugation. Note also that any element in $\mathbb{C}(\mathbf{j})$ is fixed by the bar-conjugation.

Finally, if $Z = x_1 + \mathbf{ij}y_2 \in \mathbb{D}$, that is, $y_1 = x_2 = 0$, then

$$\overline{Z} = x_1 - \mathbf{ij}y_2 = Z^\dagger, \qquad Z^* = Z,$$

thus, the bar-conjugation and the $\dagger$-conjugation restricted to $\mathbb{D}$ coincide with the intrinsic conjugation there. We will use the bar-conjugation to denote the latter. Note that any hyperbolic number is fixed by the $*$-conjugation.

Using formulas (1.4)–(1.9), the bicomplex conjugations (I)–(III) defined above can be written as

$$(\text{I'})\; \overline{Z} = \zeta_1 - \mathbf{i}\,\zeta_2 = \overline{\mathfrak{z}}_1 - \mathbf{i}\,\overline{\mathfrak{z}}_2 = \overline{\mathfrak{w}}_1 + \mathbf{j}\,\overline{\mathfrak{w}}_2 = \overline{w}_1 - \mathbf{k}\,\overline{w}_2 = \omega_1 - \mathbf{k}\,\omega_2$$

$$= x_1 - \mathbf{i}\,y_1 + \mathbf{j}\,x_2 - \mathbf{k}\,y_2;$$

$$(\text{II'}) \ \ Z^\dagger = \zeta_1^\dagger + \mathbf{i}\,\zeta_2^\dagger = \bar{\mathfrak{z}}_1 + \mathbf{i}\,\bar{\mathfrak{z}}_2 = \bar{\mathfrak{w}}_1 - \mathbf{j}\,\bar{\mathfrak{w}}_2 = w_1 - \mathbf{k}\,w_2 = \omega_1^\dagger - \mathbf{k}\,\omega_2^\dagger$$
$$= x_1 + \mathbf{i}\,y_1 - \mathbf{j}\,x_2 - \mathbf{k}\,y_2;$$

$$(\text{III'}) \ \ Z^* = \zeta_1^\dagger - \mathbf{i}\,\zeta_2^\dagger = \mathfrak{z}_1 - \mathbf{i}\,\mathfrak{z}_2 = \mathfrak{w}_1 - \mathbf{j}\,\mathfrak{w}_2 = \bar{w}_1 + \mathbf{k}\,\bar{w}_2 = \omega_1^\dagger + \mathbf{k}\,\omega_2^\dagger$$
$$= x_1 - \mathbf{i}\,y_1 - \mathbf{j}\,x_2 + \mathbf{k}\,y_2.$$

Each conjugation is an additive, involutive, and multiplicative operation on $\mathbb{BC}$:

$$\overline{(Z + W)} = \overline{Z} + \overline{W}, \quad (Z + W)^\dagger = Z^\dagger + W^\dagger, \quad (Z + W)^* = Z^* + W^* \quad (1.10)$$

$$\overline{\overline{Z}} = Z, \quad (Z^\dagger)^\dagger = Z, \quad (Z^*)^* = Z \quad\quad\quad (1.11)$$

$$\overline{(Z \cdot W)} = \overline{Z} \cdot \overline{W}, \quad (Z \cdot W)^\dagger = Z^\dagger \cdot W^\dagger, \quad (Z \cdot W)^* = Z^* \cdot W^*. \quad (1.12)$$

1.4 Moduli of bicomplex numbers

In the complex case the modulus of a complex number is intimately related with the complex conjugation: by multiplying a complex number by its conjugate one gets the square of its modulus. Applying this idea to each of the three conjugations introduced in the previous section, three possible "moduli" arise in accordance with the formulas for their squares:

- $|Z|_{\mathbf{i}}^2 := Z \cdot Z^\dagger = z_1^2 + z_2^2$

$$= \left(|\zeta_1|^2 - |\zeta_2|^2 \right) + 2\,Re\,(\zeta_1\,\zeta_2^\dagger\,)\mathbf{i}$$
$$= \left(|\mathfrak{z}_1|_{hyp}^2 + |\mathfrak{z}_2|_{hyp}^2 \right) + \left(\bar{\mathfrak{z}}_1\,\mathfrak{z}_2 - \bar{\mathfrak{z}}_1\,\mathfrak{z}_2 \right)\mathbf{j}$$
$$= \left(|\mathfrak{w}_1|_{hyp}^2 - |\mathfrak{w}_2|_{hyp}^2 \right) + \left(\mathfrak{w}_1\,\bar{\mathfrak{w}}_2 + \overline{\mathfrak{w}_1\,\bar{\mathfrak{w}}_2} \right)\mathbf{i}$$
$$= w_1^2 - w_2^2$$
$$= \left(|\omega_1|^2 - |\omega_2|^2 \right) - 2\,Im\,\left(\omega_1^\dagger \omega_2 \right)\mathbf{i} \in \mathbb{C}(\mathbf{i});$$

- $|Z|_{\mathbf{j}}^2 := Z \cdot \overline{Z} = \left(|z_1|^2 - |z_2|^2 \right) + 2\,Re\,(z_1\,\bar{z}_2\,)\mathbf{j}$

$$= \zeta_1^2 + \zeta_2^2$$
$$= \left(|\mathfrak{z}_1|_{hyp}^2 - |\mathfrak{z}_2|_{hyp}^2 \right) + \left(\mathfrak{z}_1\,\bar{\mathfrak{z}}_2 + \overline{\mathfrak{z}_1\,\bar{\mathfrak{z}}_2} \right)\mathbf{j}$$
$$= \left(|\mathfrak{w}_1|_{hyp}^2 + |\mathfrak{w}_2|_{hyp}^2 \right) + \left(\overline{\mathfrak{w}}_1\,\mathfrak{w}_2 - \overline{\overline{\mathfrak{w}}_1\,\mathfrak{w}_2} \right)\mathbf{i}$$
$$= \left(|w_1|^2 - |w_2|^2 \right) + \left(w_2\,\overline{w}_1 - w_1\,\overline{w}_2 \right)\mathbf{k}$$
$$= \omega_1^2 - \omega_2^2 \in \mathbb{C}(\mathbf{j});$$

- $|Z|_{\mathbf{k}}^2 := Z \cdot Z^* = \left(|z_1|^2 + |z_2|^2 \right) - 2\,Im\,(z_1\,\overline{z}_2\,)\mathbf{k}$

$$= \left(|\zeta_1|^2 + |\zeta_2|^2 \right) - 2\,Im\,(\zeta_1\,\zeta_2^\dagger\,)\mathbf{k}$$

$$= \mathfrak{z}_1^2 + \mathfrak{z}_2^2$$

$$= \mathfrak{w}_1^2 + \mathfrak{w}_2^2$$

$$= \left(|w_1|^2 + |w_2|^2 \right) + (w_2\,\overline{w}_1 + w_1\,\overline{w}_2\,)\,\mathbf{k}$$

$$= \left(|\omega_1|^2 + |\omega_2|^2 \right) + (\omega_1\,\omega_2^* + \omega_2\,\omega_1^*\,)\,\mathbf{k} \ \in \mathbb{D},$$

where for a complex number z (in $\mathbb{C}(\mathbf{i})$ or $\mathbb{C}(\mathbf{j})$) we denote by $|z|$ its usual modulus and for a hyperbolic number $\mathfrak{z} = a + \mathbf{k}b$ we use the notation $|\mathfrak{z}|_{hyp}^2 = a^2 - b^2$.

Unlike what happens in the complex case, these moduli are not $\mathbb{R}^+$-valued. The first two moduli are complex-valued (in $\mathbb{C}(\mathbf{i})$ and $\mathbb{C}(\mathbf{j})$ respectively), while the last one is hyperbolic-valued.

The value of $|Z|_{\mathbf{i}} = \sqrt{Z \cdot Z^\dagger}$, being the square root of a complex number, is determined by the following convention: for the complex number $z = Z \cdot Z^\dagger$, if z is a non-negative real number, then $\sqrt{z}$ denotes its non-negative value; otherwise, the $\sqrt{z}$ denotes the value of the square root of z in the upper half-plane. In many standard references, this latter one is also called the "principal" square root of z.

Although in general $|Z|_{\mathbf{i}}$ is a $\mathbb{C}(\mathbf{i})$-complex number, nevertheless if Z is in $\mathbb{C}(\mathbf{j})$, then its $\mathbb{C}(\mathbf{i})$-complex modulus $|Z|_{\mathbf{i}}$ coincides with the usual modulus of the complex number $\zeta_1 = x_1 + \mathbf{j}x_2$: since $z_1 = x_1 + \mathbf{i}0$, $z_2 = x_2 + \mathbf{i}0$, then

$$|Z|_{\mathbf{i}} = \sqrt{x_1^2 + x_2^2} = |\zeta_1|\,.$$

Hence the restriction of the quadratic form $z_1^2 + z_2^2$ onto the real two-dimensional plane $\mathbb{C}(\mathbf{j})$ determines the usual Euclidean structure on this plane.

We make similar conventions for the $\mathbb{C}(\mathbf{j})$-valued modulus

$$|Z|_{\mathbf{j}} = \sqrt{Z \cdot \overline{Z}}.$$

We note again that in the special case when $Z = z_1 = x_i + \mathbf{i}y_1$, we get:

$$|Z|_{\mathbf{j}} = \sqrt{x_1^2 + y_1^2} = |z_1|\,,$$

hence the restriction of the quadratic form

$$Z\overline{Z} = \zeta_1^2 + \zeta_2^2$$

onto the real two-dimensional plane $\mathbb{C}(\mathbf{i})$ determines the usual Euclidean structure on this plane.

We observe here a kind of a "dual" relation between the two types of complex moduli and the respective complex numbers: if $Z = z_1 \in \mathbb{C}(\mathbf{i})$, then

$$|Z|_{\mathbf{i}} = |z_1|_{\mathbf{i}} = \sqrt{z_1^2}\,,$$

which, in general, is not equal to $|z_1|$ but is equal to z_1 or $-z_1$; but somewhat paradoxically, if $Z = \zeta_1 \in \mathbb{C}(\mathbf{j})$, then $|Z|_{\mathbf{i}} = |\zeta_1|$.

Similarly, the $|Z|_{\mathbf{j}}$ of $\mathbb{C}(\mathbf{j})$-numbers, $Z = \zeta_1$, is $|Z|_{\mathbf{j}} = |\zeta_1|_{\mathbf{j}} = \sqrt{\zeta_1^2}$, meanwhile if $Z = z_1 \in \mathbb{C}(\mathbf{i})$, then $|z_1|_{\mathbf{j}} = |z_1|$. We will refer to $|Z|_{\mathbf{i}}, |Z|_{\mathbf{j}}$ as the $\mathbb{C}(\mathbf{i})$- and $\mathbb{C}(\mathbf{j})$-valued moduli of the bicomplex number Z respectively.

The last modulus introduced has its square, $|\cdot|_{\mathbf{k}}^2$, which is hyperbolic-valued, and later we will show that the modulus itself can be chosen hyperbolic-valued. For its square the following holds:

$$|Z|_{\mathbf{k}}^2 = \left(|z_1|^2 + |z_2|^2\right) + \mathbf{k}\left(-2\operatorname{Im}(z_1\,\bar{z}_2)\right) =: x + \mathbf{k}\,y\,,$$

where x and y satisfy $x^2 - y^2 \geq 0$ (this is proved using the fact that $|\operatorname{Im}(z_1\,\bar{z}_2)| \leq |z_1| \cdot |z_2|$). Thus $|Z|_{\mathbf{k}}^2 \in \mathbb{D}^+$.

We will specify the value of the square root of a hyperbolic number later on.

Although these moduli are not real-valued, nevertheless they preserve, fortunately, an important property related with the multiplication; specifically, we have:

$$|Z \cdot W|_{\mathbf{i}}^2 = |Z|_{\mathbf{i}}^2 \cdot |W|_{\mathbf{i}}^2\,,$$
$$|Z \cdot W|_{\mathbf{j}}^2 = |Z|_{\mathbf{j}}^2 \cdot |W|_{\mathbf{j}}^2\,,$$
$$|Z \cdot W|_{\mathbf{k}}^2 = |Z|_{\mathbf{k}}^2 \cdot |W|_{\mathbf{k}}^2\,.$$

1.4.1 The Euclidean norm of a bicomplex number

Since all the above moduli are not real valued, we will consider also the Euclidean norm on $\mathbb{BC}$ when it is seen as

$$\mathbb{C}^2(\mathbf{i}) := \mathbb{C}(\mathbf{i}) \times \mathbb{C}(\mathbf{i}) = \{\,(z_1, z_2) \mid z_1 + \mathbf{j}\,z_2 \in \mathbb{BC}\,\}\,,$$

or as

$$\mathbb{C}^2(\mathbf{j}) = \{\,(\zeta_1, \zeta_2) \mid \zeta_1 + \mathbf{i}\,\zeta_2 \in \mathbb{BC}\,\}\,,$$

or as

$$\mathbb{R}^4 = \{\,(x_1, y_1, x_2, y_2) \mid (x_1 + \mathbf{i}\,y_1) + \mathbf{j}(x_2 + \mathbf{i}\,y_2) \in \mathbb{BC}\,\}\,.$$

The Euclidean norm $|Z|$ is related with the properties of bicomplex numbers via the $\mathbb{D}^+$-valued modulus:

$$|Z| = \sqrt{|z_1|^2 + |z_2|^2} = \sqrt{|\zeta_1|^2 + |\zeta_2|^2} = \sqrt{\operatorname{Re}\left(|Z|_{\mathbf{k}}^2\right)} = \sqrt{x_1^2 + y_1^2 + x_2^2 + y_2^2}\,,$$

and it is again direct to prove that

$$|Z \cdot W| \leq \sqrt{2}\,|Z| \cdot |W|\,. \tag{1.13}$$

Indeed, for $Z = z_1 + \mathbf{j}z_2$ and $W = w_1 + \mathbf{j}w_2$ one has:

$$
\begin{aligned}
|Z \cdot W|^2 &= |z_1 w_1 - z_2 w_2|^2 + |z_1 w_2 + z_2 w_1|^2 \\
&\leq \left(|z_1||w_1| + |z_2||w_2| \right)^2 + \left(|z_1||w_2| + |z_2||w_1| \right)^2 \\
&= |z_1|^2|w_1|^2 + |z_2|^2|w_2|^2 + 2|z_1||w_1||z_2||w_2| \\
&\quad + |z_1|^2|w_2|^2 + |z_2|^2|w_1|^2 + 2|z_1||w_1||z_2||w_2| \\
&\leq |z_1|^2|w_1|^2 + |z_1|^2|w_2|^2 + |z_2|^2|w_2|^2 + |z_2|^2|w_1|^2 \\
&\quad + |z_1|^2|w_1|^2 + |z_2|^2|w_2|^2 + |z_1|^2|w_2|^2 + |z_2|^2|w_1|^2 \\
&= 2 \left(|z_1|^2|w_1|^2 + |z_1|^2|w_2|^2 + |z_2|^2|w_2|^2 + |z_2|^2|w_1|^2 \right) \\
&= 2 \left(|z_1|^2 + |z_2|^2 \right) \left(|w_1|^2 + |w_2|^2 \right) \\
&= 2|Z|^2|W|^2,
\end{aligned}
$$

where first, we used the triangle inequality and then we used the fact that given any two real numbers a and b, then $2ab \leq a^2 + b^2$.

We will obtain below more properties of the interplay between the Euclidean norm and the product of bicomplex numbers.

1.5   Invertibility and zero-divisors in $\mathbb{BC}$

We know already that

$$Z \cdot Z^\dagger = |Z|_{\mathbf{i}}^2 \in \mathbb{C}(\mathbf{i}) , \tag{1.14}$$

$$Z \cdot \overline{Z} = |Z|_{\mathbf{j}}^2 \in \mathbb{C}(\mathbf{j}) , \tag{1.15}$$

$$Z \cdot Z^* = |Z|_{\mathbf{k}}^2 \in \mathbb{D} \tag{1.16}$$

(compare with the complex situation where $z \cdot \overline{z} = |z|^2$).

Let us analyze (1.14). If $Z \neq 0$ but $|Z|_{\mathbf{i}} = 0$, then Z is obviously a zero-divisor since $Z^\dagger$ is also different from zero. But if $|Z|_{\mathbf{i}} \neq 0$ the number Z is invertible. Indeed, in this case, dividing both sides of (1.14) over the right-hand side one gets:

$$Z \cdot \frac{Z^\dagger}{|Z|_{\mathbf{i}}^2} = 1,$$

thus the *inverse* of an invertible bicomplex number Z is

$$Z^{-1} = \frac{Z^\dagger}{|Z|_{\mathbf{i}}^2} ,$$

similarly to what happens in the complex case. We have therefore obtained a complete description both of the invertible elements and non-invertible elements in $\mathbb{BC}$.

In a complete analogy we analyze formulas (1.15) and (1.16) arriving at the following conclusions:

1. A bicomplex number $Z \neq 0$ is invertible if and only if $|Z|_\mathbf{j} \neq 0$ or, equivalently, $|Z|_\mathbf{k}$ is not a zero-divisor and in this case the inverse of Z is

$$Z^{-1} = \frac{\overline{Z}}{|Z|_\mathbf{j}^2} = \frac{Z^*}{|Z|_\mathbf{k}^2}.$$

2. A bicomplex number $Z \neq 0$ is a zero-divisor if and only if $|Z|_\mathbf{j} = 0$ or, equivalently, $|Z|_\mathbf{k}$ is a zero-divisor.

Let us see what all this means working with specific representations of a bicomplex number. Assume that Z is given as $Z = z_1 + \mathbf{j}z_2$, then $|Z|_\mathbf{i}^2 = z_1^2 + z_2^2$. In this case Z is invertible if and only if $z_1^2 + z_2^2 \neq 0$ and the inverse of Z is

$$Z^{-1} = \frac{Z^\dagger}{z_1^2 + z_2^2} = \frac{z_1 - \mathbf{j}z_2}{z_1^2 + z_2^2}.$$

If both z_1 and z_2 are non-zero but the sum $z_1^2 + z_2^2 = 0$, then the corresponding bicomplex number $Z = z_1 + \mathbf{j}z_2$ is a *zero-divisor*. This is equivalent to $z_1^2 = -z_2^2$, i.e.,

$$z_1 = \pm \mathbf{i}z_2, \qquad (1.17)$$

and thus all zero-divisors in $\mathbb{BC}$ are of the form:

$$Z = \lambda(1 \pm \mathbf{ij}), \qquad (1.18)$$

where λ runs the whole set $\mathbb{C}(\mathbf{i}) \setminus \{0\}$.

One wonders if the description (1.18) of zero-divisors depends on the form of writing Z and what happens if $Z = \zeta_1 + \mathbf{i}\zeta_2$ with $\zeta_1, \zeta_2 \in \mathbb{C}(\mathbf{j})$. In this case

$$Z \cdot Z^\dagger = 0 \iff |\zeta_1| = |\zeta_2| \quad \text{and} \quad Re(\zeta_1 \zeta_2^\dagger) = 0. \qquad (1.19)$$

At first sight, we have something quite different from (1.17). Note however that $Re(\zeta_1 \zeta_2^\dagger)$ is the Euclidean inner product in $\mathbb{R}^2$, hence (1.19) means that ζ_1 and ζ_2 are orthogonal in $\mathbb{C}(\mathbf{j})$ and with the same magnitude (i.e., with the same modulus of complex numbers), and thus

$$\zeta_1 = \pm \mathbf{j}\zeta_2.$$

Hence, a zero-divisor $Z = \zeta_1 + \mathbf{i}\zeta_2$ becomes

$$Z = \zeta_1 \pm \mathbf{ij}\zeta_1 = \zeta_1(1 \pm \mathbf{ij}) \qquad (1.20)$$

with ζ_1 running in $\mathbb{C}(\mathbf{j}) \setminus \{0\}$.

Observe that (1.20) can be obtained as well by recalling that

$$Z \cdot \overline{Z} = \zeta_1^2 + \zeta_2^2 = 0$$

which uses yet another conjugation, not the †-conjugation but the *bar*-conjugation.

It is possible to give several other descriptions of the set of zero-divisors using all the three conjugations as well as formulas (1.3)–(1.9). This we leave as an exercise to the reader.

We denote the set of all zero-divisors in $\mathbb{BC}$ by $\mathfrak{S}$, and we set $\mathfrak{S}_0 := \mathfrak{S} \cup \{0\}$.

We can summarize this discussion as follows.

Theorem 1.5.1. *Let $Z \neq 0$, then the following are equivalent.*

1. *The bicomplex number Z is invertible.*

2. *Z is not a zero-divisor.*

3. *$Z \cdot Z^{\dagger} \neq 0$.*

4. *$Z \cdot \overline{Z} \neq 0$.*

5. *$Z \cdot Z^{*} \notin \mathfrak{S}_0$.*

6. *$|Z|_{\mathbf{i}} \neq 0$.*

7. *$|Z|_{\mathbf{j}} \neq 0$.*

8. *$|Z|_{\mathbf{k}} \notin \mathfrak{S}_0$.*

9. *If Z is given as $Z = z_1 + \mathbf{j} z_2$, then $z_1^2 + z_2^2 \neq 0$.*

10. *If Z is given as $Z = \zeta_1 + \mathbf{i} \zeta_2$, then $\zeta_1^2 + \zeta_2^2 \neq 0$.*

Since $\mathbb{BC}$ is a ring (we will comment on this with more detail in the next chapter) it is worth to single out the equivalence between (1) and (2) in Theorem 1.5.1. Indeed, in a general ring, the set of non-zero elements which are not zero-divisors is a different set from the set of invertible elements; from this point of view $\mathbb{BC}$ is a remarkable exception.

Of course the above Theorem allows us to give immediately a "dual" characterization of the set of zero-divisors.

Corollary 1.5.2. *Let $Z \neq 0$, then the following are equivalent.*

1. *Z is not invertible.*

2. *Z is a zero-divisor.*

3. *$Z \cdot Z^{\dagger} = 0 = Z \cdot \overline{Z}$.*

4. *$Z \cdot Z^{*} \in \mathfrak{S}_0$.*

5. *$|Z|_{\mathbf{i}} = 0 = |Z|_{\mathbf{j}}$.*

6. *$|Z|_{\mathbf{k}} \in \mathfrak{S}_0$.*

7. *If Z is given as $Z = z_1 + \mathbf{j} z_2$, then $z_1^2 + z_2^2 = 0$.*

8. *If Z is given as $Z = \zeta_1 + \mathbf{i} \zeta_2$, then $\zeta_1^2 + \zeta_2^2 = 0$.*

1.6   Idempotent representations of bicomplex numbers

It turns out that there are two very special zero-divisors.

Proposition 1.6.1. *The bicomplex numbers*

$$\mathbf{e} := \frac{1+\mathbf{ij}}{2} \quad and \quad \mathbf{e}^\dagger := \frac{1-\mathbf{ij}}{2}$$

have the properties:

$$\mathbf{e} \cdot \mathbf{e}^\dagger = 0$$

(thus, each of them is a zero-divisor);

$$\mathbf{e}^2 = \mathbf{e}, \qquad (\mathbf{e}^\dagger)^2 = \mathbf{e}^\dagger$$

(thus, they are idempotents);

$$\mathbf{e} + \mathbf{e}^\dagger = 1, \qquad \mathbf{e} - \mathbf{e}^\dagger = \mathbf{ij}.$$

The properties of the idempotents $\mathbf{e}$ and $\mathbf{e}^\dagger$ cause many strange phenomena. One of them is the following

Corollary 1.6.2. *There holds:*

$$\mathbf{i}\mathbf{e} = -\mathbf{j}\mathbf{e}, \qquad \mathbf{i}\mathbf{e}^\dagger = \mathbf{j}\mathbf{e}^\dagger,$$
$$\mathbf{k}\mathbf{e} = \mathbf{e}, \qquad \mathbf{k}\mathbf{e}^\dagger = -\mathbf{e}^\dagger. \qquad (1.21)$$

The next property has no analogs for complex numbers, and it exemplifies one of the deepest peculiarities of the set of bicomplex numbers. For any bicomplex number $Z = z_1 + \mathbf{j}z_2 \in \mathbb{BC}$ we have:

$$Z = z_1 + \mathbf{j}z_2 = \frac{z_1 - \mathbf{i}z_2 + z_1 + \mathbf{i}z_2}{2} + \mathbf{j}\frac{z_2 + \mathbf{i}z_1 + z_2 - \mathbf{i}z_1}{2}$$
$$= \frac{z_1 - \mathbf{i}z_2}{2} + \frac{z_1 + \mathbf{i}z_2}{2} + \mathbf{ij}\frac{z_1 - \mathbf{i}z_2}{2} - \mathbf{ij}\frac{z_1 + \mathbf{i}z_2}{2}$$
$$= (z_1 - \mathbf{i}z_2)\frac{1+\mathbf{ij}}{2} + (z_1 + \mathbf{i}z_2)\frac{1-\mathbf{ij}}{2},$$

that is,

$$Z = \beta_1 \mathbf{e} + \beta_2 \mathbf{e}^\dagger, \qquad (1.22)$$

where $\beta_1 := z_1 - \mathbf{i}z_2$ and $\beta_2 := z_1 + \mathbf{i}z_2$ are complex numbers in $\mathbb{C}(\mathbf{i})$. Formula (1.22) is called the $\mathbb{C}(\mathbf{i})$-*idempotent representation* of the bicomplex number Z.

It is obvious that since β_1 and β_2 are both in $\mathbb{C}(\mathbf{i})$, then $\beta_1\mathbf{e} + \beta_2\mathbf{e}^\dagger = 0$ if and only if $\beta_1 = 0 = \beta_2$. This implies that the above idempotent representation of the bicomplex number Z is unique: indeed, assume that $Z \neq 0$ has two idempotent representations, say,

$$Z = \beta_1\,\mathbf{e} + \beta_2\,\mathbf{e}^\dagger = \beta_1'\,\mathbf{e} + \beta_2'\,\mathbf{e}^\dagger,$$

then $0 = (\beta_1 - \beta_1')\,\mathbf{e} + (\beta_2 - \beta_2')\,\mathbf{e}^\dagger$ and thus $\beta_1 = \beta_1'$, $\beta_2 = \beta_2'$.

The following proposition shows the advantage of using the idempotent representation of bicomplex numbers in all algebraic operations.

Proposition 1.6.3. *The addition and multiplication of bicomplex numbers can be realized "term-by-term" in the idempotent representation* (1.22). *Specifically, if* $Z = \beta_1\,\mathbf{e} + \beta_2\,\mathbf{e}^\dagger$ *and* $W = \nu_1\,\mathbf{e} + \nu_2\,\mathbf{e}^\dagger$ *are two bicomplex numbers, then*

$$Z + W = (\beta_1 + \nu_1)\,\mathbf{e} + (\beta_2 + \nu_2)\,\mathbf{e}^\dagger,$$
$$Z \cdot W = (\beta_1\nu_1)\,\mathbf{e} + (\beta_2\nu_2)\,\mathbf{e}^\dagger,$$
$$Z^n = \beta_1^n\,\mathbf{e} + \beta_2^n\,\mathbf{e}^\dagger.$$

The proof of the formulas in the proposition above relies simply on the rather specific properties of the numbers $\mathbf{e}$ and $\mathbf{e}^\dagger$. For example, let us prove the second property:

$$
\begin{aligned}
Z \cdot W &= \left(\beta_1\mathbf{e} + \beta_2\mathbf{e}^\dagger\right) \cdot \left(\nu_1\mathbf{e} + \nu_2\mathbf{e}^\dagger\right) \\
&= \beta_1\mathbf{e} \cdot \nu_1\mathbf{e} + \beta_1\mathbf{e} \cdot \nu_2\mathbf{e}^\dagger + \beta_2\mathbf{e}^\dagger \cdot \nu_1\mathbf{e} + \beta_2\mathbf{e}^\dagger \cdot \nu_2\mathbf{e}^\dagger \\
&= \beta_1\nu_1 \cdot \mathbf{e} + \beta_1\nu_2 \cdot 0 + \beta_2\nu_1 \cdot 0 + \beta_2\nu_2 \cdot \mathbf{e}^\dagger \\
&= \beta_1\nu_1 \cdot \mathbf{e} + \beta_2\nu_2 \cdot \mathbf{e}^\dagger.
\end{aligned}
$$

We used the fact that $\mathbf{e}$ and $\mathbf{e}^\dagger$ are idempotents, i.e., each of them squares to itself, and that their product is zero.

We showed after formula (1.22) that the coefficients β_1 and β_2 of the idempotent representation are uniquely defined complex numbers. But this refers to the complex numbers in $\mathbb{C}(\mathbf{i})$, and the paradoxical nature of the idempotents $\mathbf{e}$ and $\mathbf{e}^\dagger$ manifests itself as follows.

Take a bicomplex number Z written in the form $Z = \zeta_1 + \mathbf{i}\,\zeta_2$, with $\zeta_1, \zeta_2 \in \mathbb{C}(\mathbf{j})$. Then a direct computation shows:

$$Z = \alpha_1\mathbf{e} + \alpha_2\mathbf{e}^\dagger := (\zeta_1 - \mathbf{j}\,\zeta_2)\mathbf{e} + (\zeta_1 + \mathbf{j}\,\zeta_2)\mathbf{e}^\dagger, \qquad (1.23)$$

where $\alpha_1 := \zeta_1 - \mathbf{j}\,\zeta_2$ and $\alpha_2 := \zeta_1 + \mathbf{j}\,\zeta_2$ are complex numbers in $\mathbb{C}(\mathbf{j})$. So, we see that as a matter of fact every bicomplex number has two idempotent representations with COMPLEX coefficients, one with coefficients in $\mathbb{C}(\mathbf{i})$, and the other with coefficients in $\mathbb{C}(\mathbf{j})$:

$$Z = \beta_1\,\mathbf{e} + \beta_2\,\mathbf{e}^\dagger = \alpha_1\,\mathbf{e} + \alpha_2\,\mathbf{e}^\dagger. \qquad (1.24)$$

Let us find out which is the relation between them. One has that

$$\mathbf{e}Z = \beta_1\,\mathbf{e} = \alpha_1\,\mathbf{e}$$

and

$$\mathbf{e}^\dagger Z = \beta_2\,\mathbf{e}^\dagger = \alpha_2\,\mathbf{e}^\dagger,$$

thus the authentic uniqueness consists of the fact that not the coefficients β_1 and α_1 (or β_2 and α_2) are equal, but the products $\beta_1\mathbf{e}$ and $\alpha_1\mathbf{e}$ (or $\beta_2\mathbf{e}^\dagger$ and $\alpha_2\mathbf{e}^\dagger$) are equal respectively. What is more, $\beta_1\mathbf{e} = \alpha_1\mathbf{e}$ is equivalent to $(\beta_1 - \alpha_1)\mathbf{e} = 0$, but since $\mathbf{e}$ is a zero-divisor, then $\beta_1 - \alpha_1$ is also a zero-divisor, that is, $\beta_1 - \alpha_1 = A\cdot\mathbf{e}^\dagger$, where A can be chosen either in $\mathbb{C}(\mathbf{i})$ or in $\mathbb{C}(\mathbf{j})$. The latter is justified with the following reasoning. Take β_1, β_2 to be $\beta_1 = c_1 + \mathbf{i}d_1$, $\beta_2 = c_2 + \mathbf{i}d_2$, then

$$\begin{aligned} Z = \beta_1\,\mathbf{e} + \beta_2\,\mathbf{e}^\dagger &= (c_1 + \mathbf{i}\,d_1)\,\mathbf{e} + (c_2 + \mathbf{i}\,d_2)\,\mathbf{e}^\dagger \\ &= c_1\,\mathbf{e} - \mathbf{j}\,d_1\,\mathbf{e} + c_2\,\mathbf{e}^\dagger + \mathbf{j}\,d_2\,\mathbf{e}^\dagger \\ &= \mathbf{e}\cdot(c_1 - \mathbf{j}\,d_1) + \mathbf{e}^\dagger\cdot(c_2 + \mathbf{j}\,d_2) \\ &= \mathbf{e}\cdot\alpha_1 + \mathbf{e}^\dagger\cdot\alpha_2\,, \end{aligned}$$

where $\alpha_1 = c_1 - \mathbf{j}\,d_1$, $\alpha_2 = c_2 + \mathbf{j}\,d_2$; thus

$$\begin{aligned} \beta_1 - \alpha_1 = c_1 + \mathbf{i}\,d_1 - c_1 + \mathbf{j}\,d_1 &= d_1\,(\mathbf{i} + \mathbf{j}) \\ &= \mathbf{i}\,d_1\,(1 - \mathbf{i}\,\mathbf{j}) \\ &= 2\,d_1\,\mathbf{i}\,\mathbf{e}^\dagger = 2\,d_1\,\mathbf{j}\,\mathbf{e}^\dagger. \end{aligned}$$

Example 1.6.4. Consider the bicomplex number:

$$Z = (1 + \mathbf{i}) + \mathbf{j}\,(3 - 2\mathbf{i}) =: z_1 + \mathbf{j}\,z_2\,.$$

Then $\beta_1 = z_1 - \mathbf{i}\,z_2 = -1 - 2\mathbf{i}$ and $\beta_2 = z_1 + \mathbf{i}\,z_2 = 3 + 4\mathbf{i}$, so in the first idempotent representation we have:

$$Z = (-1 - 2\mathbf{i})\mathbf{e} + (3 + 4\mathbf{i})\mathbf{e}^\dagger\,.$$

Now we write the same bicomplex number as

$$Z = (1 + 3\mathbf{j}) + \mathbf{i}\,(1 - 2\mathbf{j}) =: \zeta_1 + \mathbf{i}\,\zeta_2\,.$$

Then $\alpha_1 = \zeta_1 - \mathbf{j}\zeta_2 = -1 + 2\mathbf{j}$ and $\alpha_2 = \zeta_1 + \mathbf{j}\zeta_2 = 3 + 4\mathbf{j}$. The second idempotent representation of Z is then

$$Z = (-1 + 2\mathbf{j})\mathbf{e} + (3 + 4\mathbf{j})\mathbf{e}^\dagger\,.$$

Thus in this situation $\beta_1 = -1 - 2\mathbf{i} = c_1 + \mathbf{i}d_1$, $\beta_2 = 3 + 4\mathbf{i} = c_2 - \mathbf{i}d_2$, $\alpha_1 = -1 + 2\mathbf{j} = c_1 - \mathbf{j}d_1$, $\alpha_2 = 3 + 4\mathbf{j} = c_2 + \mathbf{j}d_2$ and as we know it should be that

$$\beta_1 - \alpha_1 = d_1(\mathbf{i} + \mathbf{j}).$$

Since
$$\beta_1 - \alpha_1 = -2\,(\mathbf{i} + \mathbf{j}) = -4\,\mathbf{i}\,\mathbf{e}^\dagger = -4\,\mathbf{j}\,\mathbf{e}^\dagger,$$

one obtains $d_1 = -2$, which coincides with the value of d_1 in this example. $\square$

Let us see now how the conjugations and moduli manifest themselves in idempotent representations. Take $Z = \beta_1 \mathbf{e} + \beta_2 \mathbf{e}^\dagger = \alpha_1 \mathbf{e} + \alpha_2 \mathbf{e}^\dagger$, with β_1 and β_2 in $\mathbb{C}(\mathbf{i})$, α_1 and α_2 in $\mathbb{C}(\mathbf{j})$. Then it is immediate to see that

$$\overline{Z} = \overline{\beta}_2 \mathbf{e} + \overline{\beta}_1 \mathbf{e}^\dagger = \alpha_2 \mathbf{e} + \alpha_1 \mathbf{e}^\dagger;$$
$$Z^\dagger = \beta_2 \mathbf{e} + \beta_1 \mathbf{e}^\dagger = \alpha_2^\dagger \mathbf{e} + \alpha_1^\dagger \mathbf{e}^\dagger;$$
$$Z^* = \overline{\beta}_1 \mathbf{e} + \overline{\beta}_2 \mathbf{e}^\dagger = \alpha_1^\dagger \mathbf{e} + \alpha_2^\dagger \mathbf{e}^\dagger.$$

Hence, the squares of all the three moduli become:

$$|Z|_{\mathbf{j}}^2 = Z \cdot \overline{Z}$$
$$= \left(\beta_1 \mathbf{e} + \beta_2 \mathbf{e}^\dagger\right) \cdot \left(\overline{\beta}_2 \mathbf{e} + \overline{\beta}_1 \mathbf{e}^\dagger\right)$$
$$= \beta_1 \overline{\beta}_2 \mathbf{e} + \overline{\beta_1 \overline{\beta}_2} \mathbf{e}^\dagger$$
$$= \left(\alpha_1 \mathbf{e} + \alpha_2 \mathbf{e}^\dagger\right) \cdot \left(\alpha_2 \mathbf{e} + \alpha_1 \mathbf{e}^\dagger\right)$$
$$= \alpha_1 \alpha_2 \mathbf{e} + \alpha_1 \alpha_2 \mathbf{e}^\dagger = \alpha_1 \alpha_2 \in \mathbb{C}(\mathbf{j});$$

$$|Z|_{\mathbf{i}}^2 = Z \cdot Z^\dagger$$
$$= \left(\beta_1 \mathbf{e} + \beta_2 \mathbf{e}^\dagger\right) \cdot \left(\beta_2 \mathbf{e} + \beta_1 \mathbf{e}^\dagger\right)$$
$$= \beta_1 \beta_2 \mathbf{e} + \beta_1 \beta_2 \mathbf{e}^\dagger = \beta_1 \beta_2$$
$$= \left(\alpha_1 \mathbf{e} + \alpha_2 \mathbf{e}^\dagger\right) \cdot \left(\alpha_2^\dagger \mathbf{e} + \alpha_1^\dagger \mathbf{e}^\dagger\right)$$
$$= \alpha_1 \alpha_2^\dagger \mathbf{e} + \left(\alpha_1 \alpha_2^\dagger\right)^\dagger \mathbf{e}^\dagger \in \mathbb{C}(\mathbf{i});$$

$$|Z|_{\mathbf{k}}^2 = Z \cdot Z^*$$
$$= \left(\beta_1 \mathbf{e} + \beta_2 \mathbf{e}^\dagger\right) \cdot \left(\overline{\beta}_1 \mathbf{e} + \overline{\beta}_2 \mathbf{e}^\dagger\right)$$
$$= \beta_1 \overline{\beta}_1 \mathbf{e} + \beta_2 \overline{\beta}_2 \mathbf{e}^\dagger = |\beta_1|^2 \mathbf{e} + |\beta_2|^2 \mathbf{e}^\dagger$$
$$= \left(\alpha_1 \mathbf{e} + \alpha_2 \mathbf{e}^\dagger\right) \cdot \left(\alpha_1^\dagger \mathbf{e} + \alpha_2^\dagger \mathbf{e}^\dagger\right)$$
$$= \alpha_1 \alpha_1^\dagger \mathbf{e} + \alpha_2 \alpha_2^\dagger \mathbf{e}^\dagger = |\alpha_1|^2 \mathbf{e} + |\alpha_2|^2 \mathbf{e}^\dagger \in \mathbb{D}^+.$$

Observe that in the formulas for $|Z|_{\mathbf{k}}^2$ the idempotent coefficients are non-negative real numbers and we will see soon that this is a characteristic property of non-negative hyperbolic numbers. Observe also that given $Z = \beta_1 \mathbf{e} + \beta_2 \mathbf{e}^\dagger = \alpha_1 \mathbf{e} + \alpha_2 \mathbf{e}^\dagger$ with β_1, β_2 in $\mathbb{C}(\mathbf{i})$ and α_1, α_2 in $\mathbb{C}(\mathbf{j})$, then

$$|Z| = \frac{1}{\sqrt{2}} \sqrt{|\beta_1|^2 + |\beta_2|^2} = \frac{1}{\sqrt{2}} \sqrt{|\alpha_1|^2 + |\alpha_2|^2}.$$

We can characterize now the invertibility of bicomplex numbers in terms of the idempotent representations.

Theorem 1.6.5. *Given a bicomplex number $Z \neq 0$, $Z = \beta_1 \mathbf{e} + \beta_2 \mathbf{e}^\dagger = \alpha_1 \mathbf{e} + \alpha_2 \mathbf{e}^\dagger$, with β_1 and β_2 in $\mathbb{C}(\mathbf{i})$, α_1 and α_2 in $\mathbb{C}(\mathbf{j})$, the following are equivalent:*

1. Z *is invertible;*

2. $\beta_1 \neq 0$ *and* $\beta_2 \neq 0$*;*

3. $\alpha_1 \neq 0$ *and* $\alpha_2 \neq 0$.

Whenever this holds the inverse of Z is given by

$$Z^{-1} = \beta_1^{-1}\mathbf{e} + \beta_2^{-1}\mathbf{e}^\dagger = \alpha_1^{-1}\mathbf{e} + \alpha_2^{-1}\mathbf{e}^\dagger.$$

Proof. It follows using items (6) and (7) from Theorem 1.5.1 together with the idempotent expressions for $|Z|_\mathbf{i}^2$ and $|Z|_\mathbf{j}^2$. $\qquad\square$

Again, we have a "dual" description of zero-divisors in terms of the idempotent decompositions.

Corollary 1.6.6. *Given a bicomplex number $Z \neq 0$, $Z = \beta_1 \mathbf{e} + \beta_2 \mathbf{e}^\dagger = \alpha_1 \mathbf{e} + \alpha_2 \mathbf{e}^\dagger$, with β_1 and β_2 in $\mathbb{C}(\mathbf{i})$, α_1 and α_2 in $\mathbb{C}(\mathbf{j})$, the following are equivalent:*

1. Z *is a zero-divisor;*

2. $\beta_1 = 0$ *and* $\beta_2 \neq 0$ *or* $\beta_1 \neq 0$ *and* $\beta_2 = 0$*;*

3. $\alpha_1 = 0$ *and* $\alpha_2 \neq 0$ *or* $\alpha_1 \neq 0$ *and* $\alpha_2 \neq 0$.

This means that any zero-divisor can be written in one of the following forms:

$$\begin{aligned} Z = \beta_1 \mathbf{e} &\quad \text{with} \quad \beta_1 \in \mathbb{C}(\mathbf{i}) \setminus \{0\}; \\ Z = \beta_2 \mathbf{e}^\dagger &\quad \text{with} \quad \beta_2 \in \mathbb{C}(\mathbf{i}) \setminus \{0\}; \\ Z = \alpha_1 \mathbf{e} &\quad \text{with} \quad \alpha_1 \in \mathbb{C}(\mathbf{j}) \setminus \{0\}; \\ Z = \alpha_2 \mathbf{e}^\dagger &\quad \text{with} \quad \alpha_2 \in \mathbb{C}(\mathbf{j}) \setminus \{0\}. \end{aligned}$$

One can ask if there are more idempotents in $\mathbb{BC}$, not only $\mathbf{e}$ and $\mathbf{e}^\dagger$ (of course the trivial idempotents 0 and 1 do not count). Assume that a bicomplex number $Z = \beta_1 \mathbf{e} + \beta_2 \mathbf{e}^\dagger$, with β_1 and β_2 being complex numbers either in $\mathbb{C}(\mathbf{i})$ or $\mathbb{C}(\mathbf{j})$, is an idempotent: $Z^2 = Z$. Then

$$\beta_1^2 \mathbf{e} + \beta_2^2 \mathbf{e}^\dagger = \beta_1 \mathbf{e} + \beta_2 \mathbf{e}^\dagger$$

and

$$\beta_1^2 = \beta_1 \quad \text{and} \quad \beta_2^2 = \beta_2,$$

which gives:

$$\beta_1 \in \{0,1\}, \qquad \beta_2 \in \{0,1\}.$$

Hence, combining all possible choices we have at most four candidates for idempotents in $\mathbb{BC}$:

$$Z_1 = 0 \cdot \mathbf{e} + 0 \cdot \mathbf{e}^\dagger = 0,$$

$$Z_2 = 1 \cdot \mathbf{e} + 1 \cdot \mathbf{e}^\dagger = 1,$$
$$Z_3 = 1 \cdot \mathbf{e} + 0 \cdot \mathbf{e}^\dagger = \mathbf{e},$$
$$Z_4 = 0 \cdot \mathbf{e} + 1 \cdot \mathbf{e}^\dagger = \mathbf{e}^\dagger.$$

Thus, one concludes that $\mathbf{e}$ and $\mathbf{e}^\dagger$ are the only non-trivial idempotents in $\mathbb{BC}$.

Remark 1.6.7. *The formulas*

$$Z = \beta_1 \mathbf{e} + \beta_2 \mathbf{e}^\dagger \qquad \text{and} \qquad Z^\dagger = \beta_2 \mathbf{e} + \beta_1 \mathbf{e}^\dagger,$$

with β_1 and β_2 in $\mathbb{C}(\mathbf{i})$, allow us to express the idempotent components of a bicomplex number in terms of the bicomplex number itself. Indeed:

$$\beta_1 = \beta_1 \mathbf{e} + \beta_1 \mathbf{e}^\dagger = Z\mathbf{e} + Z^\dagger \mathbf{e}^\dagger;$$
$$\beta_2 = \beta_2 \mathbf{e}^\dagger + \beta_2 \mathbf{e} = Z\mathbf{e}^\dagger + Z^\dagger \mathbf{e}.$$

Writing now the number Z with coefficients in $\mathbb{C}(\mathbf{j})$, $Z = \gamma_1 \mathbf{e} + \gamma_2 \mathbf{e}^\dagger$, we get a similar pair of formulas:

$$\gamma_1 = \gamma_1 \mathbf{e} + \gamma_1 \mathbf{e}^\dagger = Z\mathbf{e} + \overline{Z}\mathbf{e}^\dagger;$$
$$\gamma_2 = \gamma_2 \mathbf{e}^\dagger + \gamma_2 \mathbf{e} = \overline{Z}\mathbf{e} + Z\mathbf{e}^\dagger.$$

1.7   Hyperbolic numbers inside bicomplex numbers

Although the hyperbolic numbers had been found long ago and although we wrote about them at the beginning of the chapter, we believe that it would be instructive for the reader to have an intrinsic description of the properties of hyperbolic numbers, and only then to show how they can be obtained by appealing to bicomplex numbers.

For a hyperbolic number $\mathfrak{z} = x + \mathbf{k}y$, its (hyperbolic) *conjugate* $\mathfrak{z}^\diamond$ is defined by

$$\mathfrak{z}^\diamond := x - \mathbf{k}y.$$

The reader immediately notices that

$$\mathfrak{z} \cdot \mathfrak{z}^\diamond = x^2 - y^2 \in \mathbb{R}, \tag{1.25}$$

which yields the notion of the square of the *(intrinsic) modulus* of $\mathfrak{z}$:

$$|\mathfrak{z}|^2_{hyp} := x^2 - y^2,$$

which is a real number (it could be negative!).

If both x and y are non-zero real numbers, but $x^2 - y^2 = 0$, then the corresponding hyperbolic number $\mathfrak{z} = x + \mathbf{k}y$ is a zero-divisor, since its conjugate is non-zero, but the product is zero: $\mathfrak{z} \cdot \mathfrak{z}^\diamond = 0$. All zero-divisors in $\mathbb{D}$ are characterized by $x^2 = y^2$, i.e., $x = \pm y$, thus they are of the form

$$\mathfrak{z} = \lambda(1 \pm \mathbf{k})$$

for any $\lambda \in \mathbb{R} \setminus \{0\}$.

The idempotent representation of the hyperbolic number $\mathfrak{z} = x + \mathbf{k}y \in \mathbb{D}$ is

$$\mathfrak{z} = (x + y)\mathbf{e} + (x - y)\mathbf{e}^\diamond, \qquad (1.26)$$

where $\mathbf{e} = \dfrac{1}{2}(1 + \mathbf{k})$, $\mathbf{e}^\diamond = \dfrac{1}{2}(1 - \mathbf{k})$. We consciously use the same letter $\mathbf{e}$ that was used for the idempotent representation in $\mathbb{BC}$ since, as we will soon show, the two representations coincide in $\mathbb{BC}$. Direct analogs of Proposition 1.6.1 and Proposition 1.6.3 can be reformulated in this case.

Whenever there is no danger of confusion, we will denote the coefficients of the idempotent representation of a hyperbolic number $\mathfrak{z}$ by $s := x + y$ and $t := x - y$, so that we have:

$$\mathfrak{z} = s\mathbf{e} + t\mathbf{e}^\diamond. \qquad (1.27)$$

Observe that

$$|\mathfrak{z}|^2_{hyp} = x^2 - y^2 = (x + y)(x - y) = st.$$

Let us show now how these properties are related with their bicomplex antecedents. We are interested in bicomplex numbers $Z = z_1 + \mathbf{j}\, z_2$ with $\mathrm{Im}(z_1) = 0 = \mathrm{Re}(z_2)$, that is, our hyperbolic numbers are of the form $\mathfrak{z} = x_1 + \mathbf{ij}\, y_2$ and the hyperbolic unit is $\mathbf{k} = \mathbf{ij}$. Then the $\diamond$-conjugation operation is consistent with the bicomplex conjugations $\dagger$ and *bar* in the following way:

$$\mathfrak{z}^\diamond = ((x_1 + \mathbf{i}0) + \mathbf{j}(0 + \mathbf{i}y_2))^\dagger = \overline{((x_1 + \mathbf{i}0) + \mathbf{j}(0 + \mathbf{i}y_2))} = x_1 - \mathbf{k}y_2.$$

For this reason, from this point on we will not write the hyperbolic conjugate of $\mathbf{e}$ as $\mathbf{e}^\diamond$ anymore, but we will use the bicomplex notation $\mathbf{e}^\dagger$.

For a general bicomplex number, the three moduli have been defined in Section 1.4. Let us see what happens if they are evaluated on a generic hyperbolic number $\mathfrak{z} = x_1 + \mathbf{k}y_2$. Considering it as $\mathfrak{z} = z_1 + \mathbf{j}z_2 := (x_1 + \mathbf{i}0) + \mathbf{j}(0 + \mathbf{i}y_2) \in \mathbb{BC}$, we have:

$$|\mathfrak{z}|^2_{\mathbf{i}} = z_1^2 + z_2^2 = x_1^2 - y_2^2 = |\mathfrak{z}|^2_{hyp}.$$

Recalling that the definition of $|\cdot|_{\mathbf{i}}$ involves the $\dagger$-conjugation, the definition of $|\cdot|_{\mathbf{j}}$ involves the *bar*-conjugation and that on hyperbolic numbers both conjugations coincide, we see that on hyperbolic numbers both moduli reduce to the intrinsic modulus of hyperbolic numbers:

$$|\mathfrak{z}|^2_{\mathbf{i}} = |\mathfrak{z}|^2_{\mathbf{j}} = |\mathfrak{z}|^2_{hyp}. \qquad (1.28)$$

This is not the case of the third modulus: the hyperbolic-valued modulus of $Z = \mathfrak{z}$ is different than the intrinsic modulus of $\mathfrak{z}$. Indeed, we have:

$$|\mathfrak{z}|_{\mathbf{k}}^2 = Z \cdot Z^* = Z \cdot Z = Z^2 = \mathfrak{z}^2 . \tag{1.29}$$

In (1.28) we have a relation between the squares of the three moduli $|\mathfrak{z}|_{\mathbf{i}}$, $|\mathfrak{z}|_{\mathbf{j}}$ and $|\mathfrak{z}|_{hyp}$ for hyperbolic numbers. The question now is how to define the modulus $|\mathfrak{z}|_{hyp}$ itself, which obviously should be defined as the square root of $x_1^2 - y_2^2$. Note that some authors consider the non-negative values of $x_1^2 - y_2^2$ only.

It is instructive to analyze the situation more rigorously and to understand if we have other options for choosing an appropriate value of the intrinsic modulus. Although we work here with hyperbolic numbers, at the same time one can think about bicomplex numbers also as of possible values of the square roots of a hyperbolic number. So let us consider the solutions in $\mathbb{BC}$ of the equation $Z^2 = R$ for a given real number R. Write $Z = \beta_1 \mathbf{e} + \beta_2 \mathbf{e}^\dagger$, then the equation $Z^2 = R$ is equivalent to

$$\beta_1^2 \mathbf{e} + \beta_2^2 \mathbf{e}^\dagger = R\mathbf{e} + R\mathbf{e}^\dagger$$

which is equivalent to

$$\beta_1^2 = R; \quad \beta_2^2 = R.$$

If $R = 0$, then the only solution is $Z = 0$. If R is positive, then

$$\beta_1 = \pm\sqrt{R}; \quad \beta_2 = \pm\sqrt{R},$$

and we get four solutions:

$$\sqrt{R}; \quad -\sqrt{R}; \quad \mathbf{k}\sqrt{R}; \quad -\mathbf{k}\sqrt{R}.$$

These are all the solutions in $\mathbb{BC}$, and they are real or hyperbolic numbers.
    If R is negative, then one gets:

$$\beta_1 = \pm\mathbf{i}\sqrt{-R}, \quad \beta_2 = \pm\mathbf{i}\sqrt{-R}$$

giving the following solutions:

$$\mathbf{i}\sqrt{-R}; \quad -\mathbf{i}\sqrt{-R};$$
$$\mathbf{i}\mathbf{k}\sqrt{-R} = -\mathbf{j}\sqrt{-R};$$
$$-\mathbf{i}\mathbf{k}\sqrt{-R} = \mathbf{j}\sqrt{-R}.$$

Thus, for $R < 0$ the equation $Z^2 = R$ has four solutions none of which is a hyperbolic number; two of them are complex numbers in $\mathbb{C}(\mathbf{i})$ and the remaining two are complex numbers in $\mathbb{C}(\mathbf{j})$.

Returning to the intrinsic modulus $|\mathfrak{z}|_{hyp}$ of a hyperbolic number $\mathfrak{z}$ we see that in case $x_1^2 - y_2^2 > 0$ this modulus can be taken as a positive real number

$\sqrt{x_1^2 - y_2^2}$ or even as a hyperbolic number $\pm\mathbf{k}\sqrt{x_1^2 - y_2^2}$. But if $x_1^2 - y_2^2 < 0$, then there are no solutions in $\mathbb{D}$, the candidates should be taken as complex (in $\mathbb{C}(\mathbf{i})$ or in $\mathbb{C}(\mathbf{j})$) numbers.

It is instructive to note that in case $x_1^2 - y_2^2 > 0$ the positive real number $\sqrt{x_1^2 - y_2^2}$ coincides with the equal values of $|\mathfrak{z}|_\mathbf{i}$ and $|\mathfrak{z}|_\mathbf{j}$ as defined in Section 1.4.

When $x_1^2 - y_2^2 < 0$, then $|\mathfrak{z}|_{hyp}$ can be chosen either as $|\mathfrak{z}|_\mathbf{i} \in \mathbb{C}(\mathbf{i})$ or as $|\mathfrak{z}|_\mathbf{j} \in \mathbb{C}(\mathbf{j})$ (recall that we have agreed to take, in both cases, the value of the square root which is in the upper half plane); as formula (1.29) shows, it cannot be chosen as $|\mathfrak{z}|_\mathbf{k}$.

1.7.1 The idempotent representation of hyperbolic numbers

Recall that the "hyperbolic" idempotents $\mathbf{e}$ and $\mathbf{e}^\diamond$ in (1.26) and the "bicomplex" idempotents $\mathbf{e}$ and $\mathbf{e}^\dagger$ are the same bicomplex numbers (which are hyperbolic numbers!). Here $(x + y)$ and $(x - y)$ correspond to the idempotent "coordinates" β_1 and β_2 of a bicomplex number. Indeed, considering $\mathfrak{z} = z_1 + \mathbf{j}z_2 := (x_1 + \mathbf{i}0) + \mathbf{j}(0 + \mathbf{i}y_2) \in \mathbb{BC}$, its idempotent representation is

$$\mathfrak{z} = \beta_1\mathbf{e} + \beta_2\mathbf{e}^\dagger = (z_1 - \mathbf{i}z_2)\mathbf{e} + (z_1 + \mathbf{i}z_2)\mathbf{e}^\dagger$$
$$= (x_1 - \mathbf{i}(\mathbf{i}y_2))\mathbf{e} + (x_1 + \mathbf{i}(\mathbf{i}y_2))\mathbf{e}^\dagger = (x_1 + y_2)\mathbf{e} + (x_1 - y_2)\mathbf{e}^\dagger \,.$$

Recall also that we have defined the set $\mathbb{D}^+$ of non-negative hyperbolic numbers as

$$\mathbb{D}^+ = \left\{x + \mathbf{k}y \mid x^2 - y^2 \geq 0, \ x \geq 0\right\} \,.$$

The first of the defining inequalities gives the two systems:

$$\begin{cases} x - y \geq 0, \\ x + y \geq 0, \end{cases} \quad \text{or} \quad \begin{cases} x - y \leq 0, \\ x + y \leq 0, \end{cases}$$

but the condition $x \geq 0$ eliminates the second system; hence, the set $\mathbb{D}^+$ can be described as

$$\mathbb{D}^+ = \left\{x + \mathbf{k}y \mid x > 0; \ |y| \leq x\right\} \,,$$

or as

$$\mathbb{D}^+ = \left\{\nu\mathbf{e} + \mu\mathbf{e}^\dagger \mid \nu, \mu \geq 0\right\}.$$

Thus positive hyperbolic numbers are those hyperbolic numbers whose both idempotent components are non-negative, that somehow explains the origin of the name.

In Fig. 1.7.1 the points (x, y) correspond to the hyperbolic numbers $\mathfrak{z} = x + \mathbf{k}y$. One sees that, geometrically, the hyperbolic positive numbers are situated in the quarter plane denoted by $\mathbb{D}^+$. The quarter plane symmetric to it with respect to the origin corresponds to the negative hyperbolic numbers. The other points correspond to those hyperbolic numbers which cannot be called either positive or negative.

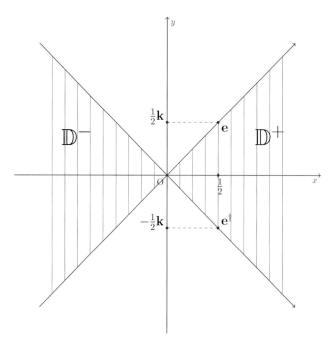

Figure 1.7.1: THE POSITIVE AND NEGATIVE HYPERBOLIC NUMBERS.

Analogously, the non-positive hyperbolic numbers form the set

$$\mathbb{D}^- = \left\{ x + \mathbf{k}y \mid x \leq 0; \ |y| \leq |x| \right\},$$

or equivalently

$$\mathbb{D}^- = \{ \nu\mathbf{e} + \mu\mathbf{e}^\dagger \mid \nu, \mu \leq 0 \}.$$

We will say sometimes that the hyperbolic number $\mathfrak{z} = \nu\mathbf{e} + \mu\mathbf{e}^\dagger$ is semi-positive if one of the coefficients μ and ν is positive and the other is zero.

We mentioned already that $\mathbb{D}^+$ plays an analogous role as non-negative real numbers, and now we illustrate this by computing the square roots of a hyperbolic number in $\mathbb{D}^+$. Take $\mathfrak{z} \in \mathbb{D}^+$, then $\mathfrak{z} = \mu\mathbf{e} + \nu\mathbf{e}^\dagger$ with $\mu, \nu \in \mathbb{R}^+ \cup \{0\}$, and it is easy to see that all the four hyperbolic numbers

$$\pm\sqrt{\mu}\,\mathbf{e} \pm \sqrt{\nu}\,\mathbf{e}^\dagger$$

square to $\mathfrak{z}$, but only one of them is a non-negative hyperbolic number: $\sqrt{\mu}\,\mathbf{e} + \sqrt{\nu}\,\mathbf{e}^\dagger$.

We are now in a position to define the meaning of the symbol $|Z|_\mathbf{k}$ for any bicomplex number $Z = \beta_1\mathbf{e} + \beta_2\mathbf{e}^\dagger$. Indeed, we have obtained that $|Z|_\mathbf{k}^2 = |\beta_1|^2\mathbf{e} + |\beta_2|^2\mathbf{e}^\dagger$ which is a non-negative hyperbolic number, hence the modulus $|Z|_\mathbf{k}$ can be taken as

$$|Z|_{\mathbf{k}} := |\beta_1|\mathbf{e} + |\beta_2|\mathbf{e}^\dagger \in \mathbb{D}^+ .$$

We will come back to this in the next chapter considering the notion of $\mathbb{BC}$ as a bicomplex normed module where the norm will be $\mathbb{D}^+$-valued. Meanwhile we can complement the above reasoning solving the equation

$$|\mathfrak{z}|_{\mathbf{k}} = \mathfrak{w}$$

where $\mathfrak{z}$ is an unknown hyperbolic number and $\mathfrak{w}$ is in $\mathbb{D}^+$. Writing $\mathfrak{z}$ and $\mathfrak{w}$ in the idempotent form $\mathfrak{z} = \beta_1\mathbf{e} + \beta_2\mathbf{e}^\dagger$ and $\mathfrak{w} = \gamma_1\mathbf{e} + \gamma_2\mathbf{e}^\dagger$ we infer easily a series of conclusions:

- If $\mathfrak{w} = 0$, then $\mathfrak{z} = 0$ is a unique solution.

- If $\mathfrak{w}$ is a semi-positive hyperbolic number, that is, $\mathfrak{w}$ is a positive zero-divisor: $\gamma_1 = 0$ and $\gamma_2 > 0$ or $\gamma_1 > 0$ and $\gamma_2 = 0$, then the solutions are also zero-divisors although not necessarily semi-positive:

$$\mathfrak{z} = \pm\gamma_2\mathbf{e}^\dagger \qquad \text{or} \qquad \mathfrak{z} = \pm\gamma_1\mathbf{e} ,$$

  respectively.

- If $\mathfrak{w}$ is positive but not semi-positive: $\gamma_1 > 0$ and $\gamma_2 > 0$, then all four solutions are

$$\mathfrak{z} = \pm\gamma_1\mathbf{e} \pm \gamma_2\mathbf{e}^\dagger .$$

1.8   The Euclidean norm and the product of bicomplex numbers

We know already that for any two bicomplex numbers Z, W one has:

$$|Z \cdot W| \leq \sqrt{2}\,|Z| \cdot |W|. \tag{1.30}$$

Note that this inequality is sharp since taking $Z = \mathbf{e}$, $W = \mathbf{e}$, one has:

$$|\mathbf{e} \cdot \mathbf{e}| = |\mathbf{e}| = \frac{1}{\sqrt{2}}$$

and

$$\sqrt{2}\,|\mathbf{e}| \cdot |\mathbf{e}| = \frac{1}{\sqrt{2}}.$$

But for particular bicomplex numbers we can say more.

Proposition 1.8.1. *If $U = u_1 + \mathbf{j}\,u_2 \in \mathbb{BC}$ is an arbitrary bicomplex number, but Z is a complex number in $\mathbb{C}(\mathbf{i})$ or $\mathbb{C}(\mathbf{j})$, or Z is a hyperbolic number, then*

    a) if $Z \in \mathbb{C}(\mathbf{i})$ or $\mathbb{C}(\mathbf{j})$, then $|Z \cdot U| = |Z| \cdot |U|$;

b) if $Z = x_1 + \mathbf{k}\, y_2 \in \mathbb{D}$, where $x_1 \in \mathbb{R}$ and $y_2 \in \mathbb{R}$, then in general

$$|Z \cdot U| \neq |Z| \cdot |U|.$$

More precisely,

$$|Z \cdot U|^2 = |Z|^2 \cdot |U|^2 + 4\, x_1\, y_2\, \mathrm{Re}(\mathbf{i}\, u_1\, \overline{u}_2).$$

Proof. We prove first a). Indeed, take $Z = z_1 \in \mathbb{C}(\mathbf{i})$ and $U = u_1 + \mathbf{j} u_2 = (u_1 - \mathbf{i}\, u_2)\, \mathbf{e} + (u_1 + \mathbf{i}\, u_2)\, \mathbf{e}^\dagger$, then

$$|Z \cdot U|^2 = |\, z_1\, (u_1 + \mathbf{j} u_2)\,|^2 = |\, (z_1\, u_1) + \mathbf{j}(z_1\, u_2)\,|^2$$
$$= |z_1 u_1|^2 + |z_1 u_2|^2 \;=\; |z_1|^2 \cdot |U|^2 = |Z|^2 \cdot |U|^2,$$

where we used the fact that the Euclidean norm of a complex number (both in $C(\mathbf{i})$ and in $\mathbb{C}(\mathbf{j})$), seen as a bicomplex number, coincides with its modulus.

Take now $Z = x_1 + \mathbf{j} x_2 = (x_1 - \mathbf{i}\, x_2)\, \mathbf{e} + (x_1 + \mathbf{i}\, x_2)\, \mathbf{e}^\dagger \in \mathbb{C}(\mathbf{j})$, then

$$|Z \cdot U|^2 = |\, \big((x_1 - \mathbf{i}\, x_2)\, \mathbf{e} + (x_1 + \mathbf{i}\, x_2)\, \mathbf{e}^\dagger\big)$$
$$\cdot \big((u_1 - \mathbf{i}\, u_2)\, \mathbf{e} + (u_1 + \mathbf{i}\, u_2)\, \mathbf{e}^\dagger\big)\,|$$
$$= |(x_1 - \mathbf{i}\, x_2)(u_1 - \mathbf{i}\, u_2)\mathbf{e} + (x_1 + \mathbf{i}\, x_2)(u_1 + \mathbf{i}\, u_2)\mathbf{e}^\dagger|^2$$
$$= \frac{1}{2}\left(|x_1 - \mathbf{i} x_2|^2 \cdot |u_1 - \mathbf{i} u_2|^2 + |x_1 + \mathbf{i} x_2|^2 \cdot |u_1 + \mathbf{i} u_2|^2\right)$$
$$= |Z|^2 \cdot |U|^2.$$

Finally, take $Z = x_1 + \mathbf{ij} y_2 = (x_1 + y_2)\, \mathbf{e} + (x_1 - y_2)\, \mathbf{e}^\dagger \in \mathbb{D}$, then

$$|Z \cdot U|^2 = |(x_1 + \mathbf{ij} y_2)\cdot(u_1 + \mathbf{j} u_2)|^2$$
$$= \left|(x_1 + y_2)\cdot(u_1 - \mathbf{i} u_2)\mathbf{e} + (x_1 - y_2)\cdot(u_1 + \mathbf{i} u_2)\, \mathbf{e}^\dagger\right|^2$$
$$= \frac{1}{2}\left((x_1 + y_2)^2 \cdot |u_1 - \mathbf{i} u_2|^2 + (x_1 - y_2)^2 \cdot |u_1 + \mathbf{i} u_2|^2\right)$$
$$= |Z|^2 \cdot |U|^2 + 4\, x_1\, y_2\, \mathrm{Re}(\mathbf{i}\, u_1\, \overline{u}_2),$$

and that is all. $\square$

We have described some classes of factors Z for which the Euclidean norm of the product is equal to the product of the Euclidean norms. Now let us ask the question: can we characterize all the pairs (Z, W) for which the Euclidean norm is multiplicative?

Proposition 1.8.2. *Let $Z = \beta_1\, \mathbf{e} + \beta_2\, \mathbf{e}^\dagger$ and $W = \gamma_1\, \mathbf{e} + \gamma_2\, \mathbf{e}^\dagger$ be two bicomplex numbers, then*

$$|Z \cdot W| = |Z| \cdot |W|$$

if and only if

$$|\beta_1| = |\beta_2|, \quad or \quad |\gamma_1| = |\gamma_2|, \quad or \ both.$$

Proof. One has:

$$Z \cdot W = \beta_1 \, \gamma_1 \, \mathbf{e} + \beta_2 \, \gamma_2 \, \mathbf{e}^\dagger,$$

$$|Z \cdot W|^2 = \frac{1}{2} \left(|\beta_1 \, \gamma_1|^2 + |\beta_2 \, \gamma_2|^2 \right),$$

$$|Z|^2 \cdot |W|^2 = \frac{1}{4} \left(|\beta_1|^2 + |\beta_2|^2 \right) \cdot \left(|\gamma_1|^2 + |\gamma_2|^2 \right),$$

$$\left(|\beta_1|^2 - |\beta_2|^2 \right) \cdot \left(|\gamma_1|^2 - |\gamma_2|^2 \right) = 0,$$

and that is all.   □

Remark 1.8.3. *Since $|\beta_1| = |\beta_2|$ implies that $|Z| = |\beta_1| = |\beta_2|$ we may conclude that the multiplicative property of the Euclidean norm holds if and only if the Euclidean norm of any of the factors coincides with the modulus (as a complex number) of its idempotent component. Of course the contents of Section 1.8.1 does not contradict this conclusion. Note also that the pair β_1 and β_2 as well as the pair γ_1 and γ_2 can be taken, equivalently, in $\mathbb{C}(\mathbf{i})$ or in $\mathbb{C}(\mathbf{j})$.*

It turns out that the condition $|\beta_1| = |\beta_2|$ can be usefully interpreted in terms of the cartesian components. Take Z as $Z = z_1 + \mathbf{j} \, z_2 = \beta_1 \, \mathbf{e} + \beta_2 \, \mathbf{e}^\dagger$. Assume first that $|\beta_1| = |\beta_2|$ where $\beta_1 = z_1 - \mathbf{i} \, z_2$, $\beta_2 = z_1 + \mathbf{i} \, z_2$; then

$$|z_1 - \mathbf{i} \, z_2|^2 = |z_1 + \mathbf{i} \, z_2|^2$$

which is equivalent to

$$z_1 \cdot \overline{z}_2 = \lambda \in \mathbb{R}.$$

The following cases arise:

(1) if $\lambda = 0$, then Z is in $\mathbb{C}(\mathbf{i})$, or $Z = \mathbf{j} \, z_2 \in \mathbf{j} \cdot \mathbb{C}(\mathbf{i})$, or both; the last means that $Z = 0$;

(2) if $\lambda \neq 0$, then

$$z_1 = \frac{\lambda}{|z_2|^2} \cdot z_2,$$

i.e., $Z = z_2 \left(\dfrac{\lambda}{|z_2|^2} + \mathbf{j} \right)$ and Z becomes the product of a $\mathbb{C}(\mathbf{i})$-complex number and a $\mathbb{C}(\mathbf{j})$-complex number.

Let us show that the reciprocal is also true. Take $a = a_1 + \mathbf{i} \, a_2$, $b = b_1 + \mathbf{j} \, b_2$, where a_1, a_2, b_1, b_2 are real numbers, and set $Z := a \cdot b = a \cdot b_1 + \mathbf{j} \, a \cdot b_2$, then

$$Z = (a \, b_1 - \mathbf{i} \, a \, b_2) \cdot \mathbf{e} + (a \, b_1 + \mathbf{i} \, a \, b_2) \cdot \mathbf{e}^\dagger$$

$$= a \cdot (b_1 - \mathbf{i} \, b_2) \cdot \mathbf{e} + a \, (b_1 + \mathbf{i} \, b_2) \cdot \mathbf{e}^\dagger =: \beta_1 \, \mathbf{e} + \beta_2 \, \mathbf{e}^\dagger$$

with $|\beta_1| = |a| \cdot |b_1 - \mathbf{i} \, b_2| = |a| \cdot |b| = |\beta_2|$.

We summarize this reasoning in the following two statements.

Proposition 1.8.4. *A bicomplex number Z is a product of a complex number in $\mathbb{C}(\mathbf{i})$ and of a complex number in $\mathbb{C}(\mathbf{j})$ if and only if the idempotent components of Z have the same moduli as complex numbers.*

Corollary 1.8.5. *The Euclidean norm of the product of two bicomplex numbers is equal to the product of their norms if and only if at least one of them is the product of a complex number in $\mathbb{C}(\mathbf{i})$ and of a complex number in $\mathbb{C}(\mathbf{j})$.*

The inequality

$$|Z \cdot W| \leq \sqrt{2}\,|Z| \cdot |W|$$

says that the relation between the Euclidean norm $|Z|$ of an invertible bicomplex number Z and the norm $|Z^{-1}|$ of its inverse is more complicated than the "conventional" one. Indeed,

$$1 = |Z \cdot Z^{-1}| \leq \sqrt{2} \cdot |Z| \cdot |Z^{-1}|$$

and thus

$$\frac{1}{|Z|} \leq \sqrt{2} \cdot |Z^{-1}|.$$

We ask now for which class of bicomplex numbers the "conventional" formula

$$|Z^{-1}| = \frac{1}{|Z|} \tag{1.31}$$

holds? The answer follows from the conditions ensuring the equality $|Z \cdot W| = |Z| \cdot |W|$, in which we can take $W = Z^{-1}$, thus obtaining that (1.31) holds if and only if Z is a product of a complex number in $\mathbb{C}(\mathbf{i})$ by a complex number in $\mathbb{C}(\mathbf{j})$ or, equivalently, if and only if the Euclidean norm of Z coincides with the modulus of any of its idempotent components.

Chapter 2

Algebraic Structures of the Set of Bicomplex Numbers

2.1 The ring of bicomplex numbers

The operations of addition and multiplication of bicomplex numbers imply directly

Proposition 2.1.1. $(\mathbb{BC}, +, \cdot)$ *is a commutative ring, i.e.,*

1. *The addition is associative, commutative, with identity element $0 = 0 + \mathbf{j}0$, and each bicomplex number has an additive inverse. This is to say that $(\mathbb{BC}, +)$ is an Abelian group.*

2. *The multiplication is associative, commutative, with identity element $1 = 1 + \mathbf{j}0$.*

3. *The multiplication is distributive with respect to the addition, i.e., for any $Z, Z_1, Z_2 \in \mathbb{BC}$, we have:*

$$Z(Z_1 + Z_2) = Z Z_1 + Z Z_2. \tag{2.1}$$

This is a very particular ring, with a number of specific features which we will explain in what follows.

Using the language of ring theory, one says that the invertible elements are called "units", and generally speaking the sets of non-units and of zero-divisors do not necessarily coincide. As we saw before, in the ring of bicomplex numbers the non-units "almost" coincide with the zero-divisors. More exactly, if we exclude the zero from non-units, the rest of the elements are zero-divisors. In other words, if $\mathbb{BC}^{-1}$ denotes the set of invertible elements in $\mathbb{BC}$, then we have a partition of $\mathbb{BC}$:

$$\mathbb{BC} = \mathbb{BC}^{-1} \cup \mathfrak{S} \cup \{0\},$$

where $\mathfrak{S}$ is the set of zero-divisors.

As it happens in any non-trivial ring that is not a field, the ring of bicomplex numbers has many ideals, but we want to single out two of them, they are $\mathbb{BC}_{\mathbf{e}} := \mathbb{BC} \cdot \mathbf{e}$ and $\mathbb{BC}_{\mathbf{e}^\dagger} := \mathbb{BC} \cdot \mathbf{e}^\dagger$. These ideals are principal in terms of the ring theory. The peculiarities of these ideals are:

$$\mathbb{BC}_{\mathbf{e}} \cap \mathbb{BC}_{\mathbf{e}^\dagger} = \{0\},$$
$$\mathbb{BC}_{\mathbf{e}} + \mathbb{BC}_{\mathbf{e}^\dagger} = \mathbb{BC},$$
$$\mathbf{e} \cdot \mathbb{BC}_{\mathbf{e}^\dagger} = 0 \quad \text{and} \quad \mathbf{e}^\dagger \cdot \mathbb{BC}_{\mathbf{e}} = 0.$$

2.2   Linear spaces and modules in $\mathbb{BC}$

It is known that if S is a subring of a ring R, then R is a module over the ring S. In our situation, the sets $\mathbb{R}$, $\mathbb{C}(\mathbf{i})$, $\mathbb{C}(\mathbf{j})$ and $\mathbb{D}$ are subrings of the ring $\mathbb{BC}$, thus $\mathbb{BC}$ can be seen as a module over each one of these subrings, and of course, it is a module over itself. Now, since $\mathbb{R}$, $\mathbb{C}(\mathbf{i})$ and $\mathbb{C}(\mathbf{j})$ are fields, $\mathbb{BC}$ is a real linear space, a $\mathbb{C}(\mathbf{i})$-complex linear space and a $\mathbb{C}(\mathbf{j})$-complex linear space.

Recalling formula (1.9), we see that the mapping

$$\mathbb{BC} \ni Z = x_1 + \mathbf{i}\,y_1 + \mathbf{j}\,x_2 + \mathbf{k}\,y_2 \longmapsto (x_1, y_1, x_2, y_2) \in \mathbb{R}^4 \qquad (2.2)$$

is an isomorphism of real spaces, which maps the bicomplex numbers 1, $\mathbf{i}$, $\mathbf{j}$, $\mathbf{k}$ into the canonical basis of $\mathbb{R}^4$. We will widely use this identification.

Equation (1.3) suggests the following isomorphism between $\mathbb{BC}$ as a $\mathbb{C}(\mathbf{i})$-linear space and $\mathbb{C}^2(\mathbf{i})$:

$$\mathbb{BC} \ni Z = z_1 + \mathbf{j}\,z_2 \longmapsto (z_1, z_2) \in \mathbb{C}^2(\mathbf{i}). \qquad (2.3)$$

In this case the bicomplex numbers 1 and $\mathbf{j}$ are mapped into the canonical basis of $\mathbb{C}^2(\mathbf{i})$. Composing this isomorphism and the inverse of the previous one, we have the following isomorphism between $\mathbb{R}^4$ and $\mathbb{C}^2(\mathbf{i})$:

$$\mathbb{R}^4 \ni (x_1, y_1, x_2, y_2) \longmapsto (x_1 + \mathbf{i}\,y_1, \ x_2 + \mathbf{i}\,y_2) \in \mathbb{C}^2(\mathbf{i}). \qquad (2.4)$$

Seeing now $\mathbb{BC}$ as a $\mathbb{C}(\mathbf{j})$-linear space and using (1.4), we have the following isomorphism:

$$\mathbb{BC} \ni Z = \zeta_1 + \mathbf{i}\zeta_2 \longmapsto (\zeta_1, \zeta_2) \in \mathbb{C}^2(\mathbf{j}). \qquad (2.5)$$

This isomorphism sends the bicomplex numbers 1 and $\mathbf{i}$ into the canonical basis in $\mathbb{C}^2(\mathbf{j})$ and it induces the following isomorphism (of real linear spaces) between $\mathbb{R}^4$ and $\mathbb{C}^2(\mathbf{j})$:

$$\mathbb{R}^4 \ni (x_1, y_1, x_2, y_2) \longmapsto (x_1 + \mathbf{j}\,x_2, \ y_1 + \mathbf{j}\,y_2) \in \mathbb{C}^2(\mathbf{j}). \qquad (2.6)$$

Obviously the isomorphisms (2.4) and (2.6) are different: this shows once again that inside $\mathbb{BC}$ the "complex sets" $\mathbb{C}^2(\mathbf{i})$ and $\mathbb{C}^2(\mathbf{j})$ play distinct roles.

One more difference that one notes considering $\mathbb{BC}$ as a $\mathbb{C}(\mathbf{i})$- or a $\mathbb{C}(\mathbf{j})$-linear space is, for example, that the set $\{1, \mathbf{i}\}$ is linearly independent when $\mathbb{BC}$ is seen as a $\mathbb{C}(\mathbf{j})$-linear space, but the same set is linearly dependent in the $\mathbb{C}(\mathbf{i})$-linear space $\mathbb{BC}$.

The reader may note that equation (1.7) suggests another $\mathbb{C}(\mathbf{i})$-linear isomorphism between $\mathbb{BC}$ and $\mathbb{C}^2(\mathbf{i})$:

$$\mathbb{BC} \ni Z = (x_1 + \mathbf{i}\, y_1) + \mathbf{k}\,(y_2 - \mathbf{i}\, x_2) = w_1 + \mathbf{k}\, w_2 \longmapsto (w_1,\, w_2) \in \mathbb{C}^2(\mathbf{i}). \quad (2.7)$$

The relation between the isomorphisms (2.3) and (2.7) is the following. Since under the isomorphism (2.3) the bicomplex numbers $1, \mathbf{j}$ are mapped to the canonical basis of $\mathbb{C}^2(\mathbf{i})$, then the bicomplex number $\mathbf{k} = \mathbf{i}\,\mathbf{j}$ is mapped to $(0, \mathbf{i}) \in \mathbb{C}^2(\mathbf{i})$. Thus we have made a change of basis from the canonical one to the basis $\{(1,0), (0, \mathbf{i})\}$. The matrix of this change of basis is

$$\begin{bmatrix} 1 & 0 \\ 0 & -\mathbf{i} \end{bmatrix},$$

that is, the pair $(z_1, z_2) \in \mathbb{C}^2(\mathbf{i})$ is mapped to the pair $(w_1, w_2) \in \mathbb{C}^2(\mathbf{i})$ by the rule

$$\begin{bmatrix} 1 & 0 \\ 0 & -\mathbf{i} \end{bmatrix} \cdot \begin{bmatrix} z_1 \\ z_2 \end{bmatrix} = \begin{bmatrix} z_1 \\ -\mathbf{i}\, z_2 \end{bmatrix} = \begin{bmatrix} w_1 \\ w_2 \end{bmatrix}. \quad (2.8)$$

This equality gives the precise relation between (2.3) and (2.7).

A similar reasoning applies to $\mathbb{C}^2(\mathbf{j})$, that is, inspired by equation (1.8), one defines the isomorphism

$$\mathbb{BC} \ni Z = (x_1 + \mathbf{j}\, x_2) + \mathbf{k}\,(y_2 - \mathbf{j}\, y_1) = w_1 + \mathbf{k}\, w_2 \longmapsto (w_1,\, w_2) \in \mathbb{C}^2(\mathbf{j}). \quad (2.9)$$

The relation between (2.5) and (2.9) is given by a change of basis from the canonical one to the basis $\{(1,0), (0, \mathbf{j})\}$. The matrix of this change of basis is

$$\begin{bmatrix} 1 & 0 \\ 0 & -\mathbf{j} \end{bmatrix},$$

and thus the pairs $(\zeta_1, \zeta_2) \in \mathbb{C}^2(\mathbf{j})$ and $(w_1, w_2) \in \mathbb{C}^2(\mathbf{j})$ are related as follows:

$$\begin{bmatrix} 1 & 0 \\ 0 & -\mathbf{j} \end{bmatrix} \cdot \begin{bmatrix} \zeta_1 \\ \zeta_2 \end{bmatrix} = \begin{bmatrix} \zeta_1 \\ -\mathbf{j}\, \zeta_2 \end{bmatrix} = \begin{bmatrix} w_1 \\ w_2 \end{bmatrix}.$$

The reader may note that we are "playing" with the different linear structures in $\mathbb{BC}$. We have pointed out some differences between $\mathbb{C}(\mathbf{i})$ and $\mathbb{C}(\mathbf{j})$, although existence of the following isomorphism of fields is evident:

$$\varphi : \mathbb{C}(\mathbf{i}) \to \mathbb{C}(\mathbf{j}),$$
$$\varphi(x + \mathbf{i}\, y) := x + \mathbf{j}\, y. \quad (2.10)$$

This isomorphism arose implicitly in the first chapter when we compared the two idempotent representations, one with coefficients in $\mathbb{C}(\mathbf{i})$ and another with coefficients in $\mathbb{C}(\mathbf{j})$:

$$Z = \beta_1\,\mathbf{e} + \beta_2\,\mathbf{e}^\dagger = (c_1 + \mathbf{i}\,d_1)\,\mathbf{e} + (c_2 + \mathbf{i}\,d_2)\,\mathbf{e}^\dagger$$

$$= \alpha_1\,\mathbf{e} + \alpha_2\,\mathbf{e}^\dagger = (c_1 - \mathbf{j}\,d_1)\,\mathbf{e} + (c_2 + \mathbf{j}\,d_2)\,\mathbf{e}^\dagger, \tag{2.11}$$

which shows that $\alpha_1 = (\varphi(\beta_1))^\dagger$ and $\alpha_2 = \varphi(\beta_2)$.

Following the line of defining isomorphisms between $\mathbb{BC}$ and the complex spaces $\mathbb{C}^2(\mathbf{i})$ and $\mathbb{C}^2(\mathbf{j})$, we will see that the idempotent representations suggest two more complex linear spaces isomorphisms. We need first the following proposition.

Proposition 2.2.1. *The zero-divisors*

$$\mathbf{e} = \frac{1 + \mathbf{i}\,\mathbf{j}}{2} \qquad and \qquad \mathbf{e}^\dagger = \frac{1 - \mathbf{i}\,\mathbf{j}}{2}$$

are linearly independent in $\mathbb{BC}$ when it is seen as a $\mathbb{C}(\mathbf{i})$-linear space or as a $\mathbb{C}(\mathbf{j})$-linear space.

Proof. Using the isomorphism (2.3), the bicomplex numbers $\mathbf{e}$ and $\mathbf{e}^\dagger$ are mapped as follows:

$$\mathbf{e} \longmapsto \left(\frac{1}{2}, \frac{\mathbf{i}}{2}\right) = \frac{1}{2}\,(1,\,\mathbf{i}),$$

$$\mathbf{e}^\dagger \longmapsto \left(\frac{1}{2}, -\frac{\mathbf{i}}{2}\right) = \frac{1}{2}\,(1,\,-\mathbf{i}),$$

and considering the equation

$$\lambda_1(1,\,\mathbf{i}) + \lambda_2\,(1,\,-\mathbf{i}) = 0,$$

with $\lambda_1\,\lambda_2 \in \mathbb{C}(\mathbf{i})$, we infer immediately that $\lambda_1 = \lambda_2 = 0$.

The same for $\mathbb{C}^2(\mathbf{j})$.   □

Using now (2.11), define the isomorphisms of complex linear spaces:

$$\mathbb{BC} \ni Z = \beta_1\,\mathbf{e} + \beta_2\,\mathbf{e}^\dagger \longmapsto (\beta_1, \beta_2) \in \mathbb{C}^2(\mathbf{i}), \tag{2.12}$$

$$\mathbb{BC} \ni Z = \alpha_1\,\mathbf{e} + \alpha_2\,\mathbf{e}^\dagger \longmapsto (\alpha_1, \alpha_2) \in \mathbb{C}^2(\mathbf{j}). \tag{2.13}$$

Again, the relations between (2.3) and (2.12) as well as between (2.5) and (2.13) are given through the change of basis from the canonical ones to the basis

$$\left\{\left(\frac{1}{2}, \frac{\mathbf{i}}{2}\right), \left(\frac{1}{2}, \frac{-\mathbf{i}}{2}\right)\right\}$$

in $\mathbb{C}^2(\mathbf{i})$ and to the basis

$$\left\{ \left(\frac{1}{2}, \frac{\mathbf{j}}{2} \right), \left(\frac{1}{2}, \frac{-\mathbf{j}}{2} \right) \right\}$$

in $\mathbb{C}^2(\mathbf{j})$. Using the matrices of change of basis we get the relations:

$$\begin{bmatrix} 1 & -\mathbf{i} \\ 1 & \mathbf{i} \end{bmatrix} \cdot \begin{bmatrix} z_1 \\ z_2 \end{bmatrix} = \begin{bmatrix} z_1 - \mathbf{i}\, z_2 \\ z_1 + \mathbf{i}\, z_2 \end{bmatrix} = \begin{bmatrix} \beta_1 \\ \beta_2 \end{bmatrix} \tag{2.14}$$

in the space $\mathbb{C}^2(\mathbf{i})$, and

$$\begin{bmatrix} 1 & -\mathbf{j} \\ 1 & \mathbf{j} \end{bmatrix} \cdot \begin{bmatrix} \zeta_1 \\ \zeta_2 \end{bmatrix} = \begin{bmatrix} \zeta_1 - \mathbf{j}\, \zeta_2 \\ \zeta_1 + \mathbf{j}\, \zeta_2 \end{bmatrix} = \begin{bmatrix} \alpha_1 \\ \alpha_2 \end{bmatrix}$$

in the space $\mathbb{C}^2(\mathbf{j})$.

Consider now the set $\mathbb{D}^2 = \mathbb{D} \times \mathbb{D}$. With the component-wise addition inherited from $\mathbb{D}$ it is an additive abelian group. Defining also the component-wise multiplication by the scalars from $\mathbb{D}$, $\mathbb{D}^2$ becomes a hyperbolic, or $\mathbb{D}$-, module. Now, formulas (1.5) and (1.6) suggest the following isomorphisms of $\mathbb{D}$-modules:

$$\mathbb{BC} \ni Z = (x_1 + \mathbf{k}\, y_2) + \mathbf{i}\, (y_1 - \mathbf{k}\, x_2) = \mathfrak{z}_1 + \mathbf{i}\, \mathfrak{z}_2 \longmapsto (\mathfrak{z}_1, \mathfrak{z}_2) \in \mathbb{D}^2 \tag{2.15}$$

and

$$\mathbb{BC} \ni Z = (x_1 + \mathbf{k}\, y_2) + \mathbf{j}\, (x_2 - \mathbf{k}\, y_1) = \mathfrak{w}_1 + \mathbf{j}\, \mathfrak{w}_2 \longmapsto (\mathfrak{w}_1, \mathfrak{w}_2) \in \mathbb{D}^2. \tag{2.16}$$

2.3 Algebra structures in $\mathbb{BC}$

We have endowed the set $\mathbb{BC}$ with three structures of linear spaces, namely, real linear space, a $\mathbb{C}(\mathbf{i})$-linear space and $\mathbb{C}(\mathbf{j})$-linear space. Since the set $\mathbb{BC}$ is a ring all these spaces generate the corresponding algebras, one is a real algebra and the others are complex algebras.

Let us begin with the real algebra. First of all, the real spaces isomorphism (2.2) induces the following multiplication in $\mathbb{R}^4$ (which is of course the "real form of the bicomplex multiplication"):

$$
\begin{aligned}
(x_1, & y_1, x_2, y_2) \cdot (s_1, t_1, s_2, t_2) \\
& := (x_1 s_1 - y_1 t_1 - x_2 s_2 + y_2 t_2, \ x_1 t_1 + y_1 s_1 - x_2 t_2 - y_2 s_2, \\
& \quad\ x_1 s_2 - y_1 t_2 + x_2 s_1 - y_2 t_1, \ x_1 t_2 + y_1 s_2 + x_2 t_1 + y_2 s_1).
\end{aligned}
\tag{2.17}
$$

It follows directly from the properties of the bicomplex multiplication that (2.17) endows $\mathbb{R}^4$ with the structure of a commutative real algebra. Moreover, the isomorphism (2.2) extends up to a real algebras isomorphism.

It is well known that there are not too many "reasonably good" multiplications in $\mathbb{R}^4$. The bicomplex multiplication given by (2.17) is one of these few options and one can compare it with the multiplication generated by quaternions.

The next step is to consider both complex algebras. The isomorphisms (2.3), (2.7) and (2.12) give us three candidates for introducing a multiplication on the $\mathbb{C}(\mathbf{i})$-linear space $\mathbb{C}^2(\mathbf{i})$:

$$(z_1, z_2) \cdot (p_1, p_2) := (z_1 p_1 - z_2 p_2,\ z_1 p_2 + z_2 p_1);$$
$$(w_1, w_2) \cdot (\theta_1, \theta_2) := (w_1 \theta_1 + w_2 \theta_2,\ w_1 \theta_2 + w_2 \theta_1);$$
$$(a_1, a_2) \cdot (c_1, c_2) := (a_1 c_1,\ a_2 c_2).$$

Again, the properties of the bicomplex multiplication guarantee that each of these three formulas defines a commutative multiplication on $\mathbb{C}^2(\mathbf{i})$. Looking at $\mathbb{C}^2(\mathbf{i})$ as just a linear space, with no basis fixed, the three formulas define indeed three different multiplications. But endowing the $\mathbb{C}^2(\mathbf{i})$ subsequently with the bases $\{(1,0),\ (0,1)\}$, $\{(1,0),\ (0,\mathbf{i})\}$, $\left\{ \left(\dfrac{1}{2}, \dfrac{\mathbf{i}}{2} \right),\ \left(\dfrac{1}{2}, -\dfrac{\mathbf{i}}{2} \right) \right\}$, and using (2.8) and (2.14), one can see that the three formulas define the same, unique multiplication on $\mathbb{C}^2(\mathbf{i})$ although expressed in terms of the three bases of $\mathbb{C}^2(\mathbf{i})$.

For example, from (2.8), one has that

$$w_1 = z_1, \qquad\qquad \theta_1 = p_1,$$
$$\text{and}$$
$$w_2 = -\mathbf{i}\, z_2, \qquad\qquad \theta_2 = -\mathbf{i}\, p_2,$$

thus,

$$\begin{aligned}
(w_1, w_2) \cdot (\theta_1, \theta_2) &= (w_1 \theta_1 + w_2 \theta_2,\ w_1 \theta_2 + w_2 \theta_1) \\
&= (z_1 p_1 - z_2 p_2,\ -\mathbf{i}\,(z_1 p_2 + z_2 p_1)) \\
&= \begin{bmatrix} 1 & 0 \\ 0 & -\mathbf{i} \end{bmatrix} \cdot \begin{bmatrix} z_1 p_1 - z_2 p_2 \\ z_1 p_2 + z_2 p_1 \end{bmatrix} \\
&= \begin{bmatrix} w_1 \theta_1 + w_2 \theta_2 \\ w_1 \theta_2 + w_2 \theta_1 \end{bmatrix}.
\end{aligned}$$

The case of the complex linear space $\mathbb{C}^2(\mathbf{j})$ is treated in exactly the same way.

Now taking into account that $\mathbb{BC}$ is a module over $\mathbb{D}$ and over itself, and that it is also a ring, we conclude that the set $\mathbb{BC}$ is a $\mathbb{D}$-algebra and a $\mathbb{BC}$-algebra. As such, $\mathbb{BC}$ is isomorphic to the $\mathbb{D}$-algebra $\mathbb{D}^2$ and the isomorphisms are given by formulas (2.15) and (2.16).

2.4 Matrix representations of bicomplex numbers

It is well known that the field $\mathbb{C}$ of complex numbers is isomorphic to the set of real 2×2 matrices of the form

$$\begin{pmatrix} x & -y \\ y & x \end{pmatrix}.$$

In other words, the mapping

$$\phi_{\mathbb{C}} : z = x + \mathbf{i}y \in \mathbb{C} \mapsto \begin{pmatrix} x & -y \\ y & x \end{pmatrix}$$

is an isomorphism of fields between $\mathbb{C}$ and

$$\left\{ \begin{pmatrix} x & -y \\ y & x \end{pmatrix} \,\middle|\, x, y \in \mathbb{R} \right\} =: \mathcal{A}_{\mathbb{C}}.$$

In particular, the matrices of this form commute under multiplication and any of them, but zero, has its inverse. Under this isomorphism the imaginary unit $\mathbf{i}$ is represented by the matrix

$$\mathcal{I} := \begin{pmatrix} 0 & -1 \\ 1 & 0 \end{pmatrix}.$$

Take $z = x + \mathbf{i}y$, then

$$\phi_{\mathbb{C}}(z) = \begin{pmatrix} x & -y \\ y & x \end{pmatrix} = x \begin{pmatrix} 1 & 0 \\ 0 & 1 \end{pmatrix} + y \begin{pmatrix} 0 & -1 \\ 1 & 0 \end{pmatrix} = x I_2 + y\mathcal{I}.$$

The square of the modulus of a complex number z coincides with $\det \phi_{\mathbb{C}}(z)$.

In terms of representation theory the mapping $\phi_{\mathbb{C}}$ is called a representation of the field $\mathbb{C}$ into a subset of the set of 2×2 real matrices.

A similar reasoning applies to the ring $\mathbb{BC}$. The mapping

$$\phi_{\mathbb{C}(\mathbf{i})} : Z = z_1 + \mathbf{j}z_2 \in \mathbb{BC} \longmapsto \begin{pmatrix} z_1 & -z_2 \\ z_2 & z_1 \end{pmatrix} \tag{2.18}$$

turns out to be an isomorphism (of rings) between $\mathbb{BC}$ and the set of matrices

$$\left\{ \begin{pmatrix} z_1 & -z_2 \\ z_2 & z_1 \end{pmatrix} \,\middle|\, z_1, z_2 \in \mathbb{C}(\mathbf{i}) \right\}.$$

Note the following identifications:

$$\phi_{\mathbb{C}(\mathbf{i})}(\mathbf{i}) = \begin{pmatrix} \mathbf{i} & 0 \\ 0 & \mathbf{i} \end{pmatrix} = \mathbf{i}I_2, \qquad \phi_{\mathbb{C}(\mathbf{i})}(\mathbf{j}) = \begin{pmatrix} 0 & -1 \\ 1 & 0 \end{pmatrix} =: \mathcal{J},$$

$$\phi_{\mathbb{C}(\mathbf{i})}(\mathbf{e}) = \frac{1}{2}\begin{pmatrix} 1 & -\mathbf{i} \\ \mathbf{i} & 1 \end{pmatrix} =: \mathcal{E}, \qquad \phi_{\mathbb{C}(\mathbf{i})}(\mathbf{e}^{\dagger}) = \frac{1}{2}\begin{pmatrix} 1 & \mathbf{i} \\ -\mathbf{i} & 1 \end{pmatrix} =: \mathcal{E}^{\dagger}.$$

Thus, for a bicomplex number $Z = z_1 + \mathbf{j}z_2$ its image $\phi_{\mathbb{C}(\mathbf{i})}(Z)$ is

$$\phi_{\mathbb{C}(\mathbf{i})}(Z) = z_1 I_2 + z_2 J = \beta_1 \mathcal{E} + \beta_2 \mathcal{E}^t \,.$$

As one can expect, there are two analogous mappings which allow us to identify the bicomplex numbers with matrices having $\mathbb{C}(\mathbf{j})$-complex entries, or hyperbolic entries:

$$\phi_{\mathbb{C}(\mathbf{j})} : Z = \zeta_1 + \mathbf{i}\zeta_2 \in \mathbb{BC} \longmapsto \begin{pmatrix} \zeta_1 & -\zeta_2 \\ \zeta_2 & \zeta_1 \end{pmatrix} ,$$

$$\phi_{\mathbb{D}} : Z = \mathfrak{z}_1 + \mathbf{i}\mathfrak{z}_2 \in \mathbb{BC} \longmapsto \begin{pmatrix} \mathfrak{z}_1 & -\mathfrak{z}_2 \\ \mathfrak{z}_2 & \mathfrak{z}_1 \end{pmatrix} .$$

It is instructive to repeat the above computations with these new mappings.

Finally, the bicomplex numbers may be identified with real 4×4 matrices:

$$\phi_{\mathbb{R}} : Z = x_1 + \mathbf{i}y_1 + \mathbf{j}x_2 + \mathbf{k}y_2 \in \mathbb{BC} \longmapsto \begin{pmatrix} x_1 & -y_1 & -x_2 & y_2 \\ y_1 & x_1 & -y_2 & -x_2 \\ x_2 & -y_2 & x_1 & -y_1 \\ y_2 & x_2 & y_1 & x_1 \end{pmatrix} .$$

Every 4×4 matrix determines a linear (more exactly, a real linear) transformation on $\mathbb{R}^4$. Of course, not all of them remain $\mathbb{BC}$-linear when $\mathbb{R}^4$ is seen as $\mathbb{BC}$. Those matrices which represent $\mathbb{BC}$-linear mappings are of the form $\phi_{\mathbb{R}}(Z)$.

Notice also that at this stage, we are considering again the following identification between $\mathbb{BC}$ and $\mathbb{R}^4$:

$$Z = x_1 + \mathbf{i}y_1 + \mathbf{j}x_2 + \mathbf{k}y_2 \longleftrightarrow (x_1, y_1, x_2, y_2) \,.$$

This "forces" the following two identifications between $\mathbb{BC}$ and $\mathbb{C}^2$:

$$Z = z_1 + \mathbf{j}z_2 \longleftrightarrow (z_1, z_2) = (x_1 + \mathbf{i}y_1, x_2 + \mathbf{i}y_2) \in \mathbb{C}^2(\mathbf{i}) \longleftrightarrow (x_1, y_1, x_2, y_2)$$

and

$$Z = \zeta_1 + \mathbf{i}\zeta_2 \longleftrightarrow (\zeta_1, \zeta_2) = (x_1 + \mathbf{j}x_2, y_1 + \mathbf{j}y_2) \in \mathbb{C}^2(\mathbf{j}) \longleftrightarrow (x_1, y_1, x_2, y_2) \,.$$

Consider now an $\mathbb{R}$-linear mapping $T : \mathbb{R}^4 \to \mathbb{R}^4$ which represents also a $\mathbb{C}(\mathbf{i})$-linear mapping, then the matrix of T is of the form

$$\begin{pmatrix} a & -b & c & -d \\ b & a & d & c \\ \ell & -m & u & -v \\ m & \ell & v & u \end{pmatrix} ,$$

meanwhile, if T represents a $\mathbb{C}(\mathbf{j})$-linear mapping, then its matrix is of the form

$$\begin{pmatrix} A & B & -E & -F \\ C & D & -G & -H \\ E & F & A & B \\ G & H & C & D \end{pmatrix} .$$

It is clear that, as one should expect, the matrix $\phi_{\mathbb{R}}(Z)$ represents both a $\mathbb{C}(\mathbf{i})$-linear mapping and a $\mathbb{C}(\mathbf{j})$-linear one.

2.5 Bilinear forms and inner products

On the real linear space $\mathbb{R}$ the following formula defines a bilinear form which serves simultaneously as an inner product: if $x, y \in \mathbb{R}$, then

$$\mathcal{B}_{\mathbb{R}}(x, y) := x \cdot y.$$

The corresponding (real) quadratic form is

$$\mathcal{Q} := \mathcal{B}_{\mathbb{R}}(x, x) = x^2$$

and it defines the (square of the) Euclidean metric on $\mathbb{R}$.

The set $\mathbb{C}$ can be seen both as a real and as a complex linear space and each of these structures generates its own analogue of what we described above.

When $\mathbb{C}$ is considered as a real linear space, that is, $\mathbb{C} = \mathbb{R}^2$, then the bilinear form is given, for $z = x + \mathbf{i}y$, $w = u + \mathbf{i}v$, by

$$\mathcal{B}_{\mathbb{C},\mathbb{R}}(z, w) := xu + yv,$$

which is exactly the canonical inner product on $\mathbb{R}^2$. The corresponding quadratic form is

$$\mathcal{Q}_{\mathbb{C},\mathbb{R}}(z) := \mathcal{B}_{\mathbb{C},\mathbb{R}}(z, z) = x^2 + y^2$$

and it defines the (square of the) Euclidean metric on $\mathbb{C}$.

When $\mathbb{C}$ is considered as a complex linear space, then it has both a bilinear and a sesquilinear form:

$$\mathcal{B}_{\mathbb{C},1}(z, w) := z \cdot w$$

and

$$\mathcal{B}_{\mathbb{C},2}(z, w) := z \cdot \overline{w},$$

the second of them being the canonical complex-valued inner product on $\mathbb{C}$. They generate the quadratic forms

$$\mathcal{Q}_{\mathbb{C},1}(z) := \mathcal{B}_{\mathbb{C},1}(z, z) := z^2$$

and

$$\mathcal{Q}_{\mathbb{C},2}(z) := \mathcal{B}_{\mathbb{C},2}(z, z) := |z|^2 = x^2 + y^2,$$

where again $\mathcal{Q}_{\mathbb{C},2}(z) = \mathcal{Q}_{\mathbb{C},\mathbb{R}}(z)$ is the square of the Euclidean metric on $\mathbb{C}$. The forms $\mathcal{B}_{\mathbb{C},1}$ and $\mathcal{Q}_{\mathbb{C},1}$ are employed widely in different areas of mathematics, but $\mathcal{B}_{\mathbb{C},1}$ is not called usually an inner product and $\mathcal{Q}_{\mathbb{C},1}$ does not define any metric in the classical sense.

Let us extend these ideas onto the bicomplex context. Starting with the real structure on $\mathbb{BC}$, we see it as $\mathbb{R}^4 = \{(x_1, y_1, x_2, y_2) = Z\}$ and thus we endow it with the (real) bilinear form

$$\mathcal{B}_{\mathbb{BC},\mathbb{R}}(Z, W) := x_1 u_1 + y_1 v_1 + x_2 u_2 + y_2 v_2 \,,$$

which is the canonical inner product on $\mathbb{R}^4$. The corresponding quadratic form is

$$\mathcal{Q}_{\mathbb{BC},\mathbb{R}}(Z) := \mathcal{B}_{\mathbb{BC},\mathbb{R}}(Z, Z) := x_1^2 + y_1^2 + x_2^2 + y_2^2,$$

and it defines the (square of the) Euclidean metric on $\mathbb{BC}$.

When $\mathbb{BC}$ is considered as a $\mathbb{C}(\mathbf{i})$-complex linear space, $\mathbb{BC} = \mathbb{C}^2(\mathbf{i})$, then it has both a $\mathbb{C}(\mathbf{i})$-bilinear and a $\mathbb{C}(\mathbf{i})$-sesquilinear forms:

$$\mathcal{B}_{\mathbb{BC};\mathbf{i},1}(Z, W) := z_1 \cdot w_1 + z_2 \cdot w_2 = \frac{1}{2}(ZW^\dagger + Z^\dagger W)$$

and

$$\mathcal{B}_{\mathbb{BC};\mathbf{i},2}(Z, W) := z_1 \cdot \overline{w}_1 + z_2 \cdot \overline{w}_2 = \frac{1}{2}(ZW^* + Z^\dagger \overline{W}) \,,$$

the second of them being the canonical $\mathbb{C}(\mathbf{i})$-valued inner product on $\mathbb{C}^2(\mathbf{i})$. They generate the respective quadratic forms

$$\mathcal{Q}_{\mathbb{BC};\mathbf{i},1}(Z) := \mathcal{B}_{\mathbb{BC};\mathbf{i},1}(Z, Z) := z_1^2 + z_2^2 = Z \cdot Z^\dagger = |Z|_\mathbf{i}^2$$

and

$$\mathcal{Q}_{\mathbb{BC};\mathbf{i},2}(Z) := \mathcal{B}_{\mathbb{BC};\mathbf{i},2}(Z, Z) := |z_1|^2 + |z_2|^2 = \frac{1}{2}(ZZ^* + Z^\dagger \overline{Z})$$
$$= \frac{1}{2}(|Z|_\mathbf{k}^2 + |Z^\dagger|_\mathbf{k}^2) = \frac{1}{2}(|Z|_\mathbf{k}^2 + |\overline{Z}|_\mathbf{k}^2) \,,$$

where, again, $\mathcal{Q}_{\mathbb{BC};\mathbf{i},2}(Z) = \mathcal{Q}_{\mathbb{BC},\mathbb{R}}(Z)$ is the square of the canonical Euclidean metric on $\mathbb{C}^2$. The forms $\mathcal{B}_{\mathbb{BC};\mathbf{i},1}$ and $\mathcal{Q}_{\mathbb{BC};\mathbf{i},1}$ are also widely known and used (for instance, in the theory of complex Laplacian and its solutions called complex harmonic functions), but $\mathcal{B}_{\mathbb{BC};\mathbf{i},1}$ is not called, usually, an inner product on $\mathbb{BC}$, and $\mathcal{Q}_{\mathbb{BC};\mathbf{i},1}$ does not define any metric in the classical sense.

The other complex structure on $\mathbb{BC}$, $\mathbb{BC} = \mathbb{C}^2(\mathbf{j})$, is dealt with in the same way. In this context, it is worth writing the corresponding quadratic forms:

$$\mathcal{Q}_{\mathbb{BC};\mathbf{j},1}(Z) := \zeta_1^2 + \zeta_2^2 = Z \cdot \overline{Z} = |Z|_\mathbf{j}^2$$

and

$$\mathcal{Q}_{\mathbb{BC};\mathbf{j},2}(Z) := |\zeta_1|^2 + |\zeta_2|^2 = \frac{1}{2}(ZZ^* + \overline{Z}Z^\dagger)$$
$$= \frac{1}{2}(|Z|_\mathbf{k}^2 + |Z^\dagger|_\mathbf{k}^2) = \frac{1}{2}(|Z|_\mathbf{k}^2 + |\overline{Z}|_\mathbf{k}^2) \,,$$

where we notice that the last one coincides with $\mathcal{Q}_{\mathbb{BC};i,2}(Z)$, and both are equal to the square of the Euclidean metric.

When $\mathbb{BC}$ is interpreted as $\mathbb{D}^2$, the situation is different. Of course, one sets:

$$\mathcal{B}_{\mathbb{BC};\mathbb{D},1}(Z,W) := \mathfrak{z}_1\mathfrak{w}_1 + \mathfrak{z}_2\mathfrak{w}_2 = \frac{1}{2}(ZW^* + Z^*W)$$

and

$$\mathcal{B}_{\mathbb{BC};\mathbb{D},2}(Z,W) := \mathfrak{z}_1\mathfrak{w}_1^\circ + \mathfrak{z}_2\mathfrak{w}_2^\circ = \frac{1}{2}(ZW^\dagger + Z^*\overline{W}) = \frac{1}{2}(Z\overline{W} + Z^*W^\dagger),$$

imitating the previous situations, but now both forms take values in $\mathbb{D}$, not in $\mathbb{C}$ or $\mathbb{R}$. Note also that the first of them is hyperbolic bilinear, and the second one can be called hyperbolic sesquilinear; what is more, setting

$$\mathcal{Q}_{\mathbb{BC};\mathbb{D},1}(\mathfrak{z}) := \mathcal{B}_{\mathbb{BC};\mathbb{D},1}(Z,Z) = \mathfrak{z}_1^2 + \mathfrak{z}_2^2 = Z \cdot Z^* = |Z|_{\mathbf{k}}^2$$

and

$$\mathcal{Q}_{\mathbb{BC};\mathbb{D},2}(\mathfrak{z}) := \mathcal{B}_{\mathbb{BC};\mathbb{D},2}(Z,Z) = \mathfrak{z}_1\mathfrak{z}_1^\circ + \mathfrak{z}_2\mathfrak{z}_2^\circ$$

$$= \frac{1}{2}(ZZ^\dagger + Z^*\overline{Z}) = \frac{1}{2}(Z\overline{Z} + Z^*Z^\dagger)$$

$$= \frac{1}{2}(|Z|_{\mathbf{i}}^2 + |Z^*|_{\mathbf{i}}^2) = \frac{1}{2}(|Z|_{\mathbf{j}}^2 + |Z^*|_{\mathbf{j}}^2),$$

one sees that $\mathcal{Q}_{\mathbb{BC};\mathbb{D},1}$ has hyperbolic values and that $\mathcal{Q}_{\mathbb{BC};\mathbb{D},2}$ takes real values, but it is not positive definite. Thus, the geometry behind them is much more sophisticated. In the next chapter we will elaborate on this.

Finally, $\mathbb{BC}$ is a bicomplex module, i.e., a module over itself, which suggests the introduction of a bicomplex bilinear form

$$\mathcal{B}_{\mathbb{BC}}(Z,W) := Z \cdot W$$

and of three bicomplex sesquilinear-type forms:

$$\mathcal{B}_{\mathbb{BC},bar}(Z,W) := Z \cdot \overline{W}, \qquad \text{(a bar-sesquilinear form)},$$
$$\mathcal{B}_{\mathbb{BC},\dagger}(Z,W) := Z \cdot W^\dagger, \qquad \text{(a †-sesquilinear form)},$$
$$\mathcal{B}_{\mathbb{BC},*}(Z,W) := Z \cdot W^*, \qquad \text{(a *-sesquilinear form)}.$$

The corresponding quadratic forms coincide with the three "moduli" previously introduced, which take complex or hyperbolic values, making the geometric aspect even more complicated than the above described case of the $\mathbb{D}$-module $\mathbb{BC} = \mathbb{D}^2$. Of course, this makes both cases even more interesting and intriguing.

Let us consider again the $\mathbb{R}$-valued quadratic form $\mathcal{Q}_{\mathbb{C},2}(z) = x^2 + y^2$; since it coincides with $\mathcal{B}_{\mathbb{C},2}(z,z)$, then $\mathcal{Q}_{\mathbb{C},2}$ enjoys the factorization

$$\mathcal{Q}_{\mathbb{C},2} = (x + \mathbf{i}y)(x - \mathbf{i}y) = z \cdot \overline{z}.$$

This identity can be seen as one of the reasons for the necessity of introducing complex numbers: if one wants to factorize $\mathcal{Q}_{\mathbb{C},2}(z)$ (which is a real-valued and positive definite quadratic form; thus, in particular, the set of its values is $\mathbb{R}$-one-dimensional) into the product of two linear forms which should be real two-dimensional, then the imaginary unit $\mathbf{i}$ emerges forcedly and generates the whole set $\mathbb{C}$.

A very similar idea is related with the bicomplex numbers. Consider the $\mathbb{C}(\mathbf{i})$-valued quadratic form $\mathcal{Q}_{\mathbb{BC},\mathbf{i},1}(Z) = z_1^2 + z_2^2$. We know that it factorizes into

$$\mathcal{Q}_{\mathbb{BC},\mathbf{i},1} = (z_1 + \mathbf{j}z_2)(z_1 - \mathbf{j}z_2) = Z \cdot Z^\dagger,$$

where the set of the values of $\mathcal{Q}_{\mathbb{BC},\mathbf{i},1}$ is $\mathbb{C}(\mathbf{i})$-one-dimensional but the factors are already $\mathbb{C}(\mathbf{i})$-two-dimensional. Thus, the $\mathbb{C}(\mathbf{i})$-algebra $\mathbb{BC}$ arises from a complex quadratic form in the same way as the real algebra $\mathbb{C}$ arises from a real quadratic form. Notice that the requirement for the factors to be $\mathbb{C}(\mathbf{i})$-two-dimensional, not one-dimensional, is crucial since without it one has an obvious factorization

$$z_1^2 + z_2^2 = (z_1 + \mathbf{i}z_2)(z_1 - \mathbf{i}z_2) \qquad (2.19)$$

which does not serve our purposes.

Let us show that no other number system, but $\mathbb{BC}$, can play the same role. Assume that there exists a $\mathbb{C}(\mathbf{i})$-two-dimensional commutative algebra such that the four elements of it, say, a, b, c, d, ensure the identity

$$z_1^2 + z_2^2 = (az_1 + bz_2)(cz_1 + dz_2)$$

for all z_1 and z_2 in $\mathbb{C}(\mathbf{i})$. Hence, for all z_1 and z_2 it holds that

$$z_1^2 + z_2^2 = acz_1^2 + bdz_2^2 + (ad + bc)z_1z_2,$$

which is equivalent to

$$ac = 1; \qquad bd = 1; \qquad ad + bc = 0.$$

Thus, all the coefficients are invertible elements and

$$c^{-1}d + d^{-1}c = 0,$$

i.e.,

$$\left(c^{-1}d\right)^2 = -1.$$

Therefore, denoting $\mathbf{j} := c^{-1}d$ we have that $\mathbf{j}^2 = -1$ and $\mathbf{j}^{-1} = -\mathbf{j}$; the factorization becomes

$$z_1^2 + z_2^2 = (z_1 + \mathbf{j}z_2)(z_1 - \mathbf{j}z_2).$$

Thus, the complex algebra we are looking for should be generated by 1 and by a new element $\mathbf{j} \neq \pm\mathbf{i}$, and we have arrived exactly at $\mathbb{BC}$.

In the same way we can begin with the $\mathbb{C}(\mathbf{j})$-valued quadratic form $\zeta_1^2 + \zeta_2^2$ and get the same $\mathbb{BC}$ which now will be seen as a $\mathbb{C}(\mathbf{j})$-algebra.

Finally, if we begin with the $\mathbb{D}$-valued quadratic form

$$\mathfrak{z}_1^2 + \mathfrak{z}_2^2$$

acting on the $\mathbb{D}$-algebra $\mathbb{D}^2$, then any of the two imaginary units, $\mathbf{i}$ or $\mathbf{j}$, will arise giving the factorizations into two factors each of which is $\mathbb{D}$-two-dimensional.

2.6 A partial order on the set of hyperbolic numbers

2.6.1 Definition of the partial order

We have noticed already a deep similarity between the role of non-negative hyperbolic numbers inside $\mathbb{D}$ and the role of non-negative real numbers inside $\mathbb{R}$. It turns out that this similarity can be extended and the (partial) notions "greater than" and "less than" can be introduced on hyperbolic numbers. Take two hyperbolic numbers $\mathfrak{z}_1$ and $\mathfrak{z}_2$; if their difference $\mathfrak{z}_2 - \mathfrak{z}_1 \in \mathbb{D}^+$, that is, the difference is a non-negative hyperbolic number, then we write $\mathfrak{z}_2 \succeq \mathfrak{z}_1$ or $\mathfrak{z}_1 \preceq \mathfrak{z}_2$ and we say that $\mathfrak{z}_2$ is $\mathbb{D}$-greater than or equal to $\mathfrak{z}_1$, or that $\mathfrak{z}_1$ is $\mathbb{D}$-less than or equal to $\mathfrak{z}_2$.

Writing these hyperbolic numbers in their idempotent form $\mathfrak{z}_1 = \beta_1 \mathbf{e} + \beta_2 \mathbf{e}^\dagger$ and $\mathfrak{z}_2 = \gamma_1 \mathbf{e} + \gamma_2 \mathbf{e}^\dagger$, with real numbers β_1, β_2, γ_1 and γ_2, we have that

$$\mathfrak{z}_1 \preceq \mathfrak{z}_2 \quad \text{if and only if} \quad \gamma_1 \geq \beta_1 \quad \text{and} \quad \gamma_2 \geq \beta_2.$$

On Figure 2.6.1, $\mathfrak{z}_0 = x_0 + \mathbf{k} y_0$ is an arbitrary hyperbolic number, and one can see that the entire plane is divided into four quarters: the quarter plane of hyperbolic numbers which are $\mathbb{D}$-greater than or equal to $\mathfrak{z}_0$ ($\mathfrak{z} \succeq \mathfrak{z}_0$); the quarter plane of hyperbolic numbers which are $\mathbb{D}$-less than or equal to $\mathfrak{z}_0$ ($\mathfrak{z} \preceq \mathfrak{z}_0$); and the two quarter planes where the hyperbolic numbers are not $\mathbb{D}$-comparable with $\mathfrak{z}_0$ (neither $\mathfrak{z} \succeq \mathfrak{z}_0$ nor $\mathfrak{z} \preceq \mathfrak{z}_0$ holds).

Thus we have introduced a binary relation on $\mathbb{D}$ which is, obviously, reflexive: for any $\mathfrak{z} \in \mathbb{D}$, $\mathfrak{z} \preceq \mathfrak{z}$; transitive: if $\mathfrak{z}_1 \preceq \mathfrak{z}_2$ and $\mathfrak{z}_2 \preceq \mathfrak{z}_3$, then $\mathfrak{z}_1 \preceq \mathfrak{z}_3$; and antisymmetric: if $\mathfrak{z}_1 \preceq \mathfrak{z}_2$ and $\mathfrak{z}_2 \preceq \mathfrak{z}_1$, then $\mathfrak{z}_1 = \mathfrak{z}_2$. Since this relation is applicable not for any pair of elements in $\mathbb{D}$, then the relation $\preceq$ defines a partial order on $\mathbb{D}$. As can be expected, this partial order extends the total order $\leq$ on $\mathbb{R}$: if one takes two real numbers x_1 and x_2, $x_1 \leq x_2$, and considers them as hyperbolic numbers with zero imaginary parts, then $x_1 \preceq x_2$.

In case $\mathfrak{z}_2 - \mathfrak{z}_1 \in \mathbb{D}^+ \setminus \{0\}$ we write $\mathfrak{z}_2 \succ \mathfrak{z}_1$ and we say that $\mathfrak{z}_2$ is $\mathbb{D}$-greater than $\mathfrak{z}_1$, or we write $\mathfrak{z}_1 \prec \mathfrak{z}_2$ and say that $\mathfrak{z}_1$ is $\mathbb{D}$-less than $\mathfrak{z}_2$. This implies that $\mathfrak{z} \in \mathbb{D}^+$ is equivalent to $\mathfrak{z} \succeq 0$ and that $\mathfrak{z} \in \mathbb{D}^+ \setminus \{0\}$ is equivalent to $\mathfrak{z} \succ 0$; $\mathfrak{z} \in \mathbb{D}^-$ is equivalent to $\mathfrak{z} \preceq 0$ and $\mathfrak{z} \in \mathbb{D}^- \setminus \{0\}$ is equivalent to $\mathfrak{z} \prec 0$.

As a matter of fact, we can trace an analogy with the future and past cones in a two-dimensional space with the Minkowski metric. Specifically, let us identify a

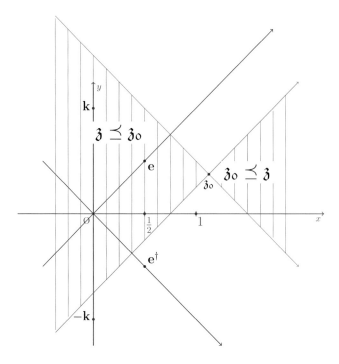

Figure 2.6.1: A PARTIAL ORDER ON $\mathbb{D}$

one-dimensional time with the x-axis and a one-dimensional space with the y-axis, both axes being embedded into $\mathbb{D}$ and assume the speed of light to be equal to one; then the set of zero-divisors with positive real part, $x > 0$ is nothing more than the future of a ray of light which was sent from the origin towards either direction. Analogously, the set of zero-divisors with negative real part, $x < 0$, represent the past of the same ray of light.

Thus, positive hyperbolic numbers represent the future cone, i. e., they correspond to the events which are in the future of the origin; the negative hyperbolic numbers, that is, those which after multiplying by (-1) become positive, represent the past cone, i.e., they correspond to the events in the past of the origin.

Accepting this interpretation, we see how the set of hyperbolic numbers which are greater than a given hyperbolic number Z represents the events in the future of the event which is represented by Z.

2.6.2   Properties of the partial order

Let us describe here some consequences of the the definition of the partial order $\preceq$. We combine them into three groups. Let $\mathfrak{z}$, $\mathfrak{w}$ and $\mathfrak{y}$ be hyperbolic numbers.

(I) Consequences of the definition itself

- If $\mathfrak{z}$ and $\mathfrak{y}$ are comparable with respect to $\preceq$, then precisely one of the following relations holds:

$$\mathfrak{z} \prec \mathfrak{y} \quad \text{or} \quad \mathfrak{z} \succ \mathfrak{y} \quad \text{or} \quad \mathfrak{z} = \mathfrak{y}.$$

- The inequalities $\mathfrak{z} \prec \mathfrak{y}$ and $\mathfrak{y} \preceq \mathfrak{w}$ imply that $\mathfrak{z} \prec \mathfrak{w}$.

- The inequalities $\mathfrak{z} \preceq \mathfrak{y}$ and $\mathfrak{y} \prec \mathfrak{w}$ imply that $\mathfrak{z} \prec \mathfrak{w}$.

(II) Connections between the addition and the order on $\mathbb{D}$

- $\mathfrak{z} \prec \mathfrak{w}$ implies that $\mathfrak{z} + \mathfrak{y} \prec \mathfrak{w} + \mathfrak{y}$.

- $\mathfrak{z} \preceq \mathfrak{w}$ implies that $\mathfrak{z} + \mathfrak{y} \preceq \mathfrak{w} + \mathfrak{y}$.

- $0 \prec \mathfrak{z}$ implies that $-\mathfrak{z} \prec 0$.

- $\mathfrak{z}_1 \preceq \mathfrak{z}_2$ and $\mathfrak{y} \preceq \mathfrak{w}$ imply that $\mathfrak{z}_1 + \mathfrak{y} \preceq \mathfrak{z}_2 + \mathfrak{w}$.

- $\mathfrak{z}_1 \preceq \mathfrak{z}_2$ and $\mathfrak{y} \prec \mathfrak{w}$ imply that $\mathfrak{z}_1 + \mathfrak{y} \prec \mathfrak{z}_2 + \mathfrak{w}$.

(III) Connections between multiplication and partial order on $\mathbb{D}$

- If $\mathfrak{z}$ and $\mathfrak{y}$ are non-negative hyperbolic numbers, then so is their product:

$$\mathfrak{z} \cdot \mathfrak{y} \in \mathbb{D}^{+}.$$

- If $\mathfrak{z}$ and $\mathfrak{y}$ are strictly positive hyperbolic numbers, then so is their product:

$$\mathfrak{z} \cdot \mathfrak{y} \in \mathbb{D}^{+} \setminus \{0\}.$$

- If $\mathfrak{z}$ and $\mathfrak{y}$ are strictly negative hyperbolic numbers, then their product is strictly positive:

$$\mathfrak{z} \cdot \mathfrak{y} \succ 0.$$

- If one of $\mathfrak{z}$ and $\mathfrak{y}$ is strictly positive and another is strictly negative, then their product is strictly negative:

$$\mathfrak{z} \cdot \mathfrak{y} \prec 0.$$

- If $\mathfrak{z} \prec \mathfrak{y}$ and $\mathfrak{w} \succ 0$, then $\mathfrak{z} \cdot \mathfrak{w} \prec \mathfrak{y} \cdot \mathfrak{w}$.

- If $\mathfrak{z} \prec \mathfrak{y}$ and $\mathfrak{w} \prec 0$, then $\mathfrak{z} \cdot \mathfrak{w} \succ \mathfrak{y} \cdot \mathfrak{w}$.

- If $\mathfrak{z}$ is a (strictly) positive hyperbolic number, then it is invertible and its inverse is also positive: if $\mathfrak{z} \succ 0$ and $\mathfrak{z} \prec \mathfrak{y}$, then $\mathfrak{y}^{-1} \succ 0$ and $\mathfrak{y}^{-1} \prec \mathfrak{z}^{-1}$.

Example 2.6.2.1. Let us illustrate the above properties solving for $\mathfrak{z} = \beta_1 \mathbf{e} + \beta_2 \mathbf{e}^{\dagger}$ in $\mathbb{D}$ the inequality

$$|\mathfrak{z}|_{\mathbf{k}} \preceq \mathfrak{w}, \tag{2.20}$$

where $\mathfrak{w} = \gamma_1 \mathbf{e} + \gamma_2 \mathbf{e}^\dagger$ is in $\mathbb{D}^+$ and $|\cdot|_\mathbf{k}$ is the $\mathbb{D}$-valued modulus. If $\mathfrak{w} = 0$ the unique solution is $\mathfrak{z} = 0$. Hence, consider $\mathfrak{w} \in \mathbb{D}^+ \setminus \{0\}$. The inequality (2.20) is equivalent to

$$|\beta_1| \mathbf{e} + |\beta_2| \mathbf{e}^\dagger \preceq \gamma_1 \mathbf{e} + \gamma_2 \mathbf{e}^\dagger$$

which in turn is equivalent to the system

$$\begin{cases} |\beta_1| \leq \gamma_1, \\ |\beta_2| \leq \gamma_2. \end{cases}$$

Thus the solutions of (2.20) are hyperbolic numbers $\mathfrak{z} = \beta_1 \mathbf{e} + \beta_2 \mathbf{e}^\dagger$ with $-\gamma_1 \leq \beta_1 \leq \gamma_1$ and $-\gamma_2 \leq \beta_2 \leq \gamma_2$. This means that the inequality (2.20) is equivalent to the double hyperbolic inequality $-\mathfrak{w} \preceq \mathfrak{z} \preceq \mathfrak{w}$.   □

We introduce now the notion of a hyperbolic interval (or hyperbolic segment). Given two hyperbolic numbers $\mathfrak{a}$ and $\mathfrak{b}$, $\mathfrak{a} \preceq \mathfrak{b}$, we set

$$[\mathfrak{a}, \mathfrak{b}]_\mathbb{D} := \big\{ \mathfrak{z} \in \mathbb{D} \,\big|\, \mathfrak{a} \preceq \mathfrak{z} \preceq \mathfrak{b} \big\}.$$

Consider now two particular cases:

- Let $\mathfrak{a} = \mathbf{k}$ and $\mathfrak{b} = 1$. Since obviously $\mathbf{k} = \mathbf{e} - \mathbf{e}^\dagger \preceq 1 = \mathbf{e} + \mathbf{e}^\dagger$, then the interval $[\mathbf{k}, 1]_\mathbb{D}$ is well defined. The inequality

$$\mathbf{k} \preceq \mathfrak{z} = \beta_1 \mathbf{e} + \beta_2 \mathbf{e}^\dagger \preceq 1$$

  gives:

$$1 \leq \beta_1 \leq 1, \qquad -1 \leq \beta_2 \leq 1.$$

  It turns out that in this case the hyperbolic interval is a one-dimensional set. See the Figure 2.6.2.

- Take now $\mathfrak{a} = \mathbf{k}$ and $\mathfrak{b} = 2$ (obviously $\mathbf{k} \prec 2$). In this case the hyperbolic interval is given by

$$[\mathbf{k}, 2]_\mathbb{D} = \big\{ \mathfrak{z} = \beta_1 \mathbf{e} + \beta_2 \mathbf{e}^\dagger \,\big|\, 1 \leq \beta_1 \leq 2 \ \text{ and } \ -1 \leq \beta_2 \leq 2 \big\},$$

  and it is now a two-dimensional set. See Figure 2.6.3.

2.6.3 $\mathbb{D}$-bounded subsets in $\mathbb{D}$.

Given a subset $\mathcal{A}$ in $\mathbb{D}$, we define as usual the notion of $\mathbb{D}$-upper and $\mathbb{D}$-lower bounds, as well as the notions of a set being $\mathbb{D}$-bounded from above, from below, and finally of a $\mathbb{D}$-bounded set. There are some fine points here. If $\mathcal{A}$ has a $\mathbb{D}$-upper or a $\mathbb{D}$-lower bound α, then this means that for any $a \in \mathcal{A}$ there holds that a is comparable with α and $a \preceq \alpha$ or $\alpha \preceq a$. But this does not mean that the elements of $\mathcal{A}$ are necessarily comparable between them; the same happens taking two $\mathbb{D}$-upper of $\mathbb{D}$-lower bounds α and β they are not always comparable.

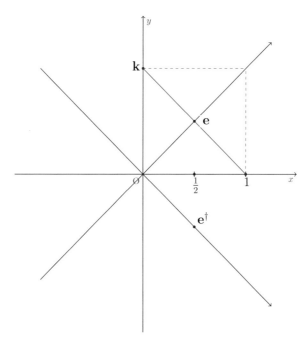

Figure 2.6.2: THE HYPERBOLIC SEGMENT $[\mathbf{k}, 1]_{\mathbb{D}}$.

If $\mathcal{A} \subset \mathbb{D}$ is a set $\mathbb{D}$-bounded from above, we define the notion of its $\mathbb{D}$-supremum, denoted by $\sup_{\mathbb{D}} \mathcal{A}$, to be the least upper bound for $\mathcal{A}$, and its $\mathbb{D}$-infimum $\inf_{\mathbb{D}} \mathcal{A}$ to be the greatest lower bound for $\mathcal{A}$. The "least" upper bound here means that $\sup_{\mathbb{D}} \mathcal{A} \preceq \alpha$ for any $\mathbb{D}$-upper bound α even if not all of the $\mathbb{D}$-upper bounds are comparable. Similarly the meaning of the "greatest" lower bound is understood. Of course, every non-empty set of hyperbolic numbers which is $\mathbb{D}$-bounded from above has its $\mathbb{D}$-supremum, and if it is $\mathbb{D}$-bounded from below, then it has its $\mathbb{D}$-infimum. This can be seen immediately if one notes that there are more convenient expressions for these notions. Given a set $\mathcal{A} \subset \mathbb{D}$, consider the sets $\mathcal{A}_1 := \{\, a_1 \mid a_1\mathbf{e} + a_2\mathbf{e}^{\dagger} \in \mathcal{A} \,\}$ and $\mathcal{A}_2 := \{\, a_2 \mid a_1\mathbf{e} + a_2\mathbf{e}^{\dagger} \in \mathcal{A} \,\}$. If $\mathcal{A}$ is $\mathbb{D}$-bounded from above, then the $\sup_{\mathbb{D}} \mathcal{A}$ can be computed by the formula

$$\sup_{\mathbb{D}}\mathcal{A} = \sup \mathcal{A}_1 \cdot \mathbf{e} + \sup \mathcal{A}_2 \cdot \mathbf{e}^{\dagger}.$$

If $\mathcal{A}$ is $\mathbb{D}$-bounded from below, then the $\inf_{\mathbb{D}} \mathcal{A}$ can be computed by the formula

$$\inf_{\mathbb{D}}\mathcal{A} = \inf \mathcal{A}_1 \cdot \mathbf{e} + \inf \mathcal{A}_2 \cdot \mathbf{e}^{\dagger}.$$

The above formulas explain a very peculiar character of the partial order on $\mathbb{D}$. Note only that although two $\mathbb{D}$-upper (or $\mathbb{D}$-lower) bounds can be incomparable nevertheless they are always comparable with $\sup_{\mathbb{D}} \mathcal{A}$ (or with $\inf_{\mathbb{D}} \mathcal{A}$).

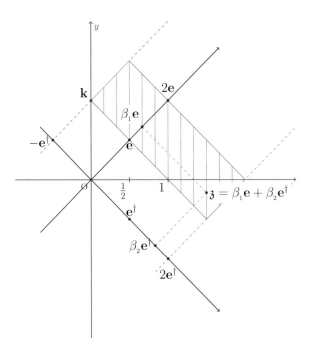

Figure 2.6.3: THE HYPERBOLIC INTERVAL $[\mathbf{k}, 2]_\mathbb{D}$.

Let $\mathcal{A}$ and $\mathcal{B}$ be two subsets of $\mathbb{D}$, denote by $-\mathcal{A}$ the set of all the elements in $\mathcal{A}$ multiplied by -1, and denote by $\mathcal{A} + \mathcal{B}$ the set of all sums $\mathfrak{z} + \mathfrak{y}$ with $\mathfrak{z} \in \mathcal{A}$ and $\mathfrak{y} \in \mathcal{B}$; define in the same fashion the set $\mathcal{A} \cdot \mathcal{B}$. The reader is invited to prove the following properties.

- A set $\mathcal{A}$ is $\mathbb{D}$-bounded from above (or from below) if and only if the set $-\mathcal{A}$ is $\mathbb{D}$-bounded from below (or from above); for such sets it holds that

$$\inf_\mathbb{D}(-\mathcal{A}) = -\sup_\mathbb{D}\mathcal{A}; \qquad \sup_\mathbb{D}(-\mathcal{A}) = -\inf_\mathbb{D}\mathcal{A}.$$

- If $\mathcal{A}$ and $\mathcal{B}$ are $\mathbb{D}$-bounded from below, then so is $\mathcal{A} + \mathcal{B}$ and for such sets one has:
$$\inf_\mathbb{D}(\mathcal{A}) + \inf_\mathbb{D}(\mathcal{B}) = \inf_\mathbb{D}(\mathcal{A} + \mathcal{B}).$$

If $\mathcal{A}$ and $\mathcal{B}$ are $\mathbb{D}$-bounded from above, then so is $\mathcal{A} + \mathcal{B}$ and for such sets one has:
$$\sup_\mathbb{D}(\mathcal{A}) + \sup_\mathbb{D}(\mathcal{B}) = \sup_\mathbb{D}(\mathcal{A} + \mathcal{B}).$$

- Assume that $\mathcal{A}$ and $\mathcal{B}$ are subsets of the set $\mathbb{D}^+$ of non-negative hyperbolic numbers. If $\mathcal{A}$ and $\mathcal{B}$ are $\mathbb{D}$-bounded from below, then so is $\mathcal{A} \cdot \mathcal{B}$ and for such

sets one has:
$$\inf_{\mathbb{D}}(\mathcal{A} \cdot \mathcal{B}) = \inf_{\mathbb{D}}(\mathcal{A}) \cdot \inf_{\mathbb{D}}(\mathcal{B}).$$

If $\mathcal{A}$ and $\mathcal{B}$ are $\mathbb{D}$-bounded from above, then so is $\mathcal{A} \cdot \mathcal{B}$ and for such sets one has:
$$\sup_{\mathbb{D}}(\mathcal{A} \cdot \mathcal{B}) = \sup_{\mathbb{D}}(\mathcal{A}) \cdot \sup_{\mathbb{D}}(\mathcal{B}).$$

2.7 The hyperbolic norm on $\mathbb{BC}$

We know already, for any bicomplex $Z = \beta_1 \mathbf{e} + \beta_2 \mathbf{e}^\dagger$, the formula
$$|Z|_{\mathbf{k}} := |\beta_1| \mathbf{e} + |\beta_2| \mathbf{e}^\dagger.$$

That is, we have the map
$$|\cdot|_{\mathbf{k}} : \mathbb{BC} \longrightarrow \mathbb{D}^+$$

with the properties:

(I) $|Z|_{\mathbf{k}} = 0$ if and only if $Z = 0$;

(II) $|Z \cdot W|_{\mathbf{k}} = |Z|_{\mathbf{k}} \cdot |W|_{\mathbf{k}}$ for any $Z, W \in \mathbb{BC}$;

(III) $|Z + W|_{\mathbf{k}} \preceq |Z|_{\mathbf{k}} + |W|_{\mathbf{k}}$.

The first two properties are clear. Let us prove (III).

$$\begin{aligned}
|Z + W|_{\mathbf{k}} &= |(\beta_1 + \nu_1) \cdot \mathbf{e} + (\beta_2 + \nu_2) \cdot \mathbf{e}^\dagger|_{\mathbf{k}} \\
&= |\beta_1 + \nu_1| \cdot \mathbf{e} + |\beta_2 + \nu_2| \cdot \mathbf{e}^\dagger \\
&\preceq (|\beta_1| + |\nu_1|) \cdot \mathbf{e} + (|\beta_2| + |\nu_2|) \cdot \mathbf{e}^\dagger \\
&= |Z|_{\mathbf{k}} + |W|_{\mathbf{k}}.
\end{aligned}$$

The three properties (I)–(III) manifest again the analogy between real positive numbers and hyperbolic positive numbers; now we see that the hyperbolic modulus of a bicomplex number has exactly the same properties as the real modulus of a complex number whenever the partial order $\preceq$ is used instead of $\leq$.

Because of properties (I)–(III) we will say that $|\cdot|_{\mathbf{k}}$ is the hyperbolic-valued ($\mathbb{D}$-valued) norm on the $\mathbb{BC}$-module $\mathbb{BC}$.

It is instructive to compare (II) with (1.13) where the norm of the product and the product of the norms are related with an inequality. We believe that one could say that the hyperbolic norm of bicomplex numbers is better suited to the algebraic structure of the latter although, of course, one has to allow hyperbolic values for the norm.

Remark 2.7.1. (1) *Since for any $Z \in \mathbb{BC}$ it holds that*

$$|Z|_{\mathbf{k}} \preceq \sqrt{2} \cdot |Z|, \tag{2.21}$$

with $|Z|$ the Euclidean norm of Z, then one has:

$$|Z \cdot W|_{\mathbf{k}} \preceq \sqrt{2} \cdot |Z| \cdot |W|_{\mathbf{k}}.$$

In contrast with property (II) above, this inequality involves both the Euclidean and the hyperbolic norms.

(2) *Take $\mathfrak{z}_1$ and $\mathfrak{z}_2$ in $\mathbb{D}^+$, then clearly*

$$\mathfrak{z}_1 \preceq \mathfrak{z}_2 \quad \text{implies that} \quad |\mathfrak{z}_1| \leq |\mathfrak{z}_2|. \tag{2.22}$$

(3) *Note that the definition of hyperbolic norm for a bicomplex number Z does not depend on the choice of its idempotent representation. We have used, for $Z \in \mathbb{BC}$, the idempotent representation $Z = \beta_1 \mathbf{e} + \beta_2 \mathbf{e}^\dagger$, with β_1 and β_2 in $\mathbb{C}(\mathbf{i})$. If we had started with the idempotent representation $Z = \gamma_1 \mathbf{e} + \gamma_2 \mathbf{e}^\dagger$, with γ_1 and γ_2 in $\mathbb{C}(\mathbf{j})$, then we would have arrived at the same definition of the hyperbolic norm since $|\beta_1| = |\gamma_1|$ and $|\beta_2| = |\gamma_2|$.*

(4) *The comparison of the Euclidean norm $|Z|$ and the $\mathbb{D}$-valued norm $|Z|_{\mathbf{k}}$ of a bicomplex number Z gives:*

$$\||Z|_{\mathbf{k}}\| = \frac{1}{\sqrt{2}} \sqrt{|\beta_1|^2 + |\beta_2|^2} = |Z| \tag{2.23}$$

where the left-hand side is the Euclidean norm of a hyperbolic number.

2.7.1   Multiplicative groups of hyperbolic and bicomplex numbers

The set $\mathbb{R}^+$ of strictly positive real numbers is a multiplicative group. For the set $\mathbb{D}^+$ of non-negative hyperbolic numbers the situation is more delicate although it preserves some analogies.

Introduce the set $\mathbb{D}^+_{inv} := \mathbb{D}^+ \setminus \mathfrak{S}_0$ of all strictly positive hyperbolic numbers. This set has the following properties:

- if λ_1 and λ_2 are in $\mathbb{D}^+_{inv}$, then $\lambda_1 \cdot \lambda_2 \in \mathbb{D}^+_{inv}$;

- $1 \in \mathbb{D}^+_{inv}$;

- if $\lambda \in \mathbb{D}^+_{inv}$, then λ is invertible in $\mathbb{D}$ and $\lambda^{-1} \in \mathbb{D}^+_{inv}$.

Hence, $\mathbb{D}^+_{inv}$ is a multiplicative group with respect to the hyperbolic multiplication and thus it is an exact analogue of $\mathbb{R}^+$; what is more, $\mathbb{R}^+$ is a subgroup of $\mathbb{D}^+_{inv}$.

The zero-divisors in $\mathbb{D}^+$ are either of the form $\mathfrak{z} = \lambda \mathbf{e}$ or of the form $\mathfrak{z} = \mu \mathbf{e}^\dagger$ where λ and μ are positive real numbers; so we will use the notations $\mathbb{D}^+_{\mathbf{e}} := \{\lambda \mathbf{e} \mid \lambda > 0\}$ and $\mathbb{D}^+_{\mathbf{e}^\dagger} := \{\mu \mathbf{e}^\dagger \mid \mu > 0\}$. Both sets of semi-positive hyperbolic numbers are closed under hyperbolic multiplication but neither of them contains the number one.

It turns out that, anyway, both $\mathbb{D}_{\mathbf{e}}^{+}$ and $\mathbb{D}_{\mathbf{e}^{\dagger}}^{+}$ can be endowed with the structure of a multiplicative group. Beginning with $\mathbb{D}_{\mathbf{e}}^{+}$ one observes that for any $\lambda > 0$ one has that $\lambda\mathbf{e}\cdot\mathbf{e} = \lambda\mathbf{e}$, and $\frac{1}{\lambda}\mathbf{e}\cdot\lambda\mathbf{e} = 1\cdot\mathbf{e} = \mathbf{e}$. Hence, if we endow $\mathbb{D}_{\mathbf{e}}^{+}$ with the multiplication $\star$ which is the restriction of the hyperbolic multiplication, then on $(\mathbb{D}_{\mathbf{e}}^{+}, \star)$ we have that

1. if $\lambda_1\mathbf{e}$ and $\lambda_2\mathbf{e}$ are in $\mathbb{D}_{\mathbf{e}}^{+}$, then their product $\lambda_1\mathbf{e}\star\lambda_2\mathbf{e} = \lambda_1\lambda_2\mathbf{e}$ is in $(\mathbb{D}_{\mathbf{e}}^{+}, \star)$;

2. the element $1\cdot\mathbf{e} = \mathbf{e}$ serves as the unit $1_\star$ for the multiplication $\star$:

$$1_\star \star \lambda\mathbf{e} = \mathbf{e}\cdot\lambda\mathbf{e} = \lambda\mathbf{e};$$

3. if $\lambda\mathbf{e} \in (\mathbb{D}_{\mathbf{e}}^{+}, \star)$, then $\frac{1}{\lambda}\mathbf{e}$ is its $\star$-inverse:

$$\lambda\mathbf{e} \star \frac{1}{\lambda}\mathbf{e} = 1\cdot\mathbf{e} = \mathbf{e} = 1_\star.$$

Thus, we conclude that $(\mathbb{D}_{\mathbf{e}}^{+}, \star)$ is a multiplicative group. The same reasoning applies to the set $\mathbb{D}_{\mathbf{e}^{\dagger}}^{+}$ with obvious changes. Each of these groups is isomorphic to the group $\mathbb{R}^{+}$.

Obviously, the sets $\mathbb{D}_{inv} := \mathbb{D} \setminus \mathfrak{S}_0$ and $\mathbb{BC}_{inv} := \mathbb{BC} \setminus \mathfrak{S}_0$ are multiplicative groups with their respective multiplications, thus

$$\mathbb{D}_{inv}^{+} \subset \mathbb{D}_{inv} \subset \mathbb{BC}_{inv},$$

where $\subset$ means the embedding of group structures. At the same time we can use the same arguments on $\mathbb{D}_{\mathbf{e}} := \mathbf{e}\cdot\mathbb{D}\setminus\{0\} = \{\lambda\mathbf{e} \mid \lambda \in \mathbb{R}\setminus\{0\}\}$, $\mathbb{D}_{\mathbf{e}^{\dagger}} := \mathbf{e}^{\dagger}\cdot\mathbb{D}\setminus\{0\} = \{\mu\mathbf{e}^{\dagger} \mid \mu \in \mathbb{R}\setminus\{0\}\}$, $\mathbb{BC}_{\mathbf{e}} \setminus \{0\} := \{\lambda\mathbf{e} \mid \lambda \in \mathbb{C}\setminus\{0\}\}$, $\mathbb{BC}_{\mathbf{e}^{\dagger}} \setminus \{0\} := \{\mu\mathbf{e}^{\dagger} \mid \mu \in \mathbb{C}\setminus\{0\}\}$. The first two of them are isomorphic to the multiplicative group of real numbers, the last two are isomorphic to the multiplicative group of complex numbers.

We leave the details to the reader.

In addition to the references from the end of Chapter 1, such as [45, 56, 57, 58, 65, 85], where some of the algebraic structures that we developed in this chapter have been introduced and studied, the most important contribution in this area is the book of Alpay, Luna–Elizarrarás, Shapiro and Struppa [2]. This study introduces for the first time the notion of the hyperbolic-valued norm on the ring of bicomplex numbers, which we fully described in this chapter.

Chapter 3

Geometry and Trigonometric Representations of Bicomplex Numbers

The geometry of complex numbers coincides with the geometry of the Euclidean space $\mathbb{R}^2$, and this is because of a good compatibility between the algebraic structure of $\mathbb{C}$ and the geometry of $\mathbb{R}^2$, which is expressed by the equality

$$z \cdot \overline{z} = x^2 + y^2 = |z|^2. \tag{3.1}$$

Moreover, the algebraic operations on complex numbers can be easily interpreted in terms of the geometry of the plane $\mathbb{R}^2$; for instance, to multiply by a complex number is equivalent to realizing a composition of a rotation and a homothety.

One would want to extend this idea for the Euclidean space $\mathbb{C}^2$. But it turns out that the algebraic properties of quaternions, not those of bicomplex numbers, are compatible with the Euclidean structure of $\mathbb{C}^2$ in the same way as the complex numbers are compatible with the Euclidean structure of $\mathbb{R}^2 - \mathbb{C}$.

As for the bicomplex numbers, the situation is much more sophisticated since now we have several quadratic forms to deal with. Indeed, we know already that the analogues of (3.1) are:

$$Z \cdot Z^\dagger = z_1^2 + z_2^2 = |Z|_{\mathbf{i}}^2 \in \mathbb{C}(\mathbf{i}), \quad z_1, z_2 \in \mathbb{C}(\mathbf{i}); \tag{3.2}$$

$$Z \cdot \overline{Z} = \zeta_1^2 + \zeta_2^2 = |Z|_{\mathbf{j}}^2 \in \mathbb{C}(\mathbf{j}), \quad \zeta_1, \zeta_2 \in \mathbb{C}(\mathbf{j}); \tag{3.3}$$

$$Z \cdot Z^* = \mathfrak{z}_1^2 + \mathfrak{z}_2^2 = |Z|_{\mathbf{k}}^2 \in \mathbb{D}, \quad \mathfrak{z}_1, \mathfrak{z}_2 \in \mathbb{D}. \tag{3.4}$$

In addition, the usual Euclidean structure of $\mathbb{BC} = \mathbb{C}^2 = \mathbb{R}^4$ yields the corresponding quadratic form

$$x_1^2 + y_1^2 + x_2^2 + y_2^2.$$

This means that the "authentic" geometry of bicomplex numbers is related to two complex quadratic forms (one is $\mathbb{C}(\mathbf{i})$-valued, the other is $\mathbb{C}(\mathbf{j})$-valued), one $\mathbb{D}$-valued quadratic form and one real-valued quadratic form.

3.1   Drawing and thinking in $\mathbb{R}^4$

In this chapter we are going to analyze the geometry of the bicomplex numbers. But we begin first with the description of some usual Euclidean geometrical objects in $\mathbb{R}^4$ and we will describe them using the bicomplex numbers; this is because $\mathbb{BC}$ coincides with $\mathbb{R}^4$ when it is seen as an $\mathbb{R}$-linear space and thus we will need to imagine geometrical objects in $\mathbb{R}^4$.

In order to realize the peculiarities of the four-dimensional cube let us consider first the cubes in lower dimensions. In $\mathbb{R}^1$ the cube with a vertex at the origin and of length one is the segment $[0, 1]$ (see Figure 3.1.1).

Figure 3.1.1:   THE ONE-DIMENSIONAL CUBE.

Using this segment, one constructs the two-dimensional cube pasting perpendicularly the cube of the dimension one at each point of $[0, 1]$, i.e., pasting the translations of $[0, 1]$ and thus obtaining the square with unitary sides. See Figure 3.1.2.

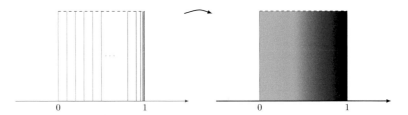

Figure 3.1.2:   THE TWO-DIMENSIONAL CUBE.

Extending this idea, we construct next the three-dimensional cube of side one by attaching at each point of $[0, 1]$ perpendicularly the cube of the previous dimension, that is, the squares of side one are attached. See Figure 3.1.3.

To construct the cube in four dimensions we repeat the procedure: the three-dimensional cube of side one is attached now, perpendicularly, to each point of $[0, 1]$. See figure 3.1.4.

In a sense, such a way of constructing contradicts our 3-dimensional intuition but since we have now an additional dimension the arising 3-dimensional cubes are parallel and they do not intersect at all.

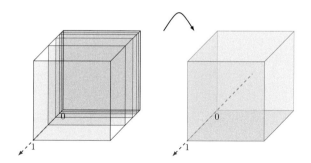

Figure 3.1.3: THE THREE-DIMENSIONAL CUBE.

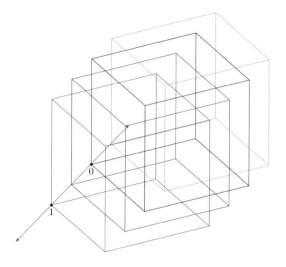

Figure 3.1.4: THE FOUR-DIMENSIONAL CUBE.

The same idea allows us to visualize the whole 4-dimensional space with the coordinate axes x_0, x_1, x_2, x_3: to each point of the axis x_0, we attach, perpendicularly a copy of $\mathbb{R}^3$. See Figure 3.1.5.

There is an alternative way of "seeing" the 4-dimensional world. Take, for instance, the real 2-dimensional plane spanned by 1 and $\mathbf{i}$; to each point of it we attach perpendicularly a copy of the 2-dimensional plane spanned by $\mathbf{j}$ and $\mathbf{k}$.

Note that whenever we mention the perpendicularity we mean, implicitly, that the two sets pass through the origin and they are the orthogonal complements of each other; if they do not pass through the origin, then they become the

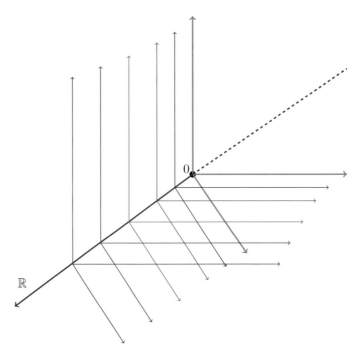

Figure 3.1.5: $\mathbb{R}^4$ AS AN INFINITE UNION OF COPIES OF $\mathbb{R}^3$.

orthogonal complements of each other if they are translated to the origin. This
is reasonably clear on the level of intuition for the 3-dimensional situation, and
for the 4-dimensional case one can appeal to analogies in three dimensions; for
instance, an analogue of the two planes which are perpendicular in $\mathbb{R}^4$ may be two
straight lines which are perpendicular in $\mathbb{R}^2$.

Traditionally, the 4-dimensional cube is presented as a solid in $\mathbb{R}^4$ whose 3-
dimensional surface consists of eight 3-dimensional cubes which are glued together
in a very particular way. We will see now that our approach does not contradict
this vision. The matter is that if instead of pasting a 3-dimensional cube at each
point of the interval $[0, 1]$ on the axis x_0, the process can be repeated pasting cubes
at each point of the interval $[0, 1]$ but now on the axis x_1, or on the axis x_2, or on
the axis x_3 (we will denote for short the corresponding intervals as $[0, \mathbf{i}]$, $[0, \mathbf{j}]$ and
$[0, \mathbf{k}]$), obtaining the same geometrical figure. With this in mind, let us introduce
the notations:

- C_r is the cube attached to the point $r \in [0, 1]$.

- I_s is the cube attached to the point $s \in [0, \mathbf{i}]$.

- J_u is the cube attached to the point $u \in [0, \mathbf{j}]$.

- K_v is the cube attached to the point $v \in [0, \mathbf{k}]$.

Proposition 3.1.1. *The (topological) boundary of the 4-dimensional cube is the union of the 3-dimensional cubes C_0, C_1, I_0, $I_\mathbf{i}$, J_0, $J_\mathbf{j}$, K_0 and $K_\mathbf{k}$.*

Proof. Follows by noting that the cubes C_0, etc., are the "extreme cubes" when the described process applies to each of the following segments: $[0, 1]$, $[0, \mathbf{i}]$, $[0, \mathbf{j}]$, $[0, \mathbf{k}]$. □

In order to see more precisely how the vertices of the cubes are glued together, we will write down the vertices explicitly.

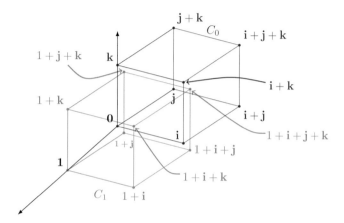

Figure 3.1.6:   THE CUBES C_0 AND C_1 AND THEIR VERTICES.

The vertices of C_0 are: 0, $\mathbf{i}$, $\mathbf{j}$, $\mathbf{k}$, $\mathbf{i}+\mathbf{j}$, $\mathbf{i}+\mathbf{k}$, $\mathbf{j}+\mathbf{k}$, $\mathbf{i}+\mathbf{j}+\mathbf{k}$. The vertices of C_1 are: 1, $1+\mathbf{i}$, $1+\mathbf{j}$, $1+\mathbf{k}$, $1+\mathbf{i}+\mathbf{j}$, $1+\mathbf{i}+\mathbf{k}$, $1+\mathbf{j}+\mathbf{k}$, $1+\mathbf{i}+\mathbf{j}+\mathbf{k}$.

The vertices of I_0 are: 0, $\mathbf{j}$, $1+\mathbf{j}$, 1, $\mathbf{k}$, $\mathbf{j}+\mathbf{k}$, $1+\mathbf{j}+\mathbf{k}$, $1+\mathbf{k}$. The vertices of $I_\mathbf{i}$ are: $\mathbf{i}$, $\mathbf{i}+\mathbf{j}$, $1+\mathbf{i}+\mathbf{j}$, $\mathbf{i}+1$, $\mathbf{i}+\mathbf{k}$, $\mathbf{i}+\mathbf{j}+\mathbf{k}$, $1+\mathbf{i}+\mathbf{j}+\mathbf{k}$, $1+\mathbf{i}+\mathbf{k}$.

As one could expect, C_0 and C_1 are disjoint sets as well as I_0 and $I_\mathbf{i}$. Nevertheless, C_0 and I_0 have the following common vertices: 0, $\mathbf{j}$, $\mathbf{k}$, $\mathbf{j}+\mathbf{k}$. This means that they are glued together along the square with these vertices. Similarly C_0 and $I_\mathbf{i}$ are glued together along the square with vertices $\mathbf{i}$, $\mathbf{i}+\mathbf{j}$, $\mathbf{i}+\mathbf{k}$, $\mathbf{i}+\mathbf{j}+\mathbf{k}$.

The common square of the cubes C_1 and I_0 has the vertices 1, $1+\mathbf{j}$, $1+\mathbf{k}$, $1+\mathbf{j}+\mathbf{k}$; the common square of the cubes C_1 and $I_\mathbf{i}$ has the vertices $1+\mathbf{i}$, $1+\mathbf{i}+\mathbf{j}$, $1+\mathbf{i}+\mathbf{j}$, $1+\mathbf{i}+\mathbf{j}+\mathbf{k}$.

The vertices of J_0 are: 0, 1, $\mathbf{i}$, $1+\mathbf{i}$, $\mathbf{k}$, $1+\mathbf{k}$, $\mathbf{i}+\mathbf{k}$ and $1+\mathbf{i}+\mathbf{k}$. The vertices of $J_\mathbf{j}$ are: $\mathbf{j}$, $1+\mathbf{j}$, $\mathbf{i}+\mathbf{j}$, $1+\mathbf{i}+\mathbf{j}$, $\mathbf{j}+\mathbf{k}$, $1+\mathbf{j}+\mathbf{k}$, $\mathbf{i}+\mathbf{j}+\mathbf{k}$ and $1+\mathbf{i}+\mathbf{j}+\mathbf{k}$.

The vertices of K_0 are: 0, 1, $\mathbf{i}$, $\mathbf{j}$, $1+\mathbf{i}$, $\mathbf{i}+\mathbf{j}$, $1+\mathbf{i}+\mathbf{j}$, $1+\mathbf{j}$; and the vertices of $K_\mathbf{k}$ are: $\mathbf{k}$, $1+\mathbf{k}$, $\mathbf{i}+\mathbf{k}$, $\mathbf{j}+\mathbf{k}$, $1+\mathbf{i}+\mathbf{k}$, $\mathbf{i}+\mathbf{j}+\mathbf{k}$, $1+\mathbf{i}+\mathbf{j}+\mathbf{k}$, $1+\mathbf{j}+\mathbf{k}$.

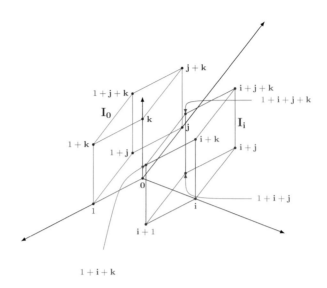

Figure 3.1.7: THE CUBES I_0 AND $I_{\mathbf{i}}$ AND THEIR VERTICES.

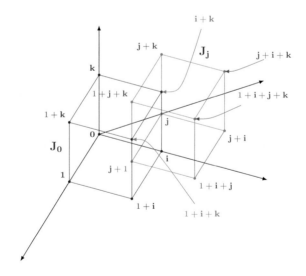

Figure 3.1.8: THE CUBES J_0 AND $J_{\mathbf{j}}$ AND THEIR VERTICES.

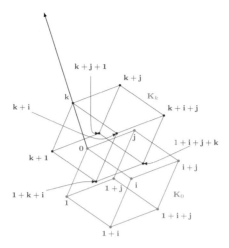

Figure 3.1.9: THE CUBES K_0 AND $K_\mathbf{k}$ AND THEIR VERTICES.

Table 3.1 shows:

- the vertices of the 4-dimensional cube (there are sixteen of them).

- The vertices of each of its 3-dimensional sides.

- At which vertices the 3-dimensional sides intersect.

Observe that the first column gives all the vertices and the other columns show which cube corresponds to each vertex; besides, each row shows the cubes with a fixed vertex.

3.2 Trigonometric representation in complex terms

Any complex number $z \neq 0$ has a trigonometric form which involves the modulus (a real non-negative number) and the argument (also a real number). In this section we will show that bicomplex numbers which are neither zero nor zero-divisors have a trigonometric representation, where the modulus and the argument are complex numbers. We will see in the next sections that there exists also a trigonometric representation where the modulus and the argument are hyperbolic numbers: note that such a representation is available as well for zero-divisors.

Thus, take $Z \in \mathbb{BC} \setminus \mathfrak{S}_0$, that is, $|Z|_\mathbf{i} \neq 0$. Then we write:

$$Z = z_1 + \mathbf{j}z_2 = |Z|_\mathbf{i} \left(\frac{z_1}{|Z|_\mathbf{i}} + \mathbf{j}\frac{z_2}{|Z|_\mathbf{i}} \right) . \tag{3.5}$$

Table 3.1:   THE VERTICES OF THE FOUR-DIMENSIONAL CUBE.

0	C_0		I_0		J_0		K_0	
1		C_1	I_0		J_0		K_0	
i	C_0			$I_\mathbf{i}$	J_0		K_0	
j	C_0		I_0			$J_\mathbf{j}$	K_0	
k	C_0		I_0		J_0			$K_\mathbf{k}$
$1+\mathbf{i}$		C_1		$I_\mathbf{i}$	J_0		K_0	
$1+\mathbf{j}$		C_1	I_0			$J_\mathbf{j}$	K_0	
$1+\mathbf{k}$		C_1	I_0		J_0			$K_\mathbf{k}$
$\mathbf{i}+\mathbf{j}$	C_0			$I_\mathbf{i}$		$J_\mathbf{j}$	K_0	
$\mathbf{i}+\mathbf{k}$	C_0			$I_\mathbf{i}$	J_0			$K_\mathbf{k}$
$\mathbf{j}+\mathbf{k}$	C_0		I_0			$J_\mathbf{j}$		$K_\mathbf{k}$
$1+\mathbf{i}+\mathbf{j}$		C_1		$I_\mathbf{i}$		$J_\mathbf{j}$	K_0	
$1+\mathbf{i}+\mathbf{k}$		C_1		$I_\mathbf{i}$	J_0			$K_\mathbf{k}$
$1+\mathbf{j}+\mathbf{k}$		C_1	I_0			$J_\mathbf{j}$		$K_\mathbf{k}$
$1+\mathbf{i}+\mathbf{j}+\mathbf{k}$		C_1		$I_\mathbf{i}$		$J_\mathbf{j}$		$K_\mathbf{k}$
$\mathbf{i}+\mathbf{j}+\mathbf{k}$	C_0			$I_\mathbf{i}$		$J_\mathbf{j}$		$K_\mathbf{k}$

Since

$$\left(\frac{z_1}{|Z|_\mathbf{i}}\right)^2 + \left(\frac{z_2}{|Z|_\mathbf{i}}\right)^2 = 1,$$

the system

$$\cos\Theta = \frac{z_1}{|Z|_\mathbf{i}}, \qquad \sin\Theta = \frac{z_2}{|Z|_\mathbf{i}}, \tag{3.6}$$

has at least a solution $\Theta \in \mathbb{C}(\mathbf{i})$. Actually there are infinitely many solutions, due to the periodicity of the complex cosine and sine functions.

Notice that if Θ_0 is a solution of the system (3.6), then the equality $\tan \Theta_0 = \frac{z_2}{z_1}$ implies that $\frac{z_2}{z_1} \neq \pm \mathbf{i}$ (it is known that the function $\tan z$ takes any value in $\mathbb{C}$ except $\pm \mathbf{i}$), but this is equivalent to saying that Z is not a zero-divisor.

In analogy with the complex case, we call any solution of the system (3.6) **a complex argument of the bicomplex number** Z, and we denote by $\mathrm{Arg}_{\mathbb{C}(\mathbf{i})}(Z)$ the set of all these solutions. If we denote by $\Theta_0 = \theta_{01} + \mathbf{i}\theta_{02} \in \mathbb{C}(\mathbf{i})$ the particular solution obtained by choosing its real part θ_{01} in the interval $[0, 2\pi)$, which is called the *principal value* of the complex argument, then

$$\mathrm{Arg}_{\mathbb{C}(\mathbf{i})}(Z) = \{\Theta_0 + 2m\pi \,|\, m \in \mathbb{Z}\}.$$

For the principal value of the complex argument, we will use the notation $\mathrm{arg}_{\mathbb{C}(\mathbf{i})}(Z)$. For a fixed $m \in \mathbb{Z}$, we denote by $\mathrm{arg}_{\mathbb{C}(\mathbf{i}),m}(Z) := \mathrm{arg}_{\mathbb{C}(\mathbf{i})}(Z) + 2m\pi$, thus

$$\mathrm{Arg}_{\mathbb{C}(\mathbf{i})}(Z) = \left\{ \mathrm{arg}_{\mathbb{C}(\mathbf{i}),m}(Z) \,\middle|\, m \in \mathbb{Z} \right\}.$$

For an invertible bicomplex number Z, we obtain its $\mathbb{BC}_{\mathbf{i}}$-*trigonometric form*:

$$\boxed{Z = |Z|_{\mathbf{i}}(\cos \Theta + \mathbf{j} \sin \Theta),} \tag{3.7}$$

where Θ is an arbitrary value of the complex argument.

For complex numbers a representation of the form

$$z = r\left(\cos \alpha + \mathbf{i} \sin \alpha\right), \quad r > 0,$$

implies that $r = |z|$ and $\alpha \in \mathrm{Arg}(z)$. The bicomplex situation is more sophisticated. If a bicomplex number has a representation of the form

$$Z = C \cdot (\cos \alpha + \mathbf{j} \sin \alpha)$$

with C and α complex numbers in $\mathbb{C}(\mathbf{i})$, then C is not necessarily the complex modulus of Z; it is so if and only if C is in the upper half-plane and in this case $\alpha \in \mathrm{Arg}_{\mathbb{C}(\mathbf{i})}(Z)$. If C is in the lower half-plane, then $|Z|_{\mathbf{i}} = -C$ and $\alpha + \pi \in \mathrm{Arg}_{\mathbb{C}(\mathbf{i})}(Z)$: this is because

$$Z = C\left(\cos \alpha + \mathbf{j} \sin \alpha\right) = -C\left(-\cos \alpha - \mathbf{j} \sin \alpha\right)$$
$$= -C\left(\cos(\alpha + \pi) + \mathbf{j} \sin(\alpha + \pi)\right).$$

We illustrate this phenomenon with the computation of the representation (3.7) for $Z = z_1 \in \mathbb{C}(\mathbf{i})$. By definition

$$|Z|_{\mathbf{i}} = |z_1|_{\mathbf{i}} = \sqrt{z_1^2} = \begin{cases} z_1 & \text{if } z_1 \text{ is in the upper half-plane;} \\ -z_1 & \text{if } z_1 \text{ is in the lower half-plane.} \end{cases}$$

Since $Z = z_1 = z_1 (\cos 0 + \mathbf{j} \sin 0)$, then

$$\mathrm{Arg}_{\mathbb{C}(\mathbf{i})}(Z) = \mathrm{Arg}_{\mathbb{C}(\mathbf{i})}(z_1)$$

$$= \begin{cases} \{0 + 2\pi m = 2\pi m \,\big|\, m \in \mathbb{Z}\} & \text{if } z_1 \text{ is in the upper half-plane;} \\ \{0 + \pi + 2\pi m = (2m+1)\pi \,\big|\, m \in \mathbb{Z}\} & \text{if } z_1 \text{ is in the lower half-plane.} \end{cases}$$

Hence the trigonometric representation of any complex number in $\mathbb{C}(\mathbf{i})$ is: if z_1 belongs to the upper half-plane, then

$$z_1 = z_1 (\cos(2\pi m) + \mathbf{j} \sin(2\pi m)), \quad m \in \mathbb{Z};$$

if z_1 belongs to the lower half-plane, then

$$z_1 = -z_1 (\cos(2m+1)\pi + \mathbf{j} \sin(2m+1)\pi), \quad m \in \mathbb{Z}.$$

Remark 3.2.1. *We used the names upper half-plane and lower half-plane as follows: the upper half-plane for us is*

$$\Pi^+ := \left\{ z \in \mathbb{C}(\mathbf{i}) \,\big|\, Im(z) > 0 \text{ or } z \in [0, \infty) \right\},$$

and the lower half-plane is

$$\Pi^- := \left\{ z \in \mathbb{C}(\mathbf{i}) \,\big|\, Im(z) < 0 \text{ or } z \in (-\infty, 0) \right\}.$$

Thus $z \in \Pi^+$ if and only if $|z|_\mathbf{i} = z$ and $z \in \Pi^-$ if and only if $|z|_\mathbf{i} = -z$. Besides if z is a real number, then $|z|_\mathbf{i} = |z|$, the usual real-valued modulus of real numbers. Hence the complex modulus of a complex number extends the modulus of real numbers, not the usual modulus of complex numbers.

Later on, we will see that the right-hand side of (3.7) is related with the bicomplex exponential function, which will be explained in Chapter 6.

Note that if $Z \notin \mathfrak{S}_0$, then the complex moduli of Z and $Z^\dagger$ are the same, which leads to

$$Z^\dagger = z_1 - \mathbf{j} z_2 = |Z|_\mathbf{i} \left(\frac{z_1}{|Z|_\mathbf{i}} - \mathbf{j} \frac{z_2}{|Z|_\mathbf{i}} \right)$$

$$= |Z|_\mathbf{i}(\cos \Theta - \mathbf{j} \sin \Theta)$$

with Θ from (3.7).

Given Z as in (3.7) and $W = |W|_\mathbf{i}(\cos \Omega + \mathbf{j} \sin \Omega)$, one has:

$$Z \cdot W = |Z|_\mathbf{i}(\cos \Theta + \mathbf{j} \sin \Theta) \cdot |W|_\mathbf{i}(\cos \Omega + \mathbf{j} \sin \Omega)$$

$$= |Z|_\mathbf{i}|W|_\mathbf{i} ((\cos \Theta \cos \Omega - \sin \Theta \sin \Omega) + \mathbf{j}(\cos \Theta \sin \Omega + \sin \Theta \cos \Omega)),$$

and therefore

$$Z \cdot W = |Z|_\mathbf{i}|W|_\mathbf{i}(\cos(\Theta + \Omega) + \mathbf{j} \sin(\Theta + \Omega)). \tag{3.8}$$

As we have previously explained, formula (3.8) does not necessarily mean that $|Z \cdot W|_\mathbf{i}$ is equal to $|Z|_\mathbf{i} \cdot |W|_\mathbf{i}$ and that $\Theta + \Omega$ belongs to $\mathrm{Arg}_{\mathbb{C}(\mathbf{i})}(Z \cdot W)$. What we can conclude is:

(a) if the complex number $|Z|_{\mathbf{i}} \cdot |W|_{\mathbf{i}}$ is in the upper half-plane, then indeed

$$|Z \cdot W|_{\mathbf{i}} = |Z|_{\mathbf{i}} \cdot |W|_{\mathbf{i}};$$

besides, in this case

$$\Theta + \Omega \in \operatorname{Arg}_{\mathbb{C}(\mathbf{i})}(Z \cdot W)$$

and

$$\operatorname{Arg}_{\mathbb{C}(\mathbf{i})}(Z \cdot W) = \operatorname{Arg}_{\mathbb{C}(\mathbf{i})}(Z) + \operatorname{Arg}_{\mathbb{C}(\mathbf{i})}(W);$$

(b) if the complex number $|Z|_{\mathbf{i}} \cdot |W|_{\mathbf{i}}$ is not in the upper half-plane, then

$$|Z \cdot W|_{\mathbf{i}} = -|Z|_{\mathbf{i}} \cdot |W|_{\mathbf{i}};$$

besides, in this case

$$\Theta + \Omega + \pi \in \operatorname{Arg}_{\mathbb{C}(\mathbf{i})}(Z \cdot W)$$

and

$$\operatorname{Arg}_{\mathbb{C}(\mathbf{i})}(Z \cdot W) + \pi = \operatorname{Arg}_{\mathbb{C}(\mathbf{i})}(Z) + \operatorname{Arg}_{\mathbb{C}(\mathbf{i})}(W).$$

The $\mathbb{BC}_{\mathbf{i}}$-analogue of De Moivre formula is a less direct generalization of its complex antecedent. On one hand, we can write, obviously, that

$$Z^n = |Z|_{\mathbf{i}}^n \left(\cos \Theta_{Z^n} + \mathbf{j} \sin \Theta_{Z^n} \right).$$

On the other hand one can obtain by induction that

$$Z^n = |Z^n|_{\mathbf{i}} \left(\cos(n\Theta) + \mathbf{j} \sin(n\Theta) \right), \tag{3.9}$$

but again not always $|Z|_{\mathbf{i}}^n$ and $|Z^n|_{\mathbf{i}}$, as well as $n\Theta$ and Θ_{Z^n}, coincide. They do coincide when the complex number $|Z|_{\mathbf{i}}^n$ is in the upper half-plane; if it is not, then

$$|Z^n|_{\mathbf{i}} = -|Z|_{\mathbf{i}}^n \quad \text{and} \quad \Theta_{Z^n} = n\Theta + \pi.$$

We will call formula (3.9) the bicomplex De Moivre formula, although understanding that it does not always coincide with the trigonometric representation in complex terms of Z^n.

Since for an invertible Z there holds:

$$0 \neq Z \cdot Z^{\dagger} = z_1^2 + z_2^2,$$

that is,

$$Z \cdot Z^{\dagger}(z_1^2 + z_2^2)^{-1} = Z \cdot Z^{\dagger}|Z|_{\mathbf{i}}^{-2} = 1,$$

then using the trigonometric form of $Z^{\dagger}$ we have:

$$Z^{-1} = Z^{\dagger} \cdot |Z|_{\mathbf{i}}^{-2} = |Z|_{\mathbf{i}}(\cos \Theta - \mathbf{j} \sin \Theta) \cdot |Z|_{\mathbf{i}}^{-2}$$

and finally

$$Z^{-1} = |Z|_{\mathbf{i}}^{-1}(\cos\Theta - \mathbf{j}\sin\Theta)\,. \tag{3.10}$$

With this, formula (3.9) becomes true for any integer number n.

This result allows us to write the quotient of two invertible bicomplex numbers in trigonometric form:

$$\frac{Z}{W} = \frac{|Z|_{\mathbf{i}}}{|W|_{\mathbf{i}}}\left(\cos(\Theta - \Omega) + \mathbf{j}\sin(\Theta - \Omega)\right). \tag{3.11}$$

In particular, if we take $W = Z^{\dagger}$ we get:

$$\frac{Z}{Z^{\dagger}} = \cos(2\Theta) + \mathbf{j}\sin(2\Theta)\,, \tag{3.12}$$

which is, obviously, a bicomplex number of complex modulus 1.

An analogous trigonometric form is obtained if we work with $|Z|_{\mathbf{j}}$. The definitions and computations are completely similar, yielding the trigonometric form

$$\boxed{Z = |Z|_{\mathbf{j}}(\cos\Psi_0 + \mathbf{i}\sin\Psi_0)}\,, \tag{3.13}$$

where the complex $\mathbb{C}(\mathbf{j})$-modulus of the bicomplex number $Z = \zeta_1 + \mathbf{i}\zeta_2$ is

$$|Z|_{\mathbf{j}} = \sqrt{\zeta_1^2 + \zeta_2^2}\,,$$

and (see the explanation above) Ψ_0 is a $\mathbb{C}(\mathbf{j})$-complex number. The notions of argument, principal argument, and all the properties such as De Moivre formulas, etc., are true and are in a complete analogy with the $\mathbb{BC}_{\mathbf{i}}$ case; it is clear, for instance, what $\arg_{\mathbb{C}(\mathbf{j})}(Z)$, $\arg_{\mathbb{C}(\mathbf{j}),m}(Z)$, $\mathrm{Arg}_{\mathbb{C}(\mathbf{j})}(Z)$ mean.

3.3   Trigonometric representation in hyperbolic terms

It turns out that for our goals it is convenient to work with bicomplex numbers given in their idempotent representation

$$Z = \beta_1\mathbf{e} + \beta_2\mathbf{e}^{\dagger}$$

with β_1 and β_2 in $\mathbb{C}(\mathbf{i})$. As we know, Z is not in $\mathfrak{S}_0$, that is, Z is neither zero nor a zero-divisor if and only if its hyperbolic modulus $|Z|_{\mathbf{k}}$ is a positive, non zero-divisor hyperbolic number. If this occurs, then we can write:

$$\begin{aligned}Z &= |Z|_{\mathbf{k}} \cdot |Z|_{\mathbf{k}}^{-1} \cdot \left(\beta_1\mathbf{e} + \beta_2\mathbf{e}^{\dagger}\right)\\ &= |Z|_{\mathbf{k}} \cdot \left(|\beta_1|^{-1} \cdot \mathbf{e} + |\beta_2|^{-1} \cdot \mathbf{e}^{\dagger}\right) \cdot \left(\beta_1\mathbf{e} + \beta_2\mathbf{e}^{\dagger}\right)\end{aligned}$$

which leads to

$$Z = |Z|_{\mathbf{k}} \cdot \left(\frac{\beta_1}{|\beta_1|} \cdot \mathbf{e} + \frac{\beta_2}{|\beta_2|} \cdot \mathbf{e}^\dagger \right), \tag{3.14}$$

where the $\mathbb{C}(\mathbf{i})$-complex numbers $\dfrac{\beta_1}{|\beta_1|}$ and $\dfrac{\beta_2}{|\beta_2|}$ are of (real) modulus one. Thus, they are of the form

$$\frac{\beta_1}{|\beta_1|} = e^{i\theta_1}, \qquad \frac{\beta_2}{|\beta_2|} = e^{i\theta_2} \tag{3.15}$$

$\{\theta_1, \theta_2\} \subset [0, 2\pi)$, hence θ_1 and θ_2 have a well-defined and well-known geometric meaning. Formula (3.14) leads us to consider the hyperbolic number

$$\Psi_Z := \theta_1 \mathbf{e} + \theta_2 \mathbf{e}^\dagger$$

which, we believe, deserves the names of *hyperbolic argument*, or *hyperbolic angle*, associated to the bicomplex number Z. We will soon justify more properly both names.

First of all, rewrite $Z = x_1 + \mathbf{i} y_1 + \mathbf{j} x_2 + \mathbf{k} y_2 \in \mathbb{BC} \setminus \mathfrak{S}_0$ as

$$Z = (x_1 + \mathbf{k} y_2) + \mathbf{i}(y_1 - \mathbf{k} x_2) \ = \ \mathfrak{z}_1 + \mathbf{i}\mathfrak{z}_2$$

with $\mathfrak{z}_1$ and $\mathfrak{z}_2$ in $\mathbb{D}$. Then $|Z|_{\mathbf{k}}^2 = Z \cdot Z^* = \mathfrak{z}_1^2 + \mathfrak{z}_2^2 \in \mathbb{D}^+ \setminus \{0\}$, hence $|Z|_{\mathbf{k}} = \sqrt{\mathfrak{z}_1^2 + \mathfrak{z}_2^2}$ is the value of the square root that belongs to $\mathbb{D}^+ \setminus \mathfrak{S}_0$. It follows that

$$Z = |Z|_{\mathbf{k}} \cdot |Z|_{\mathbf{k}}^{-1} (\mathfrak{z}_1 + \mathbf{i}\mathfrak{z}_2)$$

$$= |Z|_{\mathbf{k}} \left(\frac{\mathfrak{z}_1}{\sqrt{\mathfrak{z}_1^2 + \mathfrak{z}_2^2}} + \mathbf{i} \frac{\mathfrak{z}_2}{\sqrt{\mathfrak{z}_1^2 + \mathfrak{z}_2^2}} \right), \tag{3.16}$$

where the hyperbolic numbers $\dfrac{\mathfrak{z}_1}{\sqrt{\mathfrak{z}_1^2 + \mathfrak{z}_2^2}}$ and $\dfrac{\mathfrak{z}_2}{\sqrt{\mathfrak{z}_1^2 + \mathfrak{z}_2^2}}$ are such that the sum of their squares is equal to one.

This resembles the situation with the usual trigonometric functions cosine and sine of real or complex variables whose squares add up to one, and indeed, it will be shown in Chapter 6 that those trigonometric functions extend to all bicomplex numbers, the hyperbolic numbers included, and they still satisfy the identity

$$\sin^2 \alpha + \cos^2 \alpha = 1$$

for any $\alpha \in \mathbb{BC}$. Since this identity is true in particular for $\alpha \in \mathbb{D}$ (this is because bicomplex trigonometric functions of a hyperbolic variable take hyperbolic values!) we conclude from (3.14) that for any invertible Z there exists a hyperbolic number

$$\Psi_0 := \nu_1 \mathbf{e} + \nu_2 \mathbf{e}^\dagger$$

such that

$$\cos \Psi_0 = \frac{\mathfrak{z}_1}{|Z|_{\mathbf{k}}} \quad \text{and} \quad \sin \Psi_0 = \frac{\mathfrak{z}_2}{|Z|_{\mathbf{k}}}; \tag{3.17}$$

note that because of the periodicity of trigonometric functions the choice of ν_1 and ν_2 is not unique.

Since Ψ_0 acts somehow as a hyperbolic argument, or hyperbolic angle, of Z, we must establish the relation between Ψ_0 and Ψ_Z.

Proposition 3.3.1. *Given an invertible bicomplex number Z, it has a trigonometric representation in hyperbolic terms (or a $\mathbb{BC_k}$-trigonometric representation) given as*

$$Z = |Z|_\mathbf{k} \cdot (\cos \Psi_Z + \mathbf{i} \sin \Psi_Z)$$
$$= |Z|_\mathbf{k} \cdot \left(e^{\mathbf{i}\nu_1} \cdot \mathbf{e} + e^{\mathbf{i}\nu_2} \cdot \mathbf{e}^\dagger\right)$$

with $\Psi_Z = \nu_1 \cdot \mathbf{e} + \nu_2 \cdot \mathbf{e}^\dagger \in \mathbb{D}^+_{inv}$ being the hyperbolic angle, or hyperbolic principal argument, of Z.

Proof. It is known (see again Chapter 6 but also [45]) that

$$\cos \Psi_0 = \cos(\nu_1 \mathbf{e} + \nu_2 \mathbf{e}^\dagger) = \cos \nu_1 \cdot \mathbf{e} + \cos \nu_2 \cdot \mathbf{e}^\dagger$$

and

$$\sin \Psi_0 = \sin(\nu_1 \mathbf{e} + \nu_2 \mathbf{e}^\dagger) = \sin \nu_1 \cdot \mathbf{e} + \sin \nu_2 \cdot \mathbf{e}^\dagger;$$

hence

$$\cos \Psi_0 + \mathbf{i} \sin \Psi_0 = (\cos \nu_1 + \mathbf{i} \sin \nu_1) \cdot \mathbf{e} + (\cos \nu_2 + \mathbf{i} \sin \nu_2)\mathbf{e}^\dagger.$$

Combining this equation with (3.16) and (3.14) one gets:

$$Z = |Z|_\mathbf{k} (\cos \Psi_0 + \mathbf{i} \sin \Psi_0)$$
$$= |Z|_\mathbf{k} \left((\cos \nu_1 + \mathbf{i} \sin \nu_1)\mathbf{e} + (\cos \nu_2 + \mathbf{i} \sin \nu_2)\mathbf{e}^\dagger\right)$$
$$= |Z|_\mathbf{k} \left(\frac{\beta_1}{|\beta_1|} \cdot \mathbf{e} + \frac{\beta_2}{|\beta_2|} \cdot \mathbf{e}^\dagger\right).$$

Taking into account the periodicity of trigonometric functions and choosing the values of ν_1 and ν_2 in the interval $[0, 2\pi)$ we obtain that Ψ_0 may be taken equal to Ψ_Z. We prefer the notation Ψ_Z emphasizing that the angle is linked to $Z \in \mathbb{BC}\backslash\mathfrak{S}_0$.   □

In contrast with the complex argument of a bicomplex number which is defined for invertible bicomplex numbers only, we can define the hyperbolic argument for zero-divisors too. Consider, for instance, a zero-divisor $Z = \beta_1 \mathbf{e}$, then

$$Z \cdot Z^* = |\beta_1|^2 \mathbf{e}$$

implying that $|Z|_\mathbf{k} = |\beta_1| \cdot \mathbf{e}$, a semi-positive hyperbolic number; thus

$$Z = |Z|_\mathbf{k} \cdot \frac{\beta_1}{|\beta_1|} \cdot \mathbf{e},$$

and since $\dfrac{\beta_1}{|\beta_1|}$ is a complex number of (real) modulus one, it can be written as $e^{i\nu_1}$ with $\nu_1 \in [0, 2\pi)$. In this case the hyperbolic angle associated to Z is the hyperbolic zero-divisor

$$\nu := \nu_1 \mathbf{e}.$$

Analogously, if $Z = \beta_2 \mathbf{e}^\dagger$, then $|Z|_\mathbf{k} = |\beta_2| \mathbf{e}^\dagger$ and

$$Z = |Z|_\mathbf{k} \cdot \frac{\beta_2}{|\beta_2|} \cdot \mathbf{e}^\dagger = |Z|_\mathbf{k} e^{i\nu_2} \mathbf{e}^\dagger.$$

In this case, the hyperbolic angle associated to Z is the hyperbolic zero-divisor

$$\nu = \nu_2 \mathbf{e}^\dagger$$

with $\nu_2 \in [0, 2\pi)$.

3.3.1 Algebraic properties of the trigonometric representation of bicomplex numbers in hyperbolic terms

Our next task is to analyze the algebraic and geometric consequences of this representation. We start with the algebraic ones. First, in analogy with the (real) argument of a complex number and with the complex arguments of a bicomplex number, we introduce some notation. We denote by $\arg_\mathbb{D} Z$, for $Z \neq 0$, the principal hyperbolic argument of Z, that is, $\arg_\mathbb{D} Z = \Psi_Z$ for $Z \in \mathbb{BC}_{inv}$ and $\arg_\mathbb{D} Z = \nu_1 \mathbf{e}$ or $\arg_\mathbb{D} Z = \nu_2 \mathbf{e}^\dagger$ for a zero-divisor Z. Then, $\arg_{m,n;\mathbb{D}} Z$ is

$$\arg_{m,n;\mathbb{D}} Z := (\nu_1 + 2\pi m)\mathbf{e} + (\nu_2 + 2\pi n)\mathbf{e}^\dagger$$

for an invertible Z and similarly for a zero-divisor. Finally, $\mathrm{Arg}_\mathbb{D} Z$ denotes the set of all possible hyperbolic angles:

$$\mathrm{Arg}_\mathbb{D} Z := \big\{ \arg_{m,n;\mathbb{D}} Z \,\big|\, m, n \text{ in } \mathbb{Z} \big\}$$

for an invertible Z and similarly for a zero-divisor.

Example 3.3.1.1. Let us consider the trigonometric representation in hyperbolic terms of some particular bicomplex numbers.

(a) Given a complex number $z = x + iy \in \mathbb{C}(\mathbf{i})$, then $z = z\mathbf{e} + z\mathbf{e}^\dagger$; we have mentioned before that in this case the hyperbolic modulus coincides with the usual modulus of complex numbers, i.e.,

$$|z|_\mathbf{k} = |z|\mathbf{e} + |z|\mathbf{e}^\dagger = |z|.$$

Hence, the trigonometric form of z in hyperbolic terms is:

$$z = |z| \cdot \left(\frac{z}{|z|}\mathbf{e} + \frac{z}{|z|}\mathbf{e}^\dagger \right) = |z| \left(e^{i\theta}\mathbf{e} + e^{i\theta}\mathbf{e}^\dagger \right) = |z| e^{i\theta},$$

thus, the usual trigonometric form of the complex number z and its trigono-metric form in hyperbolic terms when seen as a bicomplex number, coincide. In particular, the hyperbolic argument of z, $\Psi_z = 0\mathbf{e} + 0\mathbf{e}^\dagger = 0$, coincides with the usual (real) argument of z. Hence

$$\mathrm{Arg}_\mathbb{D} z = \mathrm{Arg} z = \{\theta + 2m\pi \mid m \in \mathbb{Z}\}.$$

(b) The imaginary unit $\mathbf{j}$ has the idempotent form $\mathbf{j} = (-\mathrm{i})\mathbf{e} + \mathrm{i}\mathbf{e}^\dagger$, thus $|\mathbf{j}|_\mathbf{k} = 1\mathbf{e} + 1\mathbf{e}^\dagger = 1$. Hence, its trigonometric representation in hyperbolic terms is

$$\mathbf{j} = |\mathbf{j}|_\mathbf{k} \left(-\mathrm{i}\mathbf{e} + \mathrm{i}\mathbf{e}^\dagger\right) = e^{\mathrm{i}\frac{3}{2}\pi}\mathbf{e} + e^{\mathrm{i}\frac{\pi}{2}}\mathbf{e}^\dagger,$$

and its hyperbolic argument is $\Psi_\mathbf{j} = \dfrac{3}{2}\pi\mathbf{e} + \dfrac{\pi}{2}\mathbf{e}^\dagger$, thus

$$\mathrm{Arg}_\mathbb{D}\,\mathbf{j} = \left\{\left(\frac{3}{2} + 2m\right)\pi\mathbf{e} + \left(\frac{1}{2} + 2n\right)\pi\mathbf{e}^\dagger \mid m, n \in \mathbb{Z}\right\}.$$

(c) If $Z = \mathfrak{z} = a\mathbf{e} + b\mathbf{e}^\dagger$, with a, b in $\mathbb{R}$, that is, $\mathfrak{z} \in \mathbb{D}$, $|\mathfrak{z}|_\mathbf{k} = |a|\mathbf{e} + |b|\mathbf{e}^\dagger$, then its trigonometric representation in hyperbolic form is

$$\mathfrak{z} = |\mathfrak{z}|_\mathbf{k} \left(\frac{a}{|a|}\mathbf{e} + \frac{b}{|b|}\mathbf{e}^\dagger\right) = |\mathfrak{z}|_\mathbf{k}\left(\pm\mathbf{e} \pm \mathbf{e}^\dagger\right).$$

Hence there are four options for the principal value of the hyperbolic argument of a hyperbolic number as indicated in Figure 3.3.1:

$$\Psi_\mathfrak{z} = 0\mathbf{e} + 0\mathbf{e}^\dagger = 0,$$

or

$$\Psi_\mathfrak{z} = 0\mathbf{e} + \pi\mathbf{e}^\dagger = \pi\mathbf{e}^\dagger,$$

or

$$\Psi_\mathfrak{z} = \pi\mathbf{e} + 0\mathbf{e}^\dagger = \pi\mathbf{e},$$

or

$$\Psi_\mathfrak{z} = \pi\mathbf{e} + \pi\mathbf{e}^\dagger = \pi.$$

Hence the set $\mathrm{Arg}_\mathbb{D}\mathfrak{z}$ is, according to the quadrant the hyperbolic number $\mathfrak{z}$ is in, as follows:

$$\mathrm{Arg}_\mathbb{D}\mathfrak{z}$$
$$= \begin{cases} \{2\pi(m\mathbf{e} + n\mathbf{e}^\dagger) \mid m, n \in \mathbb{Z}\} & \text{if } \mathfrak{z} \text{ is in the } 1^{st} \text{ quadrant;} \\ \{2\pi m\mathbf{e} + (2n+1)\pi\mathbf{e}^\dagger \mid m, n \in \mathbb{Z}\} & \text{if } \mathfrak{z} \text{ is in the } 2^{nd} \text{ quadrant;} \\ \{(2m+1)\pi\mathbf{e} + 2n\pi\mathbf{e}^\dagger \mid m, n \in \mathbb{Z}\} & \text{if } \mathfrak{z} \text{ is in the } 3^{rd} \text{ quadrant;} \\ \{(2m+1)\pi\mathbf{e} + (2n+1)\pi\mathbf{e}^\dagger \mid m, n \in \mathbb{Z}\} & \text{if } \mathfrak{z} \text{ is in the } 4^{th} \text{ quadrant.} \quad \square \end{cases}$$

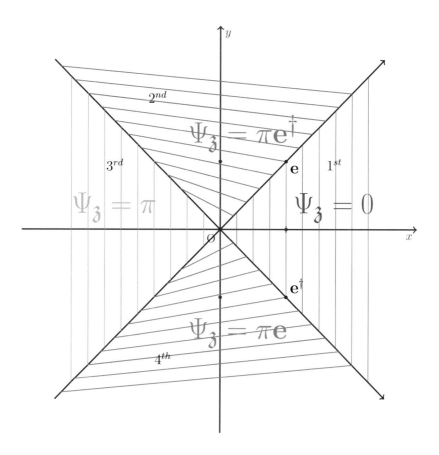

Figure 3.3.1: Hyperbolic arguments of hyperbolic numbers seen as bicomplex numbers.

Coming back to the properties of the trigonometric representation in hyperbolic terms, take two invertible bicomplex numbers Z and W in their trigonometric representations in hyperbolic terms, then

$$
\begin{aligned}
Z \cdot W &= |Z|_{\mathbf{k}} \left(\cos \Psi_Z + \mathbf{i} \sin \Psi_Z \right) \cdot |W|_{\mathbf{k}} \left(\cos \Psi_W + \mathbf{i} \sin \Psi_W \right) \\
&= |Z|_{\mathbf{k}} \cdot |W|_{\mathbf{k}} \cdot \left(e^{\mathbf{i}\nu_1} \cdot \mathbf{e} + e^{\mathbf{i}\nu_2} \cdot \mathbf{e}^\dagger \right) \cdot \left(e^{\mathbf{i}\mu_1} \cdot \mathbf{e} + e^{\mathbf{i}\mu_2} \cdot \mathbf{e}^\dagger \right) \\
&= |Z \cdot W|_{\mathbf{k}} \cdot \left(\cos(\Psi_Z + \Psi_W) + \mathbf{i} \sin(\Psi_Z + \Psi_W) \right) \\
&= |Z \cdot W|_{\mathbf{k}} \cdot \left(e^{\mathbf{i}(\nu_1 + \mu_1)} \cdot \mathbf{e} + e^{\mathbf{i}(\nu_2 + \mu_2)} \cdot \mathbf{e}^\dagger \right).
\end{aligned}
\tag{3.18}
$$

Hence, one concludes that
$$
\operatorname{Arg}_{\mathbb{D}}(Z \cdot W) = \operatorname{Arg}_{\mathbb{D}} Z + \operatorname{Arg}_{\mathbb{D}} W.
$$

Similarly for the case when one of the factors, or both, are zero-divisors. In particular, taking $Z = W$ one gets that

$$Z^2 = |Z|_\mathbf{k}^2 \cdot \left(e^{\mathbf{i}(2\nu_1)} \cdot \mathbf{e} + e^{\mathbf{i}(2\nu_2)} \cdot \mathbf{e}^\dagger\right)$$
$$= |Z|_\mathbf{k}^2 \cdot \left((\cos(2\nu_1) + \mathbf{i}\sin(2\nu_1)\mathbf{e} + (\cos(2\nu_2) + \mathbf{i}\sin(2\nu_2)\mathbf{e}^\dagger\right).$$

Using induction, we obtain an analogue of the De Moivre formula: for any $n \in \mathbb{N}$ it holds that

$$Z^n = |Z|_\mathbf{k}^n \cdot \left(e^{\mathbf{i}(n\nu_1)} \cdot \mathbf{e} + e^{\mathbf{i}(n\nu_2)} \cdot \mathbf{e}^\dagger\right)$$
$$= |Z|_\mathbf{k}^n \cdot \left((\cos(n\nu_1) + \mathbf{i}\sin(n\nu_1)\mathbf{e}) + (\cos(n\nu_2) + \mathbf{i}\sin(n\nu_2)\mathbf{e}^\dagger)\right). \tag{3.19}$$

Similarly the case of zero-divisors.

For an invertible bicomplex number its hyperbolic modulus is an invertible hyperbolic number, hence

$$Z^{-1} = |Z|_\mathbf{k}^{-1} \cdot \left(e^{-\mathbf{i}\nu_1} \cdot \mathbf{e} + e^{-\mathbf{i}\nu_2} \cdot \mathbf{e}^\dagger\right)$$

which implies immediately that the De Moivre formula is true for any $n \in \mathbb{Z}$.

Let us see what is the relation between the trigonometric representations in hyperbolic terms of Z and Z^*. Since the $*$-conjugate of $Z = \beta_1 \mathbf{e} + \beta_2 \mathbf{e}^\dagger$ is $Z^* = \overline{\beta}_1 \mathbf{e} + \overline{\beta}_2 \mathbf{e}^\dagger$, we have:

$$|Z^*|_\mathbf{k} = |\beta_1| \cdot \mathbf{e} + |\beta_2| \cdot \mathbf{e}^\dagger = |Z|_\mathbf{k},$$

thus

$$Z^* = |Z^*|_\mathbf{k} \cdot \left(\frac{\overline{\beta}_1}{|\beta_1|} \cdot \mathbf{e} + \frac{\overline{\beta}_2}{|\beta_2|} \cdot \mathbf{e}^\dagger\right)$$
$$= |Z|_\mathbf{k} \cdot \left(e^{-\mathbf{i}\nu_1} \cdot \mathbf{e} + e^{-\mathbf{i}\nu_2} \cdot \mathbf{e}^\dagger\right),$$

and we conclude that

$$\mathrm{Arg}_\mathbb{D} Z^* = -\mathrm{Arg}_\mathbb{D} Z,$$

compare again with the complex numbers case.

3.3.2 A geometric interpretation of the hyperbolic trigonometric representation.

We will see now that, in analogy with the complex numbers situation, the hyperbolic trigonometric representation of a bicomplex number provides information with clear geometrical interpretations.

Since the hyperbolic modulus of a bicomplex number $Z = \beta_1 \mathbf{e} + \beta_2 \mathbf{e}^\dagger$, with β_1 and β_2 in $\mathbb{C}(\mathbf{i})$, is the positive (not necessarily strictly positive) hyperbolic number

$$|Z|_\mathbf{k} = |\beta_1| \cdot \mathbf{e} + |\beta_2| \cdot \mathbf{e}^\dagger,$$

one can formally define the bicomplex sphere $\mathbb{S}_{\gamma_0}$ centered at the origin and with radius being a fixed positive hyperbolic number $\gamma_0 = a_0\mathbf{e} + b_0\mathbf{e}^\dagger \in \mathbb{D}^+$, i.e.,

$$\mathbb{S}_{\gamma_0} := \left\{ Z \in \mathbb{BC} \,\middle|\, |Z|_\mathbf{k} = \gamma_0 \right\}.$$

Let us show that such a set can be visualized perfectly well.

First note that if either a_0 or b_0 is zero, that is, γ_0 is a zero-divisor, say $\gamma_0 = a_0 \cdot \mathbf{e}$, then

$$\mathbb{S}_{\gamma_0} = \left\{ Z = \beta_1 \cdot \mathbf{e} \,\middle|\, |\beta_1| = a_0 \right\};$$

this set is a circumference in the real two-dimensional plane $\mathbb{BC}_\mathbf{e}$ with center at the origin and radius $\dfrac{a_0}{\sqrt{2}} = |a_0 \cdot \mathbf{e}|$. Similarly, if $\gamma_0 = b_0 \cdot \mathbf{e}^\dagger$, then the set $\mathbb{S}_{\gamma_0}$ is a circumference in $\mathbb{BC}_{\mathbf{e}^\dagger}$ with center at the origin and radius $\dfrac{b_0}{\sqrt{2}} = |b_0 \cdot \mathbf{e}^\dagger|$.

Thus, whenever the radius of a bicomplex sphere is a (positive) zero-divisor, i.e., a semi-positive hyperbolic number, then the sphere is, in fact, a usual circumference. See Figures 3.3.2 and 3.3.3.

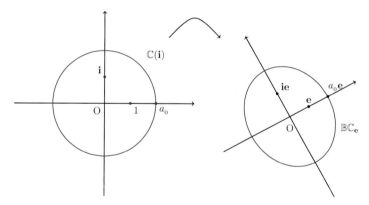

Figure 3.3.2: THE BICOMPLEX SPHERE $\mathbb{S}_{a_0\mathbf{e}}$.

If $a_0 \neq 0$ and $b_0 \neq 0$, then the intersection of the hyperbolic plane $\mathbb{D}$ and the sphere $\mathbb{S}_{\gamma_0}$ consists exactly of the four hyperbolic numbers $\pm a_0\mathbf{e} \pm b_0\mathbf{e}^\dagger$, that is, in these points the plane $\mathbb{D}$ touches the sphere tangentially. What is more, in this case the whole sphere $\mathbb{S}_{\gamma_0}$ is the surface of a three-dimensional manifold with the shape of a torus (see Figure 3.3.4).

We emphasize the fact that although the torus is three-dimensional as a manifold and its surface is a two-dimensional manifold, they both live in a four-dimensional world. This is because now the sphere $\mathbb{S}_{\gamma_0}$ is given as

$$\mathbb{S}_{\gamma_0} = \left\{ Z = \beta_1\mathbf{e} + \beta_2\mathbf{e}^\dagger \,\middle|\, |\beta_1| = a_0, \ |\beta_2| = b_0 \right\}$$

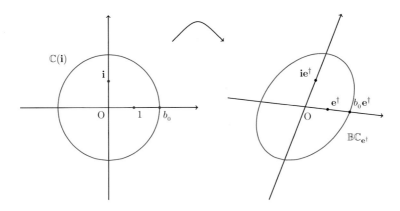

Figure 3.3.3: THE BICOMPLEX SPHERE $\mathbb{S}_{b_0 \mathbf{e}^\dagger}$.

which allows another description of this set: it is the cartesian product of the two circumferences, namely, one of them, denoted by $C_{\mathbf{e}}\left(0; \dfrac{a_0}{\sqrt{2}}\right)$, is situated in the plane $\mathbb{BC}_{\mathbf{e}}$ having the origin as its center and $\dfrac{a_0}{\sqrt{2}}$ as its radius; the other one, denoted by $C_{\mathbf{e}^\dagger}\left(0; \dfrac{b_0}{\sqrt{2}}\right)$, is situated in the plane $\mathbb{BC}_{\mathbf{e}^\dagger}$ having the origin as its center and $\dfrac{b_0}{\sqrt{2}}$ as its radius. We illustrate this with Figure 3.3.4.

These spheres of hyperbolic radius have intrinsic interest, but for the moment we will use them in order to present a geometric interpretation of the hyperbolic argument of a bicomplex number $Z \in \mathbb{BC}_{inv}$ written as

$$Z = |Z|_{\mathbf{k}}\left(e^{i\nu_1} \cdot \mathbf{e} + e^{i\nu_2} \cdot \mathbf{e}^\dagger\right)$$

with principal value hyperbolic argument $\arg_{\mathbb{D}} Z = \nu_1 \mathbf{e} + \nu_2 \mathbf{e}^\dagger \in \mathbb{D}_{inv}^+$.

Let us fix the value of $|Z|_{\mathbf{k}} =: \gamma_0 = a_0 \mathbf{e} + b_0 \mathbf{e}^\dagger$, then all the bicomplex numbers having it as their hyperbolic modulus belongs to the sphere $\mathbb{S}_{\gamma_0}$ which is the surface of a Euclidean torus; as we mentioned before, this surface can be seen as the cartesian product

$$C_{\mathbf{e}}\left(0; \frac{a_0}{\sqrt{2}}\right) \times C_{\mathbf{e}^\dagger}\left(0; \frac{b_0}{\sqrt{2}}\right).$$

Any point of the torus is in one-to-one correspondence with hyperbolic numbers of the form $\mu_1 \mathbf{e} + \mu_2 \mathbf{e}^\dagger$ where μ_1 and μ_2 are in $[0, 2\pi)$; the latter, in turn, are in one-to-one correspondence with the pairs of points each of which belongs to the

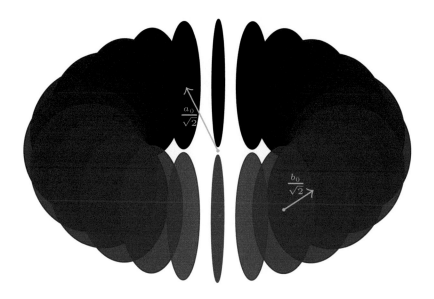

Figure 3.3.4: THE BICOMPLEX SPHERE OF HYPERBOLIC RADIUS $a_0\mathbf{e} + b_0\mathbf{e}^\dagger$.

corresponding circumference:

$$a_0 e^{\mathbf{i}\mu_1}\mathbf{e} \in C_\mathbf{e}\left(0; \frac{a_0}{\sqrt{2}}\right),$$

$$b_0 e^{\mathbf{i}\mu_2}\mathbf{e}^\dagger \in C_{\mathbf{e}^\dagger}\left(0; \frac{b_0}{\sqrt{2}}\right);$$

and moreover,

$$a_0 e^{\mathbf{i}\mu_1}\mathbf{e} + b_0 e^{\mathbf{i}\mu_2}\mathbf{e}^\dagger = \left(a_0\mathbf{e} + b_0\mathbf{e}^\dagger\right)\cdot\left(e^{\mathbf{i}\mu_1}\mathbf{e} + e^{\mathbf{i}\mu_2}\mathbf{e}^\dagger\right)$$

$$= \gamma_0 \cdot \left(e^{\mathbf{i}\mu_1}\mathbf{e} + e^{\mathbf{i}\mu_2}\mathbf{e}^\dagger\right) = Z.$$

This shows that the value of the hyperbolic modulus of Z tells us on which sphere Z is situated and its hyperbolic argument determines the exact place of Z on the sphere.

Note also that the sphere $\mathbb{S}_{\gamma_0}$ is in a bijective correspondence with the hyperbolic interval

$$[0, 2\pi)_\mathbb{D} = \left\{\mathfrak{z} \in \mathbb{D} \,\middle|\, 0 \preceq \mathfrak{z} \prec 2\pi\right\}$$

$$= \left\{\mathfrak{z} = \nu_1\mathbf{e} + \nu_2\mathbf{e}^\dagger \,\middle|\, \nu_1,\, \nu_2 \in [0, 2\pi)\right\} \subset \mathbb{D}^+$$

given in Figure 3.3.5.

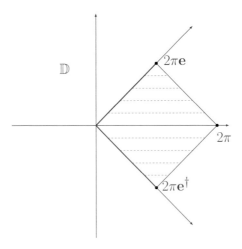

Figure 3.3.5: THE HYPERBOLIC INTERVAL $[0, 2\pi)_{\mathbb{D}}$.

The correspondence is given by

$$a_0 e^{i\nu_1}\mathbf{e} + b_0 e^{i\nu_2}\mathbf{e}^\dagger \longleftrightarrow \nu_1\mathbf{e} + \nu_2\mathbf{e}^\dagger.$$

We will see later that this is related with the exponential functions of bicomplex and hyperbolic variables.

Some of the geometric and trigonometric properties of hyperbolic and bicomplex numbers have been studied in the literature in works such as [28, 45, 56, 100]. We also mention here [87] and [11], where trigonometric representations of hyperbolic and hypercomplex numbers are studied, together with applications to Special Relativity. The authors of [11] emphasize that the hyperbolic numbers are "the mathematics of the two-dimensional Special Relativity", and adapting this two-dimensional hyperbolic geometry to "multidimensional commutative hypercomplex systems" (such as bicomplex numbers) leads to a concrete application to Special Relativity in four-dimensional Minkowski space-time. The "separation" of the four-dimensional bicomplex space into spacelike and timelike parts (events) is captured by the geometry of the complex lines $L_\mathbf{e}$ and $L_{\mathbf{e}^\dagger}$, which we carefully study in complete detail in the next chapter.

Chapter 4

Lines and curves in $\mathbb{BC}$

4.1 Straight lines in $\mathbb{BC}$

In this section we will investigate the complex straight lines in $\mathbb{BC}$ from both the algebraic and the geometric points of view. We will start by describing the well-known case of real straight lines in $\mathbb{C}(\mathbf{i})$, which we use as reference for the case of complex and hyperbolic lines in $\mathbb{BC}$.

4.1.1 Real lines in the complex plane

A *real straight line* in the plane $\mathbb{R}^2$ is given by the equation

$$a_1 x + a_2 y = b,\tag{4.1}$$

where a_1, a_2, b are real coefficients. Writing $\vec{a} = (a_1, a_2)$, $\vec{z} = (x, y)$, (4.1) is equivalent to

$$\frac{\langle \vec{a}, \vec{z} \rangle}{|\vec{a}|} = \frac{\langle (a_1, a_2), (x, y) \rangle_{\mathbb{R}^2}}{|\vec{a}|} = \frac{b}{|\vec{a}|}.\tag{4.2}$$

This formula has the following geometric description: a point (x, y) belongs to the straight line given by (4.1) if the corresponding vector has a constant projection $\frac{b}{|\vec{a}|}$ on the straight line determined by $\vec{a}$.

Setting $z := x + \mathbf{i}y$ and $a := a_1 + \mathbf{i}a_2$ and recalling that $x = \frac{1}{2}(z + \bar{z})$, $y = \frac{\mathbf{i}}{2}(\bar{z} - z)$, $a_1 = \frac{1}{2}(a + \bar{a})$ and $a_2 = \frac{\mathbf{i}}{2}(\bar{a} - a)$, we obtain the complex form of writing equation (4.1):

$$\bar{a}z + a\bar{z} = 2b.\tag{4.3}$$

Geometrically, a complex number z belongs to the line ℓ given by (4.3) if $\mathrm{Re}(\bar{a}z) = \mathrm{Re}(a\bar{z})$ is equal to the constant b (see Figure 4.1.1).

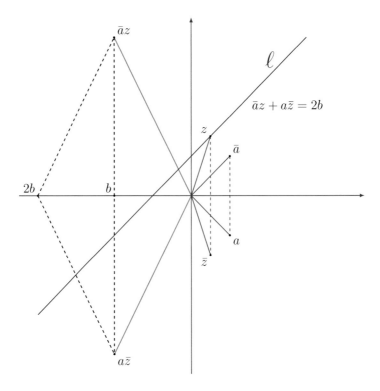

Figure 4.1.1: THE POINT z BELONG TO ℓ IF AND ONLY IF $Re(\bar{a}z) = Re(a\bar{z})$.

If we consider a and z in their trigonometric forms, i.e., $a = |a|(\cos\phi_0 + \mathbf{i}\sin\phi_0)$ and $z = |z|(\cos\phi + \mathbf{i}\sin\phi)$, then equation (4.3) becomes

$$|a||z|\cos(\phi - \phi_0) = b. \tag{4.4}$$

The latter formula can also be obtained from (4.2).

Note also that if z_1 and z_2 are two points on a real line, then the vector $\vec{z_2} - \vec{z_1}$ is a vector along the line. Then, since $\langle \vec{a}, \vec{z_2} - \vec{z_1}\rangle_{\mathbb{R}^2} = 0$, it follows that $\vec{a}$ is a *normal* vector to our line, i.e., orthogonal to it.

Thus the angle that our line makes with the positive direction of the abscissa axis is $\phi_0 + \dfrac{\pi}{2}$. The tangent of this angle is called the *slope* m of the line:

$$m = -\cot\phi_0 = \tan\left(\phi_0 + \frac{\pi}{2}\right). \tag{4.5}$$

If the line is given by equation (4.1) and $a_2 \neq 0$, then $m = -\dfrac{a_1}{a_2}$.

In complex form, the case we are considering, i.e., $a_2 \neq 0$, implies that $a \neq 0$ and $\bar{a} \neq 0$, thus if we divide by $\bar{a}$ in (4.3), we obtain:

$$z = -\frac{a}{\bar{a}}\bar{z} + \frac{2b}{\bar{a}}. \tag{4.6}$$

We introduce the notation

$$M := -\frac{a}{\bar{a}} = -e^{\mathbf{i}2\phi_0} = e^{\mathbf{i}(2\phi_0 + \pi)}.$$

Since $\bar{a} = |a|^2 a^{-1}$, then $M = -\dfrac{a^2}{|a|^2}$. Moreover, in order to capture the slope of our line, we write:

$$M = -\frac{a}{\bar{a}} = -\frac{a_1 + \mathbf{i}a_2}{a_1 - \mathbf{i}a_2} = -\frac{-\frac{a_1}{a_2} - \mathbf{i}}{-\frac{a_1}{a_2} + \mathbf{i}} = -\frac{\tan(\phi_0 + \frac{\pi}{2}) - \mathbf{i}}{\tan(\phi_0 + \frac{\pi}{2}) + \mathbf{i}} = \frac{m - \mathbf{i}}{m + \mathbf{i}},$$

thus we get the slope:

$$m = \tan\left(\phi_0 + \frac{\pi}{2}\right) = \mathbf{i}\frac{1 - M}{1 + M} = \mathbf{i}\frac{\bar{a} + a}{\bar{a} - a} = -\frac{a_1}{a_2}.$$

In conclusion, M is an intrinsic *characteristic* complex number for our real line, which determines its geometric slope. Equation (4.6) then becomes

$$z = M\bar{z} + B, \tag{4.7}$$

where $B := \dfrac{2b}{\bar{a}}$.

We note here some particular cases. If $M = -1$, then $a = \bar{a} = a_1$, a real number, i.e., $a_2 = 0$, which yields the vertical line $a_1 x = b$, which has no well-defined slope. In complex terms the equation of this vertical line is $z = -\bar{z} + B$, where $B = \dfrac{2b}{a_1}$ is a real number.

If $M = 1$, then $a = -\bar{a} = \mathbf{i}a_2$, so $a_1 = 0$, which yields the horizontal line $a_2 y = b$. In complex terms we obtain $z = \bar{z} + B$, where $B = \mathbf{i}\dfrac{2b}{a_2}$ is a purely imaginary number.

If it is known that the real line passes through a point $z_0 = x_0 + \mathbf{i}y_0$ and it has slope $m \neq 0$, then the equation can be written as

$$y - y_0 = m(x - x_0),$$

or in complex terms as

$$z + \overline{z_0} = M(\bar{z} + z_0).$$

If the real line is determined by two points $z_0 = x_0 + \mathbf{i}y_0$ and $z_1 = x_1 + \mathbf{i}y_1$, such that $x_0 \neq x_1$, it is given by equation

$$y - y_0 = \frac{y_1 - y_0}{x_1 - x_0}(x - x_0). \tag{4.8}$$

The characteristic complex number M takes the form

$$M = \frac{z_1 - z_0}{\overline{z}_1 - \overline{z}_0} \, ,$$

so the complex equation of the line is

$$z + \overline{z}_0 = \frac{z_1 - z_0}{\overline{z}_1 - \overline{z}_0}(\overline{z} + z_0) \, . \qquad (4.9)$$

If in equation (4.8) we denote $x_1 - x_0 =: \nu_1$ and $y_1 - y_0 =: \nu_2$, then, if we assume they both are non-zero real numbers, we obtain the equation

$$\frac{y - y_0}{\nu_2} = \frac{x - x_0}{\nu_1} =: t \, ,$$

where t is a real parameter. We get therefore a parametrization of this line given by

$$x(t) = x_0 + t\nu_1, \qquad y(t) = y_0 + t\nu_2 \, .$$

Note that its slope is

$$m = \frac{\nu_2}{\nu_1} \, ,$$

and its associated complex characteristic number becomes

$$M = \frac{\nu_1 + \mathbf{i}\nu_2}{\nu_1 - \mathbf{i}\nu_2} \, .$$

Then the equation of our real line is

$$\nu_2 x - \nu_1 y = \nu_2 x_0 - \nu_1 y_0 \, ,$$

thus, in complex notation, the complex coefficient $a = \nu_2 - \mathbf{i}\nu_1 = -\mathbf{i}(\nu_1 + \mathbf{i}\nu_2)$ and the parametrized equation becomes

$$z(t) = z_0 + t \cdot \mathbf{i}a \, . \qquad (4.10)$$

If we restrict the real variable t to be in an interval $[c, d] \subset \mathbb{R}$, then the equation above yields a straight line segment in $\mathbb{R}^2$.

The parametric representation of straight lines is illustrated in Figure 4.1.2.

Note that every point w on the line $\ell_{\mathbf{i}a}$ seen as a complex number is of the form $w = t \cdot \mathbf{i}a$ with a fixed, and any point of the line ℓ is obtained as the translation $t \cdot \mathbf{i}a + z_0$ of the point $t \cdot \mathbf{i}a$ by the vector z_0. Thus, any straight line which passes through the origin is the set of complex numbers

$$\ell_{\widetilde{a}} := \{w = \lambda\widetilde{a} \, | \, \lambda \in \mathbb{R}\},$$

where $\widetilde{a}$ is a fixed arbitrary complex number, and if the line does not pass through the origin but passes through a point, say z_0, then the line ℓ is the set of complex numbers of the form

$$\ell_{\widetilde{a}} + z_0 := \{\lambda\widetilde{a} + z_0 \, | \, \lambda \in \mathbb{R}\}$$

for some $\widetilde{a} \in \mathbb{C} \setminus \{0\}$.

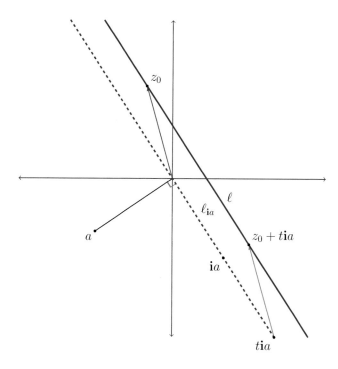

Figure 4.1.2: PARAMETRIC REPRESENTATION OF A STRAIGHT LINE.

4.1.2 Real lines in $\mathbb{BC}$

Take a non-zero bicomplex number Z_0, then the real straight line which passes through Z_0 and the origin (that is, its direction is determined by Z_0) is defined to be the set

$$\ell_{Z_0} := \{\lambda Z_0 \mid \lambda \in \mathbb{R}\}.$$

Let Z_1 be another bicomplex number different from zero, then the straight line in the direction of Z_0 and passing through Z_1 is the set

$$\ell_{Z_0} + Z_1 := \{\lambda Z_0 + Z_1 \mid \lambda \in \mathbb{R}\}.$$

4.1.3 Complex lines in $\mathbb{BC}$

In Section 4.1.1 we described real lines in $\mathbb{C}$ using both real and complex languages. In what follows we are going to extend the idea onto the complex lines, seeing them as analogues of the real lines in $\mathbb{C}$ but now inside $\mathbb{BC}$ which is seen as an analogue of the set of complex numbers. But saying "complex" in $\mathbb{BC}$ may mean any of the sets $\mathbb{C}(\mathbf{i})$ or $\mathbb{C}(\mathbf{j})$. We will start our analysis considering $\mathbb{C}(\mathbf{i})$-complex lines.

Following the analogy with the real lines, we define a *complex straight line* in $\mathbb{C}^2(\mathbf{i})$ to be the set of solutions $(z_1, z_2) \in \mathbb{C}^2(\mathbf{i})$ of the equation

$$a_1 z_1 + a_2 z_2 = b, \tag{4.11}$$

where $a_1, a_2, b \in \mathbb{C}(\mathbf{i})$ are complex coefficients. Clearly, this equation is equivalent to a system of two real linear equations with four real variables, and thus it defines, generally speaking, a 2-dimensional plane in $\mathbb{R}^4$; this happens when the matrix of coefficients has rank 2; it is clear what happens if the rank is less than 2.

If we see $\mathbb{C}^2(\mathbf{i})$ as $\mathbb{BC}$ in such a way that $Z = z_1 + \mathbf{j} z_2$, then we get: $z_1 = \dfrac{Z + Z^\dagger}{2}$ and $z_2 = -\mathbf{j}\,\dfrac{Z - Z^\dagger}{2}$, thus the complex line (4.11) will be considered as a $\mathbb{C}(\mathbf{i})$-complex line in $\mathbb{BC}$ with bicomplex equation

$$A^\dagger Z + A Z^\dagger = 2b, \tag{4.12}$$

where $A = a_1 + \mathbf{j} a_2 \neq 0$ is a bicomplex coefficient. Here we have to distinguish between the point $Z = z_1 + \mathbf{j} z_2 = x_1 + i y_1 + \mathbf{j} x_2 + \mathbf{k} y_2$, etc. (using all the possible writings for Z), and the vector $\vec{Z}$, which is a vector in $\mathbb{R}^4$ that joins the origin and the point (x_1, y_2, x_2, y_2). Thus, in what follows, when it will be necessary, we will refer to the point Z as the point $(z_1, z_2) \in \mathbb{C}^2(\mathbf{i})$ or $(w_1, w_2) \in \mathbb{C}^2(\mathbf{j})$ (writing $Z = w_1 + i w_2$) or $(\beta_1, \beta_2) \in \mathbb{C}^2(\mathbf{i})$ (writing $Z = \beta_1 \mathbf{e} + \beta_2 \mathbf{e}^\dagger$), or $(\gamma_1, \gamma_2) \in \mathbb{C}^2(\mathbf{j})$ (writing $Z = \gamma_1 \mathbf{e} + \gamma_2 \mathbf{e}^\dagger$).

Note that if $a_1 = 0$, $a_2 = 1$, $b = 0$, equation (4.11) becomes

$$0 \cdot z_1 + 1 \cdot z_2 = 0,$$

and the respective complex line is the set $\mathbb{C}(\mathbf{i}) \subset \mathbb{BC}$.

If $a_1 = 1, a_2 = 0, b = 0$, then

$$1 \cdot z_1 + 0 \cdot z_2 = 0,$$

and the complex line is the set $\mathbb{C}(\mathbf{i}) \cdot \mathbf{j} \subset \mathbb{BC}$.

4.1.4   Parametric representation of complex lines

Let us consider an alternative way of describing a $\mathbb{C}(\mathbf{i})$-complex line in $\mathbb{BC}$ which is equivalent to the one above. Take a fixed non-zero bicomplex number $Z_0 = z_1^0 + \mathbf{j} z_2^0 \in \mathbb{BC} \setminus \{0\}$. Consider the set

$$L_{Z_0} := \{\lambda Z_0 \,|\, \lambda \in \mathbb{C}(\mathbf{i})\},$$

and let us show that it is a $\mathbb{C}(\mathbf{i})$-complex line which we will call the $\mathbb{C}(\mathbf{i})$-complex line generated by Z_0 and passing through the origin. Indeed, for any element $Z = z_1 + \mathbf{j} z_2$ in L_{Z_0} there exists $\lambda \in \mathbb{C}(\mathbf{i})$ such that

$$z_1 + \mathbf{j} z_2 = Z = \lambda Z_0 = \lambda z_1^0 + \mathbf{j} \lambda z_2^0,$$

which is equivalent to the system

$$\begin{cases} z_1 = \lambda z_1^0, \\ z_2 = \lambda z_2^0. \end{cases}$$

Assuming that $z_2^0 \neq 0$, one has that $\lambda = \dfrac{z_2}{z_2^0}$ and hence $z_1 = \dfrac{z_2}{z_2^0} z_1^0$, or equivalently:

$$z_1 z_2^0 - z_2 z_1^0 = 0.$$

This equation is a particular case of (4.11) with $a_1 = z_2^0$, $a_2 = -z_1^0$ and $b = 0$.
If $z_2^0 = 0$, the set

$$L_{z_1} := \{\lambda z_1^0 \,|\, \lambda \in \mathbb{C}(\mathbf{i})\} = \mathbb{C}(\mathbf{i})$$

is the complex line $\mathbb{C}(\mathbf{i})$ with equation $z_2 = 0$.
Reciprocally, given a homogeneous equation

$$z_1 a_1 + z_2 a_2 = 0 \tag{4.13}$$

with $a_1 \neq 0$, then one can take $Z_0 := -a_2 + \mathbf{j}a_1$; the set of solutions $Z = z_1 + \mathbf{j}z_2$ of (4.13) coincides with L_{Z_0}.

Example 4.1.4.1. We know already that any zero-divisor is of the form $\beta_1 \mathbf{e}$ or $\beta_2 \mathbf{e}^\dagger$, that is, Z is a zero-divisor if and only if $Z \in L_\mathbf{e} \setminus \{0\}$ or $Z \in L_{\mathbf{e}^\dagger} \setminus \{0\}$, thus the set of zero-divisors (together with the zero) is described by

$$\mathfrak{S}_0 = L_\mathbf{e} \cup L_{\mathbf{e}^\dagger}.$$

Note also that here we have two complex lines ($L_\mathbf{e}$ and $L_{\mathbf{e}^\dagger}$) that intersect only at the origin. The reader should recall that in $\mathbb{R}^3$ a pair of two-dimensional planes cannot intersect at just one point, but in $\mathbb{BC} \cong \mathbb{R}^4$ there is an additional dimension which allows this phenomenon.
The corresponding equations of the complex lines $L_\mathbf{e}$ and $L_{\mathbf{e}^\dagger}$ are, respectively,

$$z_1 + \mathbf{i}z_2 = 0$$

and

$$z_1 - \mathbf{i}z_2 = 0. \qquad \square$$

In order to define a complex line that does not necessarily pass through the origin, take $W_0 = w_1^0 + \mathbf{j}w_2^0 \in \mathbb{BC} \setminus \{0\}$. Then the $\mathbb{C}(\mathbf{i})$-complex line passing through W_0 and parallel to L_{Z_0} (that is, a complex line with the direction determined by Z_0) is defined as the set

$$L_{Z_0} + W_0 := \{\lambda Z_0 + W_0 \,|\, \lambda \in \mathbb{C}(\mathbf{i})\}.$$

Let us show that this definition is equivalent to that of a complex line. Indeed, given a point $Z \in L_{Z_0} + W_0$, with $Z = z_1 + \mathbf{j}z_2$, there exists $\lambda \in \mathbb{C}(\mathbf{i})$ such that

$$\begin{cases} z_1 = \lambda z_1^0 + w_1^0, \\ z_2 = \lambda z_2^0 + w_2^0. \end{cases}$$

Assuming again that $z_2^0 \neq 0$, we obtain that the complex cartesian components of Z satisfy the equation

$$z_1 z_2^0 - z_2 z_1^0 = w_1^0 z_2^0 - w_2^0 z_1^0,$$

which is again a particular case of (4.11), now in its inhomogenous form.

Reciprocally, given an equation

$$z_1 a_1 + z_2 a_2 = b, \tag{4.14}$$

with $a_1 \neq 0$ and $a_2 \neq 0$, let us find Z_0 and W_0 in $\mathbb{BC}\backslash\{0\}$ such that any $Z = z_1 + \mathbf{j}z_2$ whose components satisfy (4.14) belongs to $L_{Z_0} + W_0$. From the above equation we can write z_1 as

$$z_1 = \frac{b - a_2}{a_1} + a_2 \left(\frac{1 - z_2}{a_1} \right).$$

Define $\lambda := \dfrac{z_2 - 1}{a_1}$, that is, $z_1 = \dfrac{b - a_2}{a_1} - \lambda a_2$. Substituting in (4.14), we write now z_2 as

$$z_2 = 1 + \lambda a_1.$$

Let us show that $Z_0 := -a_2 + \mathbf{j}a_1$ and $W_0 := \dfrac{b - a_2}{a_1} + \mathbf{j}$ are the bicomplex numbers which we are looking for; indeed,

$$\lambda Z_0 + W_0 = -\lambda a_2 + \mathbf{j}\lambda a_1 + \frac{b - a_2}{a_1} + \mathbf{j}$$

$$= \left(\frac{b - a_2}{a_1} - \lambda a_2 \right) + \mathbf{j}(\lambda a_1 + 1)$$

$$= z_1 + \mathbf{j}z_2 = Z,$$

thus Z is a point on the complex line $L_{Z_0} + W_0$.

What happens if either $a_1 = 0$ or $a_2 = 0$? If $a_1 = 0$, $a_2 \neq 0$, then (4.14) becomes

$$a_2 z_2 = b. \tag{4.15}$$

We have already considered the case $b = 0$. If $b \neq 0$, z_2 is the constant $z_2 = \dfrac{b}{a_2}$ and z_1 takes any value in $\mathbb{C}(\mathbf{i})$. Hence, (4.15) is the complex line $L_{\mathbf{i}} + \dfrac{b}{a_2}\mathbf{j}$, where $L_{\mathbf{i}} := \{\lambda \mathbf{i} \,|\, \lambda \in \mathbb{C}(\mathbf{i})\} = \mathbb{C}(\mathbf{i})$.

Similarly, if $a_2 = 0$ and $a_1 \neq 0$, then (4.14) becomes

$$a_1 z_1 = b,$$

and if $b \neq 0$, then it is the complex line

$$L_{\mathbf{j}} + \frac{b}{a_1}\mathbf{j} \quad \text{with} \quad L_{\mathbf{j}} := \{\lambda\mathbf{j} \,|\, \lambda \in \mathbb{C}(\mathbf{i})\}.$$

Let us come back to equation (4.12) where the coefficient A is defined to be $A = a_1 + \mathbf{j}a_2$; we know already that the bicomplex number Z_0 which determines the direction of the complex line is given as

$$Z_0 = -a_2 + \mathbf{j}a_1 = \mathbf{j}(a_1 + \mathbf{j}a_2) = \mathbf{j}A,$$

that is,

$$A = -\mathbf{j}Z_0.$$

Thus (4.12) becomes

$$Z_0 Z^\dagger - Z_0^\dagger Z = 2\mathbf{j}b. \tag{4.16}$$

4.1.5 More properties of complex lines

We continue now with some other properties of complex lines.

Proposition 4.1.5.1. *A non-zero bicomplex number U belongs to the complex line L_{Z_0} if and only if $L_{Z_0} = L_U$. Thus the complex line is generated by any non-zero bicomplex number lying on it.*

Proposition 4.1.5.2. *The non-zero bicomplex number Z_0 belongs to $\mathfrak{S}$ if and only if $L_{Z_0} = L_{\mathbf{e}}$ or $L_{Z_0} = L_{\mathbf{e}^\dagger}$.*

The proofs of these two propositions are direct applications of the definitions.

Corollary 4.1.5.3. $Z_0 \notin \mathfrak{S}_0$ *if and only if* $L_{Z_0} \cap L_{\mathbf{e}} = \{0\}$ *and* $L_{Z_0} \cap L_{\mathbf{e}^\dagger} = \{0\}$.

It is instructive to look at equation (4.16) in idempotent form. Write $Z_0 = \beta_1^0\mathbf{e} + \beta_2^0\mathbf{e}^\dagger$, $Z = \beta_1\mathbf{e} + \beta_2\mathbf{e}^\dagger$, then (4.16) becomes

$$\beta_2^0\beta_1 - \beta_1^0\beta_2 = 2\mathbf{i}b. \tag{4.17}$$

An immediate application of this formula is

Proposition 4.1.5.4. *If $Z_0 \notin \mathfrak{S}_0$, then the complex line $L_{Z_0} + W_0$ intersects each of $L_{\mathbf{e}}$ and $L_{\mathbf{e}^\dagger}$ at exactly one point.*

Proof. Since $Z_0 \notin \mathfrak{S}_0$, then $\beta_1^0 \neq 0$ and $\beta_2^0 \neq 0$. Let Z be a point in the intersection $(L_{Z_0} + W_0) \cap L_{\mathbf{e}}$, then its idempotent components satisfy (4.17) and $\beta_2 = 0$. Hence we have: $\beta_1 = \dfrac{2\mathbf{i}b}{\beta_2^0}$, i.e., $Z = \dfrac{2\mathbf{i}b}{\beta_2^0}\mathbf{e}$, and there are no other points in this intersection. Exactly in the same way, if $Z \in (L_{Z_0} + W_0) \cap L_{\mathbf{e}^\dagger}$, then $Z_2 = -\dfrac{2\mathbf{i}b}{\beta_1^0}\mathbf{e}^\dagger$. $\qquad\square$

If we perform a similar analysis in $\mathbb{C}^2(\mathbf{j})$, then all the formulas and conclusions above are valid in their corresponding analogues. For example, the $\mathbb{C}(\mathbf{j})$-complex line

$$\alpha_1 \zeta_1 + \alpha_2 \zeta_2 = \nu,$$

where all numbers and coefficients are in $\mathbb{C}(\mathbf{j})$, has the bicomplex formulation

$$\overline{A}Z + A\overline{Z} = 2\nu,$$

where $A = \alpha_1 + \mathbf{i}\alpha_2$ and $Z = \zeta_1 + \mathbf{i}\zeta_2$ are bicomplex numbers. Nevertheless, they are in the same set $\mathbb{BC}$, so that, as a matter of fact, there are two types of complex lines in $\mathbb{BC}$, the $\mathbb{C}(\mathbf{i})$-lines and $\mathbb{C}(\mathbf{j})$-lines.

Given a $\mathbb{C}(\mathbf{i})$-complex straight line, is it possible to consider it as a $\mathbb{C}(\mathbf{j})$-complex straight line or reciprocally? This question is related with the following one: given a real two-dimensional plane in $\mathbb{BC}$, is it always possible to consider it as a complex line?

Here one can find some partial answers to these questions:

- What are the $\mathbb{C}(\mathbf{i})$-complex lines containing the number 1? Assume that L is such a $\mathbb{C}(\mathbf{i})$-complex line of the form $L = L_{Z_0}$ for some $Z_0 \in \mathbb{BC} \setminus \{0\}$. Let L_1 be the complex line generated by 1, then obviously $L_1 = \mathbb{C}(\mathbf{i})$. We know also that $1 \in L_{Z_0}$ is equivalent to $L_{Z_0} = L_1$, that is, $L = \mathbb{C}(\mathbf{i})$.

  Thus, among all the two-dimensional real planes which contain the axis x_1, there is only one which has the structure of a $\mathbb{C}(\mathbf{i})$-complex line.

- In the same way, among all the two-dimensional real planes which contain the axis x_1, only $\mathbb{C}(\mathbf{j})$ is a $\mathbb{C}(\mathbf{j})$-complex line.

- Other conclusions are: $\mathbb{C}(\mathbf{j})$ is not a $\mathbb{C}(\mathbf{i})$-complex line; $\mathbb{C}(\mathbf{i})$ is not a $\mathbb{C}(\mathbf{j})$-complex line.

- As we did with the $\mathbb{C}(\mathbf{i})$-complex straight lines, it can be proved in a similar fashion that, given $Z_0 \in \mathbb{BC} \setminus \{0\}$, the $\mathbb{C}(\mathbf{j})$-complex line generated by Z_0, i.e., in the direction of Z_0, is the set

$$T_{Z_0} := \{\mu Z_0 \mid \mu \in \mathbb{C}(\mathbf{j})\}.$$

As a consequence of these observations we can prove that:

Proposition 4.1.5.5. *The $\mathbb{C}(\mathbf{i})$- and $\mathbb{C}(\mathbf{j})$-complex lines generated by a non-zero bicomplex number Z_0 coincide if and only if Z_0 is a zero-divisor.*

Proof. Assume first that $Z_0 \in \mathfrak{S}$, then $Z_0 = \beta\mathbf{e}$ or $Z_0 = \gamma\mathbf{e}^\dagger$, with $\beta, \gamma \in \mathbb{C}(\mathbf{i})$. If $Z_0 = \beta\mathbf{e}$, then

$$\begin{aligned} L_{Z_0} &= \{\lambda Z_0 = \lambda\beta\mathbf{e} \mid \lambda \in \mathbb{C}(\mathbf{i})\} = \{\nu\mathbf{e} \mid \nu \in \mathbb{C}(\mathbf{i})\} \\ &= \{(a + \mathbf{i}b)\mathbf{e} \mid a, b \in \mathbb{R}\} = \{(a - \mathbf{j}b)\mathbf{e} \mid a, b \in \mathbb{R}\} \\ &= \{\mu\mathbf{e} \mid \mu \in \mathbb{C}(\mathbf{j})\} = T_{Z_0}. \end{aligned}$$

Here we used the "peculiar" identity $\mathbf{ie} = -\mathbf{je}$. Analogously for $Z_0 = \gamma \mathbf{e}^\dagger$.

Reciprocally, assume that $T_{Z_0} = L_{Z_0}$ and suppose that $Z_0 \notin \mathfrak{S}$. The equality of the sets T_{Z_0} and L_{Z_0} is equivalent to saying that given any $\lambda \in \mathbb{C}(\mathbf{i})$, there exists $\mu \in \mathbb{C}(\mathbf{j})$ such that $\lambda Z_0 = \mu Z_0$. But since Z_0 is not a zero-divisor, the cancellation law can be applied to the latter equality, concluding that $\mathbb{C}(\mathbf{i}) \ni \lambda = \mu \in \mathbb{C}(\mathbf{j})$, for any $\lambda \in \mathbb{C}(\mathbf{i})$, which is impossible. Hence $Z_0 \in \mathfrak{S}$. $\qquad\square$

We now come back to equation (4.12) and we analyze some more geometrical aspects of it. Here we are interested only in the case where A is not a zero-divisor. We know already that on this complex line all, but at most two points, are non-zero-divisors.

We can write A and Z in their $\mathbb{BC}_\mathbf{i}$-trigonometric forms:

$$A = |A|_\mathbf{i}(\cos \Theta_0 + \mathbf{j} \sin \Theta_0), \qquad Z = |Z|_\mathbf{i}(\cos \Theta + \mathbf{j} \sin \Theta),$$

where Θ and Θ_0 are the complex arguments of A and Z, respectively. Then equation (4.12) becomes

$$|A|_\mathbf{i}|Z|_\mathbf{i} \cos(\Theta - \Theta_0) = b. \qquad (4.18)$$

We observe here the similarity between the equations of real lines (4.4) and complex lines (4.18)! Moreover, inspired by (4.18), we define the $\mathbb{C}(\mathbf{i})$-complex *projection* of Z onto A, for any invertible bicomplex numbers A and Z, by

$$\mathrm{proj}_A(Z) := |Z|_\mathbf{i} \cos(\Theta - \Theta_0),$$

where $\Theta - \Theta_0$ can be called *the complex angle between A and Z*. We will come back to this notion later.

4.1.6   Slope of complex lines

Continuing along the same path of study of real lines in $\mathbb{R}^2$, we further investigate our complex straight lines. We continue for the time being to consider the case when $A = -\mathbf{j}Z_0$ is not a zero-divisor, hence also Z_0 is not a zero-divisor. We study first the subcase when $a_1 = 0$, thus $a_2 \neq 0$ and $Z_0 = -a_2$. Hence $L_{Z_0} = L_{-a_2} = \mathbb{C}(\mathbf{i})$. Our straight line is of the form

$$L_{Z_0} + W_0 = \mathbb{C}(\mathbf{i}) + W_0,$$

with $W_0 = \dfrac{b}{a_2}\mathbf{j}$. This is a complex line which we will call parallel to the complex line $\mathbb{C}(\mathbf{i})$. Making the convention that the complex line $\mathbb{C}(\mathbf{i})$ determines the notion of $\mathbb{C}(\mathbf{i})$-*complex horizontality*, any complex line parallel to $\mathbb{C}(\mathbf{i})$ will be called $\mathbb{C}(\mathbf{i})$-*complex horizontal*.

It follows that $|A|_\mathbf{i} = \pm a_2$, where the sign should be chosen in accordance with our agreement that the imaginary part of the complex number $|A|_\mathbf{i}$ is non-negative. In this case $\Theta_0 = \dfrac{\pi}{2}$ becomes a real number.

The equation of this line in trigonometric form is

$$\pm a_2 \cdot |Z|_i \cos\left(\Theta - \frac{\pi}{2}\right) = \pm a_2 |Z|_i \sin\Theta = b.$$

Consider now the case $a_2 = 0$, thus $a_1 \neq 0$ and $Z_0 = \mathbf{j}a_1$. Hence

$$L_{Z_0} = L_{\mathbf{j}a_1} = \{\lambda a_1 \mathbf{j} \,|\, \lambda \in \mathbb{C}(\mathbf{i})\} = \{\mu \mathbf{j} \,|\, \mu \in \mathbb{C}(\mathbf{i})\} = L_{\mathbf{j}}.$$

Notice that for any $\mu = a + \mathbf{i}b \in \mathbb{C}(\mathbf{i})$ the product $\mu\mathbf{j}$ is $\mu\mathbf{j} = a\mathbf{j} + \mathbf{k}b$, hence the $\mathbb{C}(\mathbf{i})$-complex line $L_{\mathbf{j}}$ is the two-dimensional real plane generated by $\mathbf{j}$ and $\mathbf{k}$. In this case, our complex line is of the form

$$L_{Z_0} + W_0 = L_{\mathbf{j}} + W_0$$

with $W_0 = \dfrac{b}{a_1} + \mathbf{j}$.

Note that for the set $L_{\mathbf{j}} + W_0$ one has:

$$L_{\mathbf{j}} + W_0 = \{\mu \cdot \mathbf{j} + \frac{b}{a_1} + \mathbf{j} \,|\, \mu \in \mathbb{C}(\mathbf{i})\}$$

$$= \{(\mu + 1) \cdot \mathbf{j} + \frac{b}{a_1} \,|\, \mu \in \mathbb{C}(\mathbf{i})\} = \{\mu \cdot \mathbf{j} + \frac{b}{a_1} \,|\, \mu \in \mathbb{C}(\mathbf{i})\}$$

$$= L_{\mathbf{j}} + \frac{b}{a_1}.$$

In other words, one may take $W_0 = \dfrac{b}{a_1}$ and thus any point $Z \in L_{\mathbf{j}} + \dfrac{b}{a_1}$ is of the form $Z = \mu \cdot \mathbf{j} + \dfrac{b}{a_1}$, i.e., $Z = z_1 + \mathbf{j}z_2$ with a constant $z_1 = \dfrac{b}{a_1}$ and an arbitrary complex number z_2.

Making now the convention that the complex line $L_{\mathbf{j}}$ determines the notion of $\mathbb{C}(\mathbf{i})$-complex *verticality*, any $\mathbb{C}(\mathbf{i})$-complex line $L_{\mathbf{j}} + W_0$, that is parallel to $L_{\mathbf{j}}$, will be called $\mathbb{C}(\mathbf{i})$-*vertical*.

Since in this case $|A|_i = \pm a_1$ and Θ_0 can be taken to be zero, then the equation of any vertical line in trigonometric form is

$$\pm a_1 \cdot |Z|_i \cos(\Theta) = b.$$

Since $L_{\mathbf{j}} \cap \mathbb{C}(\mathbf{i}) = \{0\}$, then any horizontal line intersects any vertical line in just one point. We have seen this phenomenon before, in the sense that $\mathbb{R}^4$ is "wide" enough and a pair of two-dimensional planes may intersect at one point only.

Now some words about the case when Z_0 is a zero-divisor. By Proposition 4.1.5.2, this is equivalent to $L_{Z_0} = L_{\mathbf{e}}$ or $L_{Z_0} = L_{\mathbf{e}^\dagger}$, thus, if $Z_0 \in \mathfrak{S}$, then any $\mathbb{C}(\mathbf{i})$-complex line $L_{Z_0} + W_0$ is parallel to $L_{\mathbf{e}}$ or to $L_{\mathbf{e}^\dagger}$. We call these types of complex lines *zero-divisor complex lines*.

We will see now that it is quite appropriate to say that a complex line $L :=$ $L_{Z_0} + W_0$ has the direction along Z_0 (or that it is parallel to L_{Z_0}) since we will relate the complex line with the complex argument of Z_0.

Recall that $Z_0 = -a_2 + \mathbf{j}a_1$, with

$$a_1 z_1 + a_2 z_2 = b \tag{4.19}$$

being the equation of a complex line parallel to L_{Z_0} and passing through W_0. Let us start with the case $Z_0 \notin \mathfrak{S}_0$, which is equivalent to $Z_0 Z_0^\dagger \neq 0$; thus, we can consider the trigonometric form of Z_0:

$$Z_0 = |Z_0|_\mathbf{i} \left(-\frac{a_2}{|Z_0|_\mathbf{i}} + \mathbf{j}\frac{a_1}{|Z_0|_\mathbf{i}} \right) = |Z_0|_\mathbf{i} \left(\cos \Psi_0 + \mathbf{j} \sin \Psi_0 \right).$$

One obtains directly that

$$\tan \Psi_0 = -\frac{a_1}{a_2} =: \Lambda,$$

which we will call the *complex slope* of the line L. Comparing with equation (4.19), we have a deep analogy with the definition of slope of usual real lines in $\mathbb{R}^2$. Thus, the slope Λ of the complex line L is the tangent of the complex argument Ψ_0 of Z_0, with $Z_0 \notin \mathfrak{S}_0$ being a bicomplex vector that determines the direction of the line L.

Note that if $Z_0 = -a_2 + \mathbf{j}a_1$ is still a non-zero-divisor, but $a_2 = 0$, we have seen before that in this case we have a vertical line, and again the analogy with real lines in $\mathbb{R}^2$ is maintained, since we can define $\Lambda = \infty$ and the complex argument of Z_0 is $\dfrac{\pi}{2}$, coinciding with the usual notion of a vertical line.

What happens when Z_0 is a zero-divisor? We know already that $L_{Z_0} = L_\mathbf{e}$ or $L_{Z_0} = L_{\mathbf{e}^\dagger}$ and our line is a zero-divisor complex line. Since in this case we have also that

$$Z_0 = -a_2 + \mathbf{j}a_1 = \lambda \frac{1}{2}(1 \pm \mathbf{ij}) = \frac{1}{2}\lambda \pm \frac{1}{2}\mathbf{i}\lambda \cdot \mathbf{j}, \qquad \lambda \in \mathbb{C}(\mathbf{i}),$$

then

$$\Lambda = -\frac{a_1}{a_2} = \pm\mathbf{i}$$

and since there are no complex numbers Ψ such that $\tan \Psi = \pm\mathbf{i}$ we can define the slope of the zero-divisor complex lines as being equal to $\mathbf{i}$ or $-\mathbf{i}$.

The previous reasoning motivates the question about the relationship between complex arguments of bicomplex numbers belonging to the same complex line.

4.1.7 Complex lines and complex arguments of bicomplex numbers

Proposition 4.1.7.1. *Given a complex line $L = L_{Z_0}$ generated by $Z_0 \notin \mathfrak{S}_0$ and any $S \in L_{Z_0}$, then the complex arguments of S and Z_0 are related as follows:*

$$\mathrm{Arg}_{\mathbb{C}(\mathbf{i})}(S) = \mathrm{Arg}_{\mathbb{C}(\mathbf{i})}(Z_0) \qquad or \qquad \mathrm{Arg}_{\mathbb{C}(\mathbf{i})}(S) = \mathrm{Arg}_{\mathbb{C}(\mathbf{i})}(Z_0) + \pi.$$

In other words, the slope of L_{Z_0} does not depend on the generator Z_0.

Proof. Given $S \in L_{Z_0} \setminus \{0\}$, then there exists $\lambda \in \mathbb{C}(\mathbf{i}) \setminus \{0\}$ such that

$$S = \lambda Z_0 = \lambda \left(z_1^0 + \mathbf{j} z_2^0 \right) = \lambda z_1^0 + \mathbf{j} \lambda z_2^0;$$

hence $S \notin \mathfrak{S}_0$. We know already that $|S|_{\mathbf{i}} = \pm|\lambda|_{\mathbf{i}} \cdot |Z_0|_{\mathbf{i}}$. More precisely we have to consider the following cases.

(a) If $\lambda \in \Pi^+$, the upper half-plane, this is equivalent to $|\lambda|_{\mathbf{i}} = \lambda$, and thus $\mathrm{Arg}_{\mathbb{C}(\mathbf{i})}(\lambda) = \{2\pi m \,|\, m \in \mathbb{Z}\}$.

   (a.1) If $|\lambda|_{\mathbf{i}} \cdot |Z_0|_{\mathbf{i}} = \lambda|Z_0|_{\mathbf{i}} \in \Pi^+$, then $|S|_{\mathbf{i}} = \lambda|Z_0|_{\mathbf{i}}$ and thus

   $$\mathrm{Arg}_{\mathbb{C}(\mathbf{i})}(S) = \mathrm{Arg}_{\mathbb{C}(\mathbf{i})}(\lambda) + \mathrm{Arg}_{\mathbb{C}(\mathbf{i})}(Z_0) = \mathrm{Arg}_{\mathbb{C}(\mathbf{i})}(Z_0);$$

   (a.2) if $|\lambda|_{\mathbf{i}} \cdot |Z_0|_{\mathbf{i}} = \lambda|Z_0|_{\mathbf{i}} \in \Pi^-$, then $|S|_{\mathbf{i}} = -\lambda|Z_0|_{\mathbf{i}}$ and thus

   $$\mathrm{Arg}_{\mathbb{C}(\mathbf{i})}(S) = \mathrm{Arg}_{\mathbb{C}(\mathbf{i})}(\lambda) + \mathrm{Arg}_{\mathbb{C}(\mathbf{i})}(Z_0) + \pi = \mathrm{Arg}_{\mathbb{C}(\mathbf{i})}(Z_0) + \pi.$$

(b) Now, if $\lambda \in \Pi^-$, the lower half-plane, this is equivalent to $|\lambda|_{\mathbf{i}} = -\lambda$ and thus $\mathrm{Arg}_{\mathbb{C}(\mathbf{i})}(\lambda) = \{(2m+1)\pi \,|\, m \in \mathbb{Z}\}$.

   (b.1) If $|\lambda|_{\mathbf{i}} \cdot |Z_0|_{\mathbf{i}} = -\lambda|Z_0|_{\mathbf{i}} \in \Pi^+$, then $|S|_{\mathbf{i}} = -\lambda|Z_0|_{\mathbf{i}}$ and thus

   $$\begin{aligned} \mathrm{Arg}_{\mathbb{C}(\mathbf{i})}(S) &= \mathrm{Arg}_{\mathbb{C}(\mathbf{i})}(\lambda) + \mathrm{Arg}_{\mathbb{C}(\mathbf{i})}(Z_0) \\ &= \{(2m+1)\pi \,|\, m \in \mathbb{Z}\} + \mathrm{Arg}_{\mathbb{C}(\mathbf{i})}(Z_0) \\ &= \mathrm{Arg}_{\mathbb{C}(\mathbf{i})}(Z_0) + \pi. \end{aligned}$$

   (b.2) if $|\lambda|_{\mathbf{i}} \cdot |Z_0|_{\mathbf{i}} = -\lambda|Z_0|_{\mathbf{i}} \in \Pi^-$, then $|S|_{\mathbf{i}} = \lambda|Z_0|_{\mathbf{i}}$ and thus

   $$\mathrm{Arg}_{\mathbb{C}(\mathbf{i})}(S) = \mathrm{Arg}_{\mathbb{C}(\mathbf{i})}(\lambda) + \mathrm{Arg}_{\mathbb{C}(\mathbf{i})}(Z_0) + \pi = \mathrm{Arg}_{\mathbb{C}(\mathbf{i})}(Z_0). \qquad \square$$

The proposition above inspires the idea of complex rays.

Definition 4.1.7.2. *We define the $\mathbb{C}(\mathbf{i})$-complex ray in the direction of $Z_0 \neq 0$ to be the set*

$$R_{Z_0}^+ := \left\{ S = \lambda Z_0 \,|\, \lambda \in \mathbb{C} \setminus \{0\} \text{ and } \mathrm{Arg}_{\mathbb{C}(\mathbf{i})}(S) = \mathrm{Arg}_{\mathbb{C}(\mathbf{i})}(Z_0) \right\} \cup \{0\},$$

and the $\mathbb{C}(\mathbf{i})$-complex ray in the "opposite" direction of $Z_0 \neq 0$ to be the set

$$R_{Z_0}^- := \left\{ S = \lambda Z_0 \,|\, \lambda \in \mathbb{C} \setminus \{0\} \text{ and } \mathrm{Arg}_{\mathbb{C}(\mathbf{i})}(S) = \mathrm{Arg}_{\mathbb{C}(\mathbf{i})}(Z_0) + \pi \right\} \cup \{0\}.$$

We are able now to provide different ways of writing the equation of a complex line, according to the information that one has at hand. Consider first the case in which L is not a vertical line, thus $a_2 \neq 0$, and write the equation of L as

$$z_2 = -\frac{a_1}{a_2} z_1 + \frac{b}{a_2} = \Lambda z_1 + c, \qquad (4.20)$$

with $c = \dfrac{b}{a_2}$. Let $Z = z_1 + \mathbf{j} z_2$ be such that its cartesian components satisfy (4.20). Using also the relations

$$z_1 = \frac{1}{2}(Z + Z^\dagger), \qquad z_2 = \frac{\mathbf{j}}{2}(Z^\dagger - Z), \qquad (4.21)$$

one obtains:

$$Z = \rho Z^\dagger + C, \qquad (4.22)$$

with $\rho := -\dfrac{\Lambda - \mathbf{j}}{\Lambda + \mathbf{j}}$ and $C := \dfrac{2b}{a_2(\Lambda + \mathbf{j})}$.

Assume now that we have a line L with slope $\Lambda = \Lambda_1 + \mathbf{j}\Lambda_2 \neq 0$ and passing through the point $S_0 = s_1^0 + \mathbf{j} s_2^0$ with s_1^0 and s_2^0 in $\mathbb{C}(\mathbf{i})$. We know that L is of the form $L_{Z_0} + S_0$. Note also that if $Z = z_1 + \mathbf{j} z_2 \in L$, then $Z - S_0 \in L_{Z_0}$; since

$$Z - S_0 = (z_1 - s_1^0) + \mathbf{j}(z_2 - s_2^0)$$

one has that the slope can be written as

$$\Lambda = \frac{z_2 - s_2^0}{z_1 - s_1^0},$$

that is, in this case the equation of L is

$$z_2 - s_2^0 = \Lambda(z_1 - s_1^0). \qquad (4.23)$$

Using again (4.21) and the recently defined ρ, the previous equation can be written in terms of Z and $Z^\dagger$:

$$Z - S_0 = \rho(Z^\dagger - S_0^\dagger). \qquad (4.24)$$

Here one notes that if L passes through the origin, one may take $S_0 = 0$, and any non-zero $Z \in L$ satisfies then

$$\frac{Z}{Z^\dagger} = \rho = -\frac{\Lambda - \mathbf{j}}{\Lambda + \mathbf{j}}.$$

Following what one does for real lines in $\mathbb{R}^2$, we would like now to find the equation of a complex line passing through two given points. We need first a few results.

Proposition 4.1.7.3. *A real two-dimensional plane L containing the origin (thus L is an $\mathbb{R}$-linear subspace of dimension 2) is a $\mathbb{C}(\mathbf{i})$-complex line passing through the origin if and only if $Z \in L$ implies $\mathbf{i}Z \in L$.*

Proof. Assume first that L is a $\mathbb{C}(\mathbf{i})$-complex line. Since $0 \in L$, then $L = L_{Z_0}$ for some $Z_0 \in \mathbb{BC} \setminus \{0\}$. Given $S \in L_{Z_0}$, then $S = \lambda Z_0$ for some $\lambda \in \mathbb{C}(\mathbf{i})$, hence

$$\mathbf{i}S = (\mathbf{i}\lambda)Z_0 \in L_{Z_0} = L \,.$$

For the reciprocal statement, given any $Z_0 \in L$, the set $\{Z_0, \mathbf{i}Z_0\}$ is a linearly independent set in $\mathbb{R}^4$, thus the $\mathbb{R}$-linear subspace $\mathcal{L}\{Z_0, \mathbf{i}Z_0\}$ generated by it coincides with L. Thus $L = L_{Z_0}$ is a complex line. $\square$

Proposition 4.1.7.4. *If there are two points Z and W in a real two-dimensional plane L containing the origin, such that the difference of their complex arguments is neither zero nor π, then L is not a $\mathbb{C}(\mathbf{i})$-complex line.*

Proof. Follows directly from Proposition 4.1.7.1. $\square$

Proposition 4.1.7.5. *Given $Z_0 \in \mathbb{BC} \setminus \{0\}$, there exists a unique $\mathbb{C}(\mathbf{i})$-complex line L passing through Z_0 and the origin.*

Proof. The proof follows from the fact that L is a real two-dimensional plane containing Z_0 and $\mathbf{i}Z_0$. $\square$

Let us come back to the task of finding the equation of the complex line passing through two given points S and W. Set $Z_0 := S - W$ and let L_{Z_0} be the unique $\mathbb{C}(\mathbf{i})$-complex line that contains Z_0. If $S = s_1^0 + \mathbf{j}s_2^0$ and $W = w_1^0 + \mathbf{j}w_2^0$, then $Z_0 = (s_1^0 - w_1^0) + \mathbf{j}(s_2^0 - w_2^0)$, thus the slope of L_{Z_0} is $\Lambda = \dfrac{s_2^0 - w_2^0}{s_1^0 - w_1^0}$. Having the slope and the point S that belongs to L, and using (4.23), the equation of L is:

$$z_2 - s_2^0 = \frac{s_2^0 - w_2^0}{s_1^0 - w_1^0} \cdot (z_1 - s_2^0) \,. \tag{4.25}$$

4.2   Hyperbolic lines in $\mathbb{BC}$

In this section we will follow the scheme developed for complex lines to introduce hyperbolic lines in $\mathbb{BC}$. This means that we are interested in solutions $(\mathfrak{z}_1, \mathfrak{z}_2) \in \mathbb{D}^2$ of the equation

$$\mathfrak{a}_1\mathfrak{z}_1 + \mathfrak{a}_2\mathfrak{z}_2 = \mathfrak{b}, \tag{4.26}$$

with $\mathfrak{a}_1$, $\mathfrak{a}_2$ and $\mathfrak{b}$ in $\mathbb{D}$, but only in those solutions which form real two-dimensional planes in $\mathbb{BC}$; only such sets of solutions we will call hyperbolic lines. Thus, we have to analyze equation (4.26) for different combinations of values of its coefficients.

But first we use the bicomplex language for rewriting (4.26). Writing $\mathfrak{z}_1 := x_1 + \mathbf{k}y_2$ and $\mathfrak{z}_2 := y_1 + \mathbf{k}(-x_2)$, we can identify any solution $(\mathfrak{z}_1, \mathfrak{z}_2) \in \mathbb{D}^2$ of (4.26) with the bicomplex number $Z = x_1 + \mathbf{i}y_1 + \mathbf{j}x_2 + \mathbf{k}y_2 = (x_1 + \mathbf{k}y_2) + \mathbf{i}(y_1 - \mathbf{k}x_2) = \mathfrak{z}_1 + \mathbf{i}\mathfrak{z}_2 \in \mathbb{BC}$, thus we will say also that Z is a solution of (4.26).

Since $Z^* = \mathfrak{z}_1 - \mathbf{i}\mathfrak{z}_2$, hence

$$\mathfrak{z}_1 = \frac{1}{2}(Z + Z^*) \qquad \text{and} \qquad \mathfrak{z}_2 = \frac{\mathbf{i}}{2}(Z^* - Z)$$

and (4.26) becomes

$$A^*Z + AZ^* = 2\mathfrak{b}, \tag{4.27}$$

with the bicomplex coefficient $A := \mathfrak{a}_1 + \mathbf{i}\mathfrak{a}_2$.

Consider as an illustration some particular cases.

1) If $\mathfrak{a}_1 = 0$, $\mathfrak{a}_2 = 1$, $\mathfrak{b} = 0$, then $\mathfrak{z}_2 = 0$ and $\mathfrak{z}_1$ takes any hyperbolic value, that is, the set of solutions of the equation

$$\mathfrak{z}_2 = 0$$

is the set $\{\mathfrak{z}_1 \mid \mathfrak{z}_1 \in \mathbb{D}\} = \mathbb{D}$.

2) If $\mathfrak{a}_1 = 1$, $\mathfrak{a}_2 = 0$, $\mathfrak{b} = 0$, we have the equation

$$\mathfrak{z}_1 = 0,$$

which describes the set $\mathbf{i}\mathbb{D}$.

As it would be expected both the set $\mathbb{D}$ and the set $\mathbf{i}\mathbb{D}$ are hyperbolic lines. Recall that the (real) two-dimensional plane $\mathbb{D}$ is neither a $\mathbb{C}(\mathbf{i})$-complex line nor a $\mathbb{C}(\mathbf{j})$-complex line.

In order to analyze (4.26) we write every hyperbolic number in its idempotent form:

$$\mathfrak{a}_1 = \alpha_1^1 \cdot \mathbf{e} + \alpha_2^1 \cdot \mathbf{e}^\dagger, \quad \mathfrak{a}_2 = \alpha_1^2 \cdot \mathbf{e} + \alpha_2^2 \cdot \mathbf{e}^\dagger, \quad \mathfrak{b} = \nu_1 \cdot \mathbf{e} + \nu_2 \cdot \mathbf{e}^\dagger,$$

$$\mathfrak{z}_1 = \eta_1^1 \cdot \mathbf{e} + \eta_2^1 \cdot \mathbf{e}^\dagger, \qquad \mathfrak{z}_2 = \eta_1^2 \cdot \mathbf{e} + \eta_2^2 \cdot \mathbf{e}^\dagger,$$

where all the coefficients are real numbers, and we write

$$A = \mathfrak{a}_1 + \mathbf{i}\mathfrak{a}_2 = (\alpha_1^1 \mathbf{e} + \alpha_2^1 \mathbf{e}^\dagger) + \mathbf{i}(\alpha_1^2 \mathbf{e} + \alpha_2^2 \mathbf{e}^\dagger)$$
$$= (\alpha_1^1 + \mathbf{i}\alpha_1^2)\mathbf{e} + (\alpha_2^1 + \mathbf{i}\alpha_2^2)\mathbf{e}^\dagger$$

and

$$Z = \mathfrak{z}_1 + \mathbf{i}\mathfrak{z}_2 = (\eta_1^1 + \mathbf{i}\eta_1^2)\mathbf{e} + (\eta_2^1 + \mathbf{i}\eta_2^2)\mathbf{e}^\dagger.$$

We observe that the component $(\eta_1^1 + \mathbf{i}\eta_1^2)\mathbf{e} = \eta_1^1 \mathbf{e} + \eta_1^2 \mathbf{i}\mathbf{e}$ can be seen as a point (or vector) in the real plane $\mathbb{BC}_\mathbf{e}$ (which is a complex line) with coordinates (η_1^1, η_1^2); the same comment applies to the other component of Z as well as to both components of A.

Hence, equation (4.26) in idempotent form becomes

$$(\alpha_1^1 \eta_1^1 + \alpha_1^2 \eta_1^2)\mathbf{e} + (\alpha_2^1 \eta_2^1 + \alpha_2^2 \eta_2^2)\mathbf{e}^\dagger = \nu_1 \mathbf{e} + \nu_2 \mathbf{e}^\dagger,$$

which is equivalent to a couple (not a system) of independent real equations:

$$\begin{cases} \alpha_1^1 \eta_1^1 + \alpha_1^2 \eta_1^2 = \nu_1, \\ \alpha_2^1 \eta_2^1 + \alpha_2^2 \eta_2^2 = \nu_2. \end{cases} \tag{4.28}$$

We begin by excluding the situations in which we obtain, as the set of solutions, the empty set or the whole $\mathbb{BC}$:

(I) $\mathfrak{a}_1 = \mathfrak{a}_2 = \mathfrak{b} = 0$ (in this case equation (4.26) defines the whole $\mathbb{BC}$), i.e., $A = \mathfrak{b} = 0$.

(II) $\mathfrak{a}_1 = \mathfrak{a}_2 = 0$, $\mathfrak{b} \neq 0$ (in this case the set of solutions is the empty set), i.e., $A = 0$ and $\mathfrak{b} \neq 0$.

(III) $\mathfrak{a}_1 = \alpha_1^1 \mathbf{e}$ and $\mathfrak{a}_2 = \alpha_1^2 \mathbf{e}$ with real coefficients α_1^1 and α_1^2 which are not zero simultaneously, but $\mathfrak{b} = \nu_1 \mathbf{e} + \nu_2 \mathbf{e}^\dagger$, with $\nu_2 \neq 0$ (empty set), i.e., $A = (\alpha_1^1 + i\alpha_1^2)\mathbf{e}$ and $\mathfrak{b} \notin \mathbb{BC}_\mathbf{e}$.

(IV) $\mathfrak{a}_1 = \alpha_2^1 \mathbf{e}^\dagger$ and $\mathfrak{a}_2 = \alpha_2^2 \mathbf{e}^\dagger$ again α_2^1 and α_2^2 real numbers not zero simultaneously, but $\mathfrak{b} = \nu_1 \mathbf{e} + \nu_2 \mathbf{e}^\dagger$, with $\nu_1 \neq 0$ (empty set), i.e., $A = (\alpha_2^1 + i\alpha_2^2)\mathbf{e}^\dagger$ and $\mathfrak{b} \notin \mathbb{BC}_{\mathbf{e}^\dagger}$.

In all these situations the solutions of (4.26) do not form a hyperbolic line.

There are more sets of solutions of (4.26) which do not form a real two-dimensional plane; they arise if

(V) A and $\mathfrak{b}$ are in $\mathbb{BC}_\mathbf{e}$, i.e., $A = (\alpha_1^1 + i\alpha_1^2)\mathbf{e}$ and $\mathfrak{b} = \nu_1 \mathbf{e}$ (note that $\alpha_2^1 = \alpha_2^2 = \nu_2 = 0$).

(VI) A and $\mathfrak{b}$ are in $\mathbb{BC}_{\mathbf{e}^\dagger}$, i.e., $A = (\alpha_2^1 + i\alpha_2^2)\mathbf{e}^\dagger$ and $\mathfrak{b} = \nu_2 \mathbf{e}^\dagger$ (note that $\alpha_1^1 = \alpha_1^2 = \nu_1 = 0$).

If condition (V) holds, then the second equation in (4.28) determines the whole real two-dimensional plane $\mathbb{BC}_{\mathbf{e}^\dagger}$ and the first equation determines a real line in the real plane $\mathbb{BC}_\mathbf{e}$; hence (4.28) determines the cartesian product of the real line in $\mathbb{BC}_\mathbf{e}$ with the plane $\mathbb{BC}_{\mathbf{e}^\dagger}$, that is, a three-dimensional set in $\mathbb{BC}$. The same applies to the case (VI). We do not call these sets hyperbolic lines.

Theorem 4.2.1. *Equation (4.26) determines a hyperbolic line if and only if none of the above six restrictions is valid.*

Proof. The independent equations in (4.28) determine a two-dimensional set if and only if each of them determines a one-dimensional set, which holds if and only if none of the six conditions hold.   □

Corollary 4.2.2. *A real two-dimensional plane P in $\mathbb{R}^4 \cong \mathbb{BC}$ is a hyperbolic line if and only if its projections onto the complex lines $\mathbb{BC}_\mathbf{e}$ and $\mathbb{BC}_{\mathbf{e}^\dagger}$ are usual real lines.*

Remark 4.2.3. *Let Γ_1 be any real line in the plane $\mathbb{BC}_\mathbf{e}$ and let Γ_2 be any real line in the plane $\mathbb{BC}_{\mathbf{e}^\dagger}$; then Theorem 4.2.1 and Corollary 4.2.2 together say that the set*

$$\Gamma_1 \mathbf{e} + \Gamma_2 \mathbf{e}^\dagger$$

is a hyperbolic line in $\mathbb{BC}$ and that every hyperbolic line in $\mathbb{BC}$ is of this form.

4.2.1 Parametric representation of hyperbolic lines

As we did with complex lines, we will see now that any hyperbolic line that passes through the origin can be obtained by taking "hyperbolic multiples" of a non-zero bicomplex number. In other words, we will prove the following

Theorem 4.2.1.1. *Given a bicomplex number Z_0, the set*

$$P_{Z_0} := \left\{ \mu Z_0 \,\middle|\, \mu \in \mathbb{D} \right\}$$

is a hyperbolic line passing through the origin if and only if $Z_0 \notin \mathfrak{S}_0$. More precisely, the set P_{Z_0} is given by the solutions of the equation

$$\mathfrak{a}_1 \mathfrak{z}_1 + \mathfrak{a}_2 \mathfrak{z}_2 = 0, \tag{4.29}$$

with $A := \mathfrak{a}_1 + i\mathfrak{a}_2 = -iZ_0$.

Proof. Take $Z \in P_{Z_0}$ and write $Z = (u_1 + iv_1)\mathbf{e} + (u_2 + iv_2)\mathbf{e}^\dagger$, then there exists $\mu = m_1\mathbf{e} + m_2\mathbf{e}^\dagger \in \mathbb{D}$ such that

$$Z = \mu Z_0. \tag{4.30}$$

Indeed, write $Z_0 = \beta_1^0 \mathbf{e} + \beta_2^0 \mathbf{e}^\dagger = (u_1^0 + iv_1^0)\mathbf{e} + (u_2^0 + iv_2^0)\mathbf{e}^\dagger$, then (4.30) is equivalent to the pair of independent equations

$$\begin{cases} m_1 \beta_1^0 = \beta_1, & m_1 \in \mathbb{R}, \\ m_2 \beta_2^0 = \beta_2, & m_2 \in \mathbb{R}, \end{cases}$$

or, equivalently,

$$\begin{cases} m_1(u_1^0 + iv_1^0) = u_1 + iv_1, & m_1 \in \mathbb{R}, \\ m_2(u_2^0 + iv_2^0) = u_2 + iv_2, & m_2 \in \mathbb{R}; \end{cases}$$

this means that the pairs (u_1, v_1) and (u_2, v_2) are solutions, respectively, of the equations

$$\begin{cases} u_1^0 u_1 + (-u_1^0)v_1 = 0, \\ u_2^0 u_2 + (-u_2^0)v_2 = 0. \end{cases} \tag{4.31}$$

Hence Z belongs to P_{Z_0} if and only if its components satisfy the equations (4.31).

Comparing systems (4.31) and (4.28), we see that (4.31) determines a hyperbolic line if and only if it satisfies Corollary 4.2.2.

On the other hand the set of solutions of (4.31) admits descriptions in hyperbolic and bicomplex terms as in equations (4.26) and (4.27). To see this we compute:

$$
\begin{aligned}
Z_0 = \mathfrak{z}_1 + \mathbf{i}\mathfrak{z}_2 &= (u_1^0 + iv_1^0)\mathbf{e} + (u_2^0 + iv_2^0)\mathbf{e}^\dagger \\
&= \mathbf{i}(v_1^0 + \mathbf{i}(-u_1^0))\mathbf{e} + \mathbf{i}(v_2^0 + \mathbf{i}(-u_2^0))\mathbf{e}^\dagger \\
&= \mathbf{i}(v_1^0\mathbf{e} + v_2^0\mathbf{e}^\dagger) - (-u_1^0\mathbf{e} - u_2^0\mathbf{e}^\dagger) \\
&=: \mathbf{i}\mathfrak{a}_1 - \mathfrak{a}_2 = \mathbf{i}(\mathfrak{a}_1 + \mathbf{i}\mathfrak{a}_2) =: \mathbf{i}A,
\end{aligned}
$$

from where $\mathfrak{a}_1 = \mathfrak{z}_2^0$ and $\mathfrak{a}_2 = -\mathfrak{z}_1^0$. Thus, the set P_{Z_0} is the set of solutions of the equation

$$
\mathfrak{z}_2^0\mathfrak{z}_1 - \mathfrak{z}_1^0\mathfrak{z}_2 = 0.
$$

Finally we know that such a set determines a hyperbolic line if and only if $\mathfrak{z}_2^0$ and $\mathfrak{z}_1^0$ satisfy the restrictions (I)–(VI) and this is equivalent to saying that $Z_0 \notin \mathfrak{S}_0$.   □

4.2.2   More properties of hyperbolic lines

As it was in the case of complex lines, any hyperbolic line that passes through $W_0 \neq 0$ and that is parallel to P_{Z_0} (i.e., a hyperbolic line with the direction determined by Z_0) is defined as the set

$$
P_{Z_0} + W_0 := \left\{ \mu Z_0 + W_0 \,\middle|\, \mu \in \mathbb{D} \right\}.
$$

We leave it to the reader to show that any hyperbolic line can be described as $P_{Z_0} + W_0$ for some bicomplex numbers $Z_0,\ W_0$.

We also leave it to the reader to prove the following results, which are analogous to those given in the case of complex lines.

Proposition 4.2.2.1. *Given the hyperbolic line P_{Z_0} and given $U \in P_{Z_0}$ such that $U = \mu_0 Z_0$, with $\mu_0 \notin \mathfrak{S}_0$ (hence $U \notin \mathfrak{S}_0$), then $P_U = P_{Z_0}$. Reciprocally, if $P_U = P_{Z_0}$, then $U \in P_{Z_0}$.*

Another characterization of real two-dimensional planes passing through the origin and being a hyperbolic line is given in the following Proposition whose proof is left to the reader.

Proposition 4.2.2.2. *A real two-dimensional plane $T \subset \mathbb{BC}$ with $0 \in T$ is a hyperbolic line if and only if $U \in T$ implies $\mathbf{k}U \in T$.*

Corollary 4.2.2.3. *A hyperbolic line P through the origin, regarded as a two-dimensional subspace of $\mathbb{R}^4 \cong \mathbb{BC}$, is generated over $\mathbb{R}$ by any $U_0 \in P \setminus \mathfrak{S}_0$ and by $\mathbf{k}U_0$, i.e., it is the real span of U_0 and $\mathbf{k}U_0$.*

Now, in analogy with complex lines, we describe the relation between the hyperbolic arguments of the points of a hyperbolic line P_{Z_0}.

Theorem 4.2.2.4. *Consider the hyperbolic line P_{Z_0} for a point $Z_0 \in \mathbb{BC} \backslash \mathfrak{S}_0$. Write*

$$Z_0 = |Z_0|_\mathbf{k} \left(e^{i\nu_1^0} \mathbf{e} + e^{i\nu_2^0} \mathbf{e}^\dagger \right),$$

with $\Psi_{Z_0} = \nu_1^0 \mathbf{e} + \nu_2^0 \mathbf{e}^\dagger = \arg_\mathbb{D} Z_0 \in \mathbb{D}$ the principal hyperbolic argument of Z_0. Then the set $\mathrm{Arg}_\mathbb{D} U$ of any point $U \in P_{Z_0} \backslash \mathfrak{S}_0$ is one of the four sets:

$$\mathrm{Arg}_\mathbb{D} U = \left\{ \left(\nu_1^0 + 2m\pi \right) \mathbf{e} + \left(\nu_2^0 + 2n\pi \right) \mathbf{e}^\dagger \,\middle|\, m, n \in \mathbb{Z} \right\},$$

$$\mathrm{Arg}_\mathbb{D} U = \left\{ \left(\nu_1^0 + 2m\pi \right) \mathbf{e} + \left(\nu_2^0 + (2n+1)\pi \right) \mathbf{e}^\dagger \,\middle|\, m, n \in \mathbb{Z} \right\},$$

$$\mathrm{Arg}_\mathbb{D} U = \left\{ \left(\nu_1^0 + (2m+1)\pi \right) \mathbf{e} + \left(\nu_2^0 + 2n\pi \right) \mathbf{e}^\dagger \,\middle|\, m, n \in \mathbb{Z} \right\},$$

or

$$\mathrm{Arg}_\mathbb{D} U = \left\{ \left(\nu_1^0 + (2m+1)\pi \right) \mathbf{e} + \left(\nu_2^0 + (2n+1)\pi \right) \mathbf{e}^\dagger \,\middle|\, m, n \in \mathbb{Z} \right\}.$$

Proof. Since $U \in P_{Z_0} \backslash \mathfrak{S}_0$, then $U = \mu Z_0$ for some $\mu \in \mathbb{D} \backslash \mathfrak{S}_0$. Since the trigonometric form of μ in hyperbolic terms is

$$\mu = |\mu|_\mathbf{k} \left(\pm \mathbf{e} \pm \mathbf{e}^\dagger \right),$$

the statement follows from the fact that

$$U = \mu Z_0 = |\mu Z_0|_\mathbf{k} \left(\pm e^{i\nu_1^0} \mathbf{e} \pm e^{i\nu_2^0} \mathbf{e}^\dagger \right). \qquad \square$$

We are now ready to define the slope of hyperbolic lines; Theorem 4.2.2.4 ensures that the slope will not depend on the bicomplex number Z_0 that determines its direction.

We now need to use a result about trigonometric functions of a hyperbolic variable, a topic that is studied in detail in Chapter 6. As we will show there, the tangent of a hyperbolic number $\nu = \nu_1 \mathbf{e} + \nu_2 \mathbf{e}^\dagger$ can be given a rigorous definition and it turns out that

$$\tan \nu = \tan \nu_1 \mathbf{e} + \tan \nu_2 \mathbf{e}^\dagger.$$

Definition 4.2.2.5. *The slope N of a hyperbolic line $P_{Z_0} + W_0$, is defined by*

$$N := \tan \Psi_{Z_0} = \tan \nu_1^0 \mathbf{e} + \tan \nu_2^0 \mathbf{e}^\dagger,$$

where $\Psi_{Z_0} \in \mathrm{Arg}_\mathbb{D}(Z_0)$.

Example 4.2.2.6. (a) The hyperbolic line $\mathbb{D}$ is generated by any non zero-divisor $a\mathbf{e} + b\mathbf{e}^\dagger$ whose hyperbolic argument is $\Psi_{a\mathbf{e}+b\mathbf{e}^\dagger} = 0 = 0\mathbf{e} + 0\mathbf{e}^\dagger$. Therefore the slope of $\mathbb{D}$ is $\tan 0 = \tan 0\mathbf{e} + \tan 0\mathbf{e}^\dagger = 0$. Thus, any hyperbolic line $\mathbb{D} + W_0$ "parallel" to $\mathbb{D}$ has the slope zero, and all of them together give the notion of *hyperbolic horizontality*.

(b) Each real line L in $\mathbb{C}(\mathbf{i})$ is exactly the intersection of a hyperbolic line P and $\mathbb{C}(\mathbf{i})$ itself, and the slope of the hyperbolic line P coincides with the (usual real) slope of L. Indeed, assume that $0 \in L$, then any $z \in L$ is of the form

$$z = |z| \left(\cos \theta + \mathbf{i} \sin \theta \right) = |z|_{\mathbf{k}} \left(e^{\mathbf{i}\theta} \mathbf{e} + e^{\mathbf{i}\theta} \mathbf{e}^\dagger \right);$$

thus, the hyperbolic line $P_z = P$ which passes through z has slope

$$N = \tan \theta \mathbf{e} + \tan \theta \mathbf{e}^\dagger = \tan \theta.$$

(c) Consider the hyperbolic line of the type $P_{\mathbf{i}\mathbf{e} + \beta_2^0 \mathbf{e}^\dagger}$ or $P_{\beta_1^0 \mathbf{e} + \mathbf{i}\mathbf{e}^\dagger}$; since the hyperbolic arguments are $\Psi_{\mathbf{i}\mathbf{e} + \beta_2^0 \mathbf{e}^\dagger} = \dfrac{\pi}{2} \mathbf{e} + \nu_2^0 \mathbf{e}^\dagger$ and $\Psi_{\beta_1^0 \mathbf{e} + \mathbf{i}\mathbf{e}^\dagger} = \nu_1^0 \mathbf{e} + \dfrac{\pi}{2} \mathbf{e}^\dagger$, then the slopes are not well defined but can be symbolically represented as

$$\tan \Psi_{\mathbf{i}\mathbf{e} + \beta_2^0 \mathbf{e}^\dagger} = \infty \mathbf{e} + \tan \nu_2^0 \mathbf{e}^\dagger,$$

$$\tan \Psi_{\beta_1^0 \mathbf{e} + \mathbf{i}\mathbf{e}^\dagger} = \tan \nu_1^0 \mathbf{e} + \infty \mathbf{e}^\dagger.$$

We say in this case that these hyperbolic lines determine the notion of *hyperbolic verticality*.   □

Recalling that a hyperbolic line is a real two-dimensional plane such that its projections onto $\mathbb{BC}_\mathbf{e}$ and $\mathbb{BC}_{\mathbf{e}^\dagger}$ are usual real lines, we can give a precise geometrical notion of the hyperbolic angle between two hyperbolic lines. Clearly it is enough to define the hyperbolic angle between hyperbolic lines that pass through the origin.

Definition 4.2.2.7. *Take hyperbolic lines P_{Z_0} and P_{W_0}; if $\arg_{\mathbb{D}} Z_0 = \nu_1^0 \mathbf{e} + \nu_2^0 \mathbf{e}^\dagger$ and $\arg_{\mathbb{D}} W_0 = \mu_1^0 \mathbf{e} + \mu_2^0 \mathbf{e}^\dagger$, then the (trigonometric) hyperbolic angle α between P_{Z_0} and P_{W_0} is*

$$\alpha := \arg_{\mathbb{D}} Z_0 - \arg_{\mathbb{D}} W_0 = (\nu_1^0 - \mu_1^0)\mathbf{e} + (\nu_2^0 - \mu_2^0)\mathbf{e}^\dagger, \tag{4.32}$$

and the (geometric) hyperbolic angle between P_{Z_0} and P_{W_0} is

$$\begin{aligned} |\alpha|_{\mathbf{k}} &:= |\arg_{\mathbb{D}} Z_0 - \arg_{\mathbb{D}} W_0|_{\mathbf{k}} \\ &= |(\nu_1^0 - \mu_1^0)\mathbf{e} + (\nu_2^0 - \mu_2^0)\mathbf{e}^\dagger|_{\mathbf{k}} = |\nu_1^0 - \mu_1^0|\mathbf{e} + |\nu_2^0 - \mu_2^0|\mathbf{e}^\dagger. \end{aligned} \tag{4.33}$$

In analogy with the complex plane, if the angle α is a positive hyperbolic number, then we say that the angle between the lines is positively oriented; if the angle α is a hyperbolic negative number, then we say that the angle between the lines is negatively oriented; if the angle is neither negative nor positive, then no orientation is assigned.

Denoting by $L_{Z_0}^1$ and $L_{W_0}^1$ the real lines in $\mathbb{BC}_\mathbf{e}$ that are the projections of P_{Z_0} and P_{W_0} on $\mathbb{BC}_\mathbf{e}$ and by $L_{Z_0}^2$, $L_{W_0}^2$ the respective projections on $\mathbb{BC}_{\mathbf{e}^\dagger}$ we are not able to say which of P_{Z_0} or P_{W_0} is the initial or the final hyperbolic line that determines the orientation of the hyperbolic angle. The reason for this is that in the projections on $\mathbb{BC_e}$ and $\mathbb{BC_{e^\dagger}}$ the roles of $L^1_{Z_0}$, $L^1_{W_0}$ and $L^2_{Z_0}$, $L^2_{W_0}$ do not coincide in general; just recall that $\mathbb{D}$ is not a totally ordered set but a partially ordered one. That is why we use not the difference of the arguments but its hyperbolic modulus.

4.3 Hyperbolic and Complex Curves in $\mathbb{BC}$

4.3.1 Hyperbolic curves

We have described, quite extensively, the properties of complex and hyperbolic lines. Although their properties are rather similar, the hyperbolic lines have, in our opinion, some clear advantages from the point of view of geometrical interpretation and visualization; for example, the (hyperbolic-valued) angle between two hyperbolic lines has been defined, and the definition is not only formal but it has a precise geometrical description.

For this reason we will start with the study of hyperbolic curves which will be followed by the study of complex ones. Before giving their definition, let us point out a motivation why among all two-dimensional surfaces in $\mathbb{R}^4 \cong_{\mathbb{R}} \mathbb{BC}$ only very particular surfaces will be called hyperbolic curves. First of all, they have to be smooth two-dimensional surfaces; then the tangent (real two-dimensional) plane has to exist at any point and we will require that it should be a hyperbolic line! Since we know that a real two-dimensional plane is a hyperbolic line if and only if its projections onto the complex lines $\mathbb{BC_e}$ and $\mathbb{BC_{e^\dagger}}$ are usual real lines (see Corollary 4.2.2), this is the key idea that we will use in order to define hyperbolic curves.

Definition 4.3.1.1. *Let* Γ *be a two-dimensional surface with parametrization*

$$\phi : I := [a, b] \times [c, d] \subset \mathbb{R}^2 \to \mathbb{BC};$$

we say that Γ *is a hyperbolic curve if the idempotent representation of* ϕ *is of the form*

$$\phi(u, v) = \phi_1(u)\mathbf{e} + \phi_2(v)\mathbf{e}^\dagger$$

with $\phi_1 : [a, b] \to \mathbb{C}(\mathbf{i})$ *and* $\phi_2 : [c, d] \to \mathbb{C}(\mathbf{i})$ *being parametrizations of usual curves in* $\mathbb{C}(\mathbf{i})$.

The reader may note that a hyperbolic curve is a two-dimensional manifold in $\mathbb{R}^4$ of the simplest type. Note also that if γ_1 and γ_2 are two curves with parametrizations ϕ_1 and ϕ_2, then the hyperbolic curve Γ can be written as

$$\Gamma = \gamma_1 \mathbf{e} + \gamma_2 \mathbf{e}^\dagger.$$

$$\gamma_1 \mathbf{e} + \gamma_2 \mathbf{e}^\dagger$$

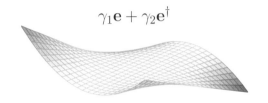

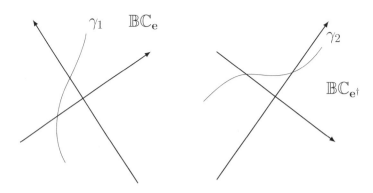

Figure 4.3.1: A HYPERBOLIC CURVE.

If additionally γ_1 and γ_2 are two smooth arcs, then Γ is a smooth two-manifold without self-intersections and with boundary

$$\partial\Gamma = \left(\phi_1(a)\mathbf{e} + \gamma_2\mathbf{e}^\dagger\right) \cup \left(\phi_1(b)\mathbf{e} + \gamma_2\mathbf{e}^\dagger\right)$$
$$\cup \left(\gamma_1\mathbf{e} + \phi_2(c)\mathbf{e}^\dagger\right) \cup \left(\gamma_1\mathbf{e} + \phi_2(d)\mathbf{e}^\dagger\right).$$

If γ_1 and γ_2 are closed Jordan curves, then the hyperbolic curve Γ is a manifold without boundary. In this case we say that Γ is a closed hyperbolic curve.

As an example, recall the definition of the bicomplex sphere of hyperbolic radius $\nu_1\mathbf{e} + \nu_2\mathbf{e}^\dagger \in \mathbb{D}^+$; it implies obviously that such a sphere is a closed hyperbolic curve.

If γ_1 and γ_2 are piece-wise smooth curves with parametrizations

$$\phi_1 : [a = a_1, a_2) \cup (a_2, a_3) \cup \cdots \cup (a_{s-1}, a_s = b] \longrightarrow \mathbb{C}(\mathbf{i})$$

and

$$\phi_2 : [c = c_1, c_2) \cup (c_2, c_3) \cup \cdots \cup (c_{t-1}, c_t = d] \longrightarrow \mathbb{C}(\mathbf{i}),$$

then Γ is called a piece-wise smooth hyperbolic curve. The smoothness is lost on the real one-dimensional curves $\phi_1(a_\ell)\mathbf{e} + \gamma_2\mathbf{e}^\dagger$ and $\gamma_1\mathbf{e} + \phi_2(c_r)\mathbf{e}^\dagger$ with $\ell \in \{1, 2, \ldots, s\}$

and $r \in \{1, 2, \ldots t\}$; note that their union is a set of measure zero (as a subset of the manifold Γ).

Of course, one may define more exotic types of hyperbolic curves eliminating other restrictions that we imposed on γ_1 and γ_2, but for our purposes the definition given above is sufficient.

4.3.2   Hyperbolic tangent lines to a hyperbolic curve

We will see now that if Γ is a smooth hyperbolic curve (with boundary or closed), then for each point $Z_0 \in \Gamma \setminus \partial\Gamma$ there exists the hyperbolic tangent line T to Γ.

Indeed, by definition of hyperbolic curve, Γ has a parametrization ϕ such that any point of Γ is of the form (see Definition 4.3.1.1):

$$\Gamma \ni Z = \phi(u, v) = \phi_1(u)\mathbf{e} + \phi_2(v)\mathbf{e}^\dagger$$

with ϕ_1 and ϕ_2 being parametrizations of smooth curves γ_1 and γ_2 in $\mathbb{C}(\mathbf{i})$. Fix a point $Z_0 \in \Gamma$, then $Z_0 = \phi_1(u_0)\mathbf{e} + \phi_2(v_0)\mathbf{e}^\dagger$, with $u_0 \in [a, b]$ and $v_0 \in [c, d]$ such that $\phi_1'(u_0) \neq 0$ and $\phi_2'(v_0) \neq 0$. Hence, there exist the tangent line T_1 to γ_1 at $\phi_1(u_0)$ and the tangent line T_2 to γ_2 at $\phi_2(v_0)$, which means that the set $T := T_1\mathbf{e} + T_2\mathbf{e}^\dagger$ is a real two-dimensional plane tangent to Γ at Z_0 and by Corollary 4.2.2, T is a hyperbolic line. See figure 4.3.2.

The above reasoning is summarized as

Theorem 4.3.2.1. *Let Γ be a smooth hyperbolic curve, then for each point Z_0 which is not on the boundary of Γ the tangent plane T to Γ at Z_0 is a hyperbolic line.*

4.3.3   Hyperbolic angle between hyperbolic curves

Let Γ and Λ be two smooth hyperbolic curves and let $Z_0 \in \Gamma \cap \Lambda$, $Z_0 \notin \partial\Gamma \cup \partial\Lambda$; let T be the hyperbolic tangent line to Γ at Z_0 and let S be the hyperbolic tangent line to Λ at Z_0.

Definition 4.3.3.1. *The hyperbolic angle between the hyperbolic tangent lines T and S at the point Z_0 is called the hyperbolic angle between the hyperbolic curves Γ and Λ at Z_0.*

Figure 4.3.3 illustrates the hyperbolic angle between the hyperbolic curves Γ and Λ that intersect at $Z_0 = \beta_1^0\mathbf{e} + \beta_2^0\mathbf{e}^\dagger$. Since Γ and Λ are smooth hyperbolic curves, they have the projections γ_1 and λ_1 on $\mathbb{BC}\mathbf{e}$ and the projections γ_2 and λ_2 on $\mathbb{BC}\mathbf{e}^\dagger$. Similarly, since T and S are hyperbolic tangent lines, their projections on $\mathbb{BC}\mathbf{e}$ are T_1 and S_1 and their projections on $\mathbb{BC}\mathbf{e}^\dagger$ are T_2 and S_2. Thus, the hyperbolic angle between the curves Γ and Λ is the hyperbolic angle between their tangents: $\theta_1\mathbf{e} + \theta_2\mathbf{e}^\dagger$.

It is possible to introduce the oriented hyperbolic angle between two hyperbolic curves; since a hyperbolic angle is, in a sense, a pair of planar angles, then there are four options for choosing an orientation of the hyperbolic angle. We will not elaborate on this here.

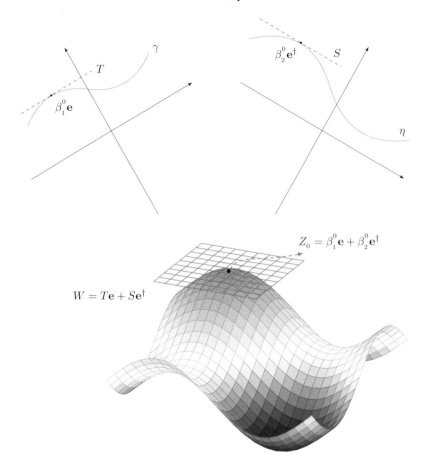

Figure 4.3.2: THE HYPERBOLIC TANGENT LINE TO A HYPERBOLIC CURVE.

4.3.4 Complex curves

While working with the notion of (straight) lines, we introduced the three types
of them: hyperbolic, $\mathbb{C}(\mathbf{i})$-complex and $\mathbb{C}(\mathbf{j})$-complex lines. Since a generalization
of the first has been given already with hyperbolic curves, then the next step is
to introduce the definition of ($\mathbb{C}(\mathbf{i})$-) complex curves, and that is what we will
develop in this section. Analogously, as it was made with hyperbolic curves, the
definition of a complex curve will be linked with complex lines.

Definition 4.3.4.1. *A smooth complex curve M is a smooth two-dimensional surface
in $\mathbb{R}^4$ such that at any point $Z_0 \in M$ the real two-dimensional tangent plane to
M is a complex line.*

We admit that this definition differs significantly from its hyperbolic coun-
terpart; in particular, it does not give an intrinsic description of the object and it

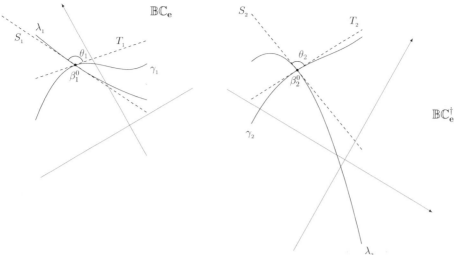

Figure 4.3.3: THE HYPERBOLIC ANGLE BETWEEN HYPERBOLIC CURVES.

is not easy to provide immediately an example of a complex curve. Thus, let us provide a sufficient condition that ensures that M is a complex curve.

Theorem 4.3.4.2. *Let* $\Gamma \subset \mathbb{R}^4 = \mathbb{C}^2(\mathbf{i})$ *be a real two-dimensional surface with a parametrization* $\varphi : \Omega \subset \mathbb{R}^2 \longrightarrow \mathbb{BC}$,

$$\varphi(u,v) = \psi_1(u,v) + \mathbf{j}\psi_2(u,v) := (\varphi_1(u,v) + \mathbf{i}\varphi_2(u,v)) + \mathbf{j}\,(\varphi_3(u,v) + \mathbf{i}\varphi_4(u,v)),$$

such that ψ_1 *and* ψ_2 *are holomorphic mappings. Then* Γ *is a complex curve.*

Proof. Take $Z_0 \in \Gamma$, then the real two-dimensional plane T tangent to Γ at Z_0 is of the form

$$T = Z_0 + T_1 := Z_0 + \left\{ a\frac{\partial \varphi}{\partial u}(Z_0) + b\frac{\partial \varphi}{\partial v}(Z_0) \,\Big|\, a,b \in R \right\}$$

with a plane T_1 passing through the origin. Thus, any point $W \in T$ can be written as

$$
\begin{aligned}
W = Z_0 + \bigg(& a\frac{\partial \varphi_1}{\partial u}(Z_0) + b\frac{\partial \varphi_1}{\partial v}(Z_0), \; a\frac{\partial \varphi_2}{\partial u}(Z_0) + b\frac{\partial \varphi_2}{\partial v}(Z_0), \\
& a\frac{\partial \varphi_3}{\partial u}(Z_0) + b\frac{\partial \varphi_3}{\partial v}(Z_0), \; a\frac{\partial \varphi_4}{\partial u}(Z_0) + b\frac{\partial \varphi_4}{\partial v}(Z_0) \bigg).
\end{aligned}
\tag{4.34}
$$

Recall (see Section 4.1.3) that T is a complex line if and only if T_1 is a complex line passing through the origin and that T_1 is a complex line if and only if it is invariant under the multiplication by $\mathbf{i}$, that is, T_1 as a subspace of dimension two in $\mathbb{R}^4$ is invariant under the action of the operator $M_{\mathbf{i}}$ of multiplication by $\mathbf{i}$ whose matrix is

$$M_{\mathbf{i}} = \begin{pmatrix} 0 & -1 & 0 & 0 \\ 1 & 0 & 0 & 0 \\ 0 & 0 & 0 & -1 \\ 0 & 0 & 1 & 0 \end{pmatrix}.$$

Here the identification $\mathbb{BC} \ni Z = x_1 + iy_1 + jx_2 + ky_2 \longleftrightarrow (x_1, y_1, x_2, y_2) \in \mathbb{R}^4$ is used, under which $\mathbf{i} \in \mathbb{BC}$ is identified with $(0, 1, 0, 0) \in \mathbb{R}^4$. Hence

$$M_{\mathbf{i}} \begin{pmatrix} x_1 \\ y_1 \\ x_2 \\ y_2 \end{pmatrix} = \begin{pmatrix} -y_1 \\ x_1 \\ -y_2 \\ x_2 \end{pmatrix} \tag{4.35}$$

and $iZ = -y_1 + ix_1 - jy_2 + kx_2$, i.e., the vector in (4.35) corresponds to iZ.

Thus, applying $M_{\mathbf{i}}$ to a point $w_1 \in T_1$ we get:

$$M_{\mathbf{i}} w_1 = \left(-a\frac{\partial\varphi_2}{\partial u}(Z_0) - b\frac{\partial\varphi_2}{\partial v}(Z_0),\; a\frac{\partial\varphi_1}{\partial u}(Z_0) + b\frac{\partial\varphi_1}{\partial v}(Z_0), \right.$$
$$\left. -a\frac{\partial\varphi_4}{\partial u}(Z_0) - b\frac{\partial\varphi_4}{\partial v}(Z_0),\; a\frac{\partial\varphi_3}{\partial u}(Z_0) + b\frac{\partial\varphi_3}{\partial v}(Z_0) \right). \tag{4.36}$$

Since ψ_1 and ψ_2 are holomorphic mappings, then the Cauchy–Riemann equations hold:

$$\begin{cases} \dfrac{\partial\varphi_1}{\partial u}(Z_0) = \dfrac{\partial\varphi_2}{\partial v}(Z_0) \\[2mm] \dfrac{\partial\varphi_1}{\partial v}(Z_0) = -\dfrac{\partial\varphi_2}{\partial u}(Z_0) \end{cases} \quad \text{and} \quad \begin{cases} \dfrac{\partial\varphi_3}{\partial u}(Z_0) = \dfrac{\partial\varphi_4}{\partial v}(Z_0) \\[2mm] \dfrac{\partial\varphi_3}{\partial v}(Z_0) = -\dfrac{\partial\varphi_4}{\partial u}(Z_0), \end{cases}$$

implying that $M_{\mathbf{i}} w_1$ becomes:

$$M_{\mathbf{i}} w_1 = \left(a\frac{\partial\varphi_1}{\partial v}(Z_0) - b\frac{\partial\varphi_1}{\partial u}(Z_0),\; a\frac{\partial\varphi_2}{\partial v}(Z_0) - b\frac{\partial\varphi_2}{\partial u}(Z_0), \right.$$
$$\left. a\frac{\partial\varphi_3}{\partial v}(Z_0) - b\frac{\partial\varphi_3}{\partial u}(Z_0),\; a\frac{\partial\varphi_4}{\partial v}(Z_0) - b\frac{\partial\varphi_4}{\partial u}(Z_0) \right)$$
$$= -b\frac{\partial\varphi}{\partial u}(Z_0) + a\frac{\partial\varphi}{\partial v}(Z_0) \in T_1.$$

Hence, T_1 is invariant under multiplication by $\mathbf{i}$ and T is a complex line. Since Z_0 was an arbitrary point of Γ, then Γ is a complex curve. $\qquad\square$

4.4 Bicomplex spheres and balls of hyperbolic radius

We have endowed the $\mathbb{BC}$-module $\mathbb{BC}$ with the hyperbolic-valued norm $|\cdot|_{\mathbf{k}}$; thus we can apply usefully the geometric language of normed spaces although normed spaces with hyperbolic-valued norm have practically no developed theory; such a norm was first introduced in [2] which covers all that is actually known about it. Given a positive hyperbolic number $\gamma_0 = a_0 \mathbf{e} + b_0 \mathbf{e}^{\dagger}$, recall that the set

$$\mathbb{S}_{\gamma_0} := \{ Z \in \mathbb{BC} \mid |Z|_{\mathbf{k}} = \gamma_0 \}$$

defines "the bicomplex sphere of hyperbolic radius γ_0 centered at the origin"; similarly, the set

$$\mathbb{B}_{\gamma_0} := \{ z \in \mathbb{BC} \mid |Z|_{\mathbf{k}} \prec \gamma_0 \}$$

is the (open) "bicomplex ball of hyperbolic radius γ_0 centered at the origin"; it is clear what is meant by a sphere $\mathbb{S}_{\gamma_0, Z_0}$ and a ball $\mathbb{B}_{\gamma_0, Z_0}$ centered at a point $Z_0 \in \mathbb{BC}$.

As a matter of fact, we have considered already a few facts tightly related with bicomplex spheres and balls. In Section 1.7 we have considered the equation $|\mathfrak{z}|_{\mathbf{k}} = R \geq 0$ in hyperbolic numbers. Such an equation defines a "hyperbolic sphere of real radius R centered at the origin" which is the slice of the bicomplex sphere $\mathbb{B}_R$ by the hyperbolic plane $\mathbb{D}$. It was shown that the intersection consists of exactly four points (for $R \geq 0$), and it is instructive to compare this fact with the sphere of radius $R > 0$ in $\mathbb{R}$: it consists of exactly two points which are the intersection of the circumference of the same radius in $\mathbb{C}$ with the real axis.

In the same Section 1.7 we have considered the equation $|\mathfrak{z}|_{\mathbf{k}} = \mathfrak{w}$ in hyperbolic numbers, with $\mathfrak{w}$ in $\mathbb{D}^{+}$. Thus, there we dealt with a "hyperbolic sphere of hyperbolic radius $\mathfrak{w}$" and it is the intersection of the bicomplex sphere $\mathbb{B}_{\mathfrak{w}}$ and of the hyperbolic plane $\mathbb{D}$. The structure of this hyperbolic sphere depends on $\mathfrak{w}$: if $\mathfrak{w}$ is a semi-positive hyperbolic number, i.e., $\mathfrak{w}$ is a positive zero-divisor, then the hyperbolic sphere consists of two points which are also zero-divisors although one of them is positive and the other is negative; if $\mathfrak{w}$ is strictly positive, then the hyperbolic sphere consists of four points.

Finally, in Section 2.6 we have considered the inequality $|\mathfrak{z}|_{\mathbf{k}} \preceq \mathfrak{w}$ in hyperbolic numbers with $\mathfrak{w}$ in $\mathbb{D}^{+}$ which is, again, about the intersection of $\mathbb{B}_{\mathfrak{w}} \cup \mathbb{S}_{\mathfrak{w}}$ and the hyperbolic plane $\mathbb{D}$, or, equivalently, about the hyperbolic ball of radius $\mathfrak{w}$ (that is, the solutions of the inequality $|\mathfrak{z}|_{\mathbf{k}} \prec \mathfrak{w}$) together with the corresponding hyperbolic sphere. It has been shown that this ball coincides with the hyperbolic interval $[-\mathfrak{w}, \mathfrak{w}]_{\mathbb{D}} := \{ \mathfrak{w} \in \mathbb{D} \mid -\mathfrak{w} \preceq \mathfrak{z} \preceq \mathfrak{w} \}$, and one may compare again with $\mathbb{R}$ where a "ball" is an interval which can be obtained by intersecting a "complex ball" in $\mathbb{C}$ (which is a disk) with the line $\mathbb{R}$.

Let us come back to the general situation, that is, let us consider the structures of arbitrary bicomplex spheres and balls.

Recall that, when the radius of the sphere is a zero-divisor of the form $\gamma_0 = a_0 \mathbf{e}$, then the sphere $\mathbb{S}_{a_0 \mathbf{e}}$ degenerates to a circumference with center at the origin and radius $\dfrac{a_0}{\sqrt{2}}$, contained in the complex line $\mathbb{BC_e}$. Similarly, if $\gamma_0 = b_0 \mathbf{e}^\dagger$,

the sphere $\mathbb{S}_{b_0 \mathbf{e}^\dagger}$ is the circumference with center at the origin and radius $\dfrac{b_0}{\sqrt{2}}$,

contained in the complex line $\mathbb{BC_{e^\dagger}}$. Finally, if γ_0 is not a zero-divisor, recall that the sphere $\mathbb{S}_{a_0 \mathbf{e} + b_0 \mathbf{e}^\dagger}$ is the surface of a torus:

$$ \mathbb{S}_{a_0 \mathbf{e} + b_0 \mathbf{e}^\dagger} = \left\{ Z = \beta_1 \cdot \mathbf{e} + \beta_2 \cdot \mathbf{e}^\dagger \,\big|\, |\beta_1| = a_0, \, , \, |\beta_2| = b_0 \right\}. $$

We are now ready to describe what a bicomplex ball is.

If γ_0 is not a zero-divisor, i.e., $a_0 \neq 0$ and $b_0 \neq 0$, then

$$ \mathbb{B}_{\gamma_0} = \left\{ Z = \beta_1 \mathbf{e} + \beta_2 \mathbf{e}^\dagger \,\big|\, |\beta_1| < a_0, \, |\beta_2| < b_0 \right\}, $$

i.e., if we are looking at $\mathbb{BC}$ as $\mathbb{C}^2(\mathbf{i})$ with the idempotent coordinates, then the bicomplex ball $\mathbb{B}_{\gamma_0}$ is the bicomplex form of writing for the usual bidisk centered at the origin and with bi-radius (a_0, b_0).

If γ_0 is a zero-divisor, then we cannot define the ball in the same way because none of the inequalities $|\beta_1| < 0$ or $|\beta_2| < 0$ has solutions. So we define in this case the ball $\mathbb{B}_{\gamma_0}$ to be one of the two disks: one is located in $\mathbb{BC_e}$ with center at the origin and radius $\dfrac{a_0}{\sqrt{2}}$, and the other is located in $\mathbb{BC_{e^\dagger}}$ with center at the origin and radius $\dfrac{b_0}{\sqrt{2}}$.

It is worth noting that for a bicomplex ball $\mathbb{B}_{\gamma_0}$, with γ_0 not a zero-divisor, the respective bicomplex sphere $\mathbb{S}_{\gamma_0}$ is not its topological boundary but it is its distinguished, or Shilov, boundary.

4.5   Multiplicative groups of bicomplex spheres

It is known that in the study of Euclidean spaces $\mathbb{R}^n$, the cases of $n = 2$ and $n = 4$ are peculiar for many reasons but in particular because the corresponding unitary spheres $\mathbb{S}^1$ in $\mathbb{R}^2$ and $\mathbb{S}^3$ in $\mathbb{R}^4$ are multiplicative groups; this is thanks to the complex numbers multiplication in $\mathbb{R}^2$ and the quaternionic multiplication in $\mathbb{R}^4$.

Let us consider some analogues of the above facts related to the bicomplex multiplication and bicomplex spheres with a hyperbolic radius:

$$ \mathbb{S}_\lambda := \left\{ Z = \beta_1 \mathbf{e} + \beta_2 \mathbf{e}^\dagger \,\big|\, |Z|_\mathbf{k} = \lambda \right\} $$

where $\lambda \in \mathbb{D}^+$. The multiplicative property of the hyperbolic modulus

$$ |Z \cdot W|_\mathbf{k} = |Z|_\mathbf{k} \cdot |W|_\mathbf{k} \tag{4.37} $$

will be crucial for the reasoning below.

Obviously, $\mathbb{S}_0 = \{0\}$. Consider the bicomplex unitary sphere $\mathbb{S}_1$; clearly

1) $1 \in \mathbb{S}_1$;

2) if $Z \in \mathbb{S}_1$, then it is invertible and $Z^{-1} \in \mathbb{S}_1$;

moreover, formula (4.37) says that

3) if $Z, W \in \mathbb{S}_1$, then $Z \cdot W \in \mathbb{S}_1$.

Thus, the bicomplex unitary sphere $\mathbb{S}_1$ is a multiplicative group, but this is an exceptional case for bicomplex spheres having a non-zero-divisor as its radius, and for other non-zero-divisor values of the parameter λ the sphere $\mathbb{S}_\lambda$ is not a multiplicative group; the same as in $\mathbb{R}^2$.

Recall what happens when λ is a positive zero-divisor, i.e., when $\lambda = \lambda_1 \mathbf{e} \in \mathbb{D}_{\mathbf{e}}^+$, with $\lambda_1 > 0$, or when $\lambda = \lambda_2 \mathbf{e}^\dagger \in \mathbb{D}_{\mathbf{e}^\dagger}^+$, with $\lambda_2 > 0$. If $\lambda \in \mathbb{D}_{\mathbf{e}}^+$, then for the points $Z = \beta_1 \mathbf{e} + \beta_2 \mathbf{e}^\dagger$ of $\mathbb{S}_\lambda$ we have that

$$|Z|_{\mathbf{k}} = |\beta_1| \cdot \mathbf{e} + |\beta_2| \cdot \mathbf{e}^\dagger = \lambda_1 \mathbf{e},$$

that is,

$$|\beta_1| = \lambda_1, \quad |\beta_2| = 0,$$

thus, the bicomplex sphere $\mathbb{S}_\lambda$ is

$$\mathbb{S}_\lambda = \{ Z \mid Z = \beta_1 \mathbf{e}, \ |\beta_1| = \lambda_1 > 0 \},$$

and it can be seen as the circumference of radius λ_1 centered at the origin which is located in the real two-dimensional plane, spanned by $\mathbf{e}$ and $\mathbf{ie}$ (which is a complex line in $\mathbb{BC}$).

Take $\lambda_1 = 1$ here, then $\mathbb{S}_\lambda$ becomes

$$\mathbb{S}_{\mathbf{e}} = \{ Z \mid Z = \beta_1 \mathbf{e}, \ |\beta_1| = 1 \}.$$

We are now in a situation very similar to that of Section 2.7.1, and we will use a similar approach in order to make $\mathbb{S}_{\mathbf{e}}$ a multiplicative group. The set $\mathbb{S}_{\mathbf{e}}$ is endowed with the multiplication $\circledast$ which is the restriction of the bicomplex multiplication and which acts invariantly on $\mathbb{S}_{\mathbf{e}}$; then the element $\mathbf{e}$ is the neutral element for $\circledast$, every element $\beta_1 \mathbf{e}$ in $\mathbb{S}_{\mathbf{e}}$ is $\circledast$-invertible and its $\circledast$-inverse is $\dfrac{1}{\beta_1} \mathbf{e} \in \mathbb{S}_{\mathbf{e}}$.

Thus, we conclude that $(\mathbb{S}_{\mathbf{e}}, \circledast)$ is a multiplicative group which is not a subgroup of the multiplicative group $(\mathbb{BC} \setminus \mathfrak{S}_0, \cdot)$. Of course, $(\mathbb{S}_{\mathbf{e}}, \circledast)$ is isomorphic to the multiplicative group of complex numbers of modulus one.

Note that if $\lambda_1 \neq 1$, then $\mathbb{S}_{\lambda_1 \mathbf{e}}$ is not a multiplicative group. In other words, among all the spheres $\mathbb{S}_{\lambda_1 \mathbf{e}}$ with $\lambda_1 > 0$, only one, $\mathbb{S}_{\mathbf{e}}$, is a multiplicative group. The situation with $\lambda = \lambda_2 \mathbf{e}^\dagger$ is similar.

In what follows we assume that

$$\lambda \in \mathbb{D}_{inv}^+, \tag{4.38}$$

that is, λ is a positive hyperbolic number; recall that $\mathbb{D}_{inv}^+$ is a multiplicative group. For any such λ the sphere $\mathbb{S}_\lambda$ enjoys the property that any Z in $\mathbb{S}_\lambda$ is invertible and

$$Z \in \mathbb{S}_\lambda \quad \text{if and only if} \quad Z^{-1} \in \mathbb{S}_{\lambda^{-1}}.$$

Notice also that if $Z \in \mathbb{S}_\lambda$ and $W \in \mathbb{S}_\mu$, then by (4.37)

$$|Z \cdot W|_{\mathbf{k}} = \lambda \cdot \mu \in \mathbb{D}_{inv}^+,$$

hence

$$Z \cdot W \in \mathbb{S}_{\lambda\mu}.$$

These properties hint that we should consider all the spheres "together" and we set:

$$\mathbf{S}_{\mathbb{D}} := \bigsqcup_{\lambda \in \mathbb{D}_{inv}^+} \mathbb{S}_\lambda$$

(i.e., $\mathbf{S}_{\mathbb{D}}$ is a disjoint union of spheres). The set $\mathbf{S}_{\mathbb{D}}$ is a multiplicative group with respect to the bicomplex multiplication.

We discussed already some specific features of bicomplex spheres with real radii, so consider the subset of $\mathbf{S}_{\mathbb{D}}$ defined by

$$\mathbf{S}_{\mathbb{R}} := \bigsqcup_{\lambda \in \mathbb{R}^+} \mathbb{S}_\lambda.$$

A sphere $\mathbb{S}_\lambda$ with $\lambda > 0$ is characterized by the condition: if $Z = \beta_1 \mathbf{e} + \beta_2 \mathbf{e}^\dagger \in \mathbb{S}_\lambda$, then

$$|\beta_1| = |\beta_2| = \lambda = |Z| = |Z|_{\mathbf{k}}.$$

It is easily seen that $\mathbf{S}_{\mathbb{R}}$ is a group also, and thus $\mathbf{S}_{\mathbb{R}}$ is a subgroup of $\mathbf{S}_{\mathbb{D}}$. Since $\mathbb{S}_1 \subset \mathbf{S}_{\mathbb{R}}$ and since $\mathbb{S}_1$ is a group, then $\mathbb{S}_1$ is a subgroup of both $\mathbf{S}_{\mathbb{R}}$ and $\mathbf{S}_{\mathbb{D}}$.

The bicomplex spheres generate another set which can be endowed with the structure of a multiplicative group. Indeed, introduce

$$\widetilde{\mathbf{S}}_{\mathbb{D}} := \{\mathbb{S}_\lambda\}_{\lambda \in \mathbb{D}_{inv}^+}$$

and define the multiplication "$\circ$" of bicomplex spheres by the formula

$$\mathbb{S}_\lambda \circ \mathbb{S}_\mu := \mathbb{S}_{\lambda\mu}.$$

The unitary sphere $\mathbb{S}_1$ will serve as the identity in $\widetilde{\mathbf{S}}_{\mathbb{D}}$ (i.e., the multiplicative neutral element), and the sphere $\mathbb{S}_{\lambda^{-1}}$ is the inverse of the sphere $\mathbb{S}_\lambda$.

The subset

$$\widetilde{\mathbf{S}}_{\mathbb{R}} := \{\mathbb{S}_\lambda\}_{\lambda \in \mathbb{R}^+}$$

of $\widetilde{\mathbf{S}}_{\mathbb{D}}$ is also a group and, thus, a subgroup of the group $\widetilde{\mathbf{S}}_{\mathbb{D}}$.

Observe that the mapping

$$\varphi : \lambda \in \mathbb{D}_{inv}^+ \longmapsto \mathbb{S}_\lambda \in \tilde{\mathbf{S}}_\mathbb{D}$$

is a group isomorphism, and its restriction $\varphi|_{\mathbb{R}^+}$ is a group isomorphism as well:

$$\varphi|_{\mathbb{R}^+} : \lambda \in \mathbb{R}^+ \longmapsto \mathbb{S}_\lambda \in \tilde{\mathbf{S}}_\mathbb{R}.$$

The sets $\mathbf{S}_\mathbb{D}$ and $\tilde{\mathbf{S}}_\mathbb{D}$ have proved to be multiplicative groups since the zero-divisors were forbidden to be radii of the spheres. Let us return to these exceptions and let us consider the sets

$$\mathcal{S}_{\mathbf{e},\mathbb{D}} := \bigsqcup_{\lambda \in \mathbb{D}_{\mathbf{e}}^+} \mathbb{S}_\lambda; \quad \mathcal{S}_{\mathbf{e}^\dagger,\mathbb{D}} := \bigsqcup_{\lambda \in \mathbb{D}_{\mathbf{e}^\dagger}^+} \mathbb{S}_\lambda.$$

Note that the sets $\mathbb{D}_{\mathbf{e}}^+$ and $\mathbb{D}_{\mathbf{e}^\dagger}^+$ have become multiplicative groups when a specific multiplication was introduced in each of them. This allows us, again, to make $\mathcal{S}_{\mathbf{e},\mathbb{D}}$ and $\mathcal{S}_{\mathbf{e}^\dagger,\mathbb{D}}$ multiplicative groups following the pattern of Section 2.7.1. If $Z \in \mathbb{S}_\lambda$ and $W \in \mathbb{S}_\mu$, i.e., $\lambda = \lambda_1 \mathbf{e}$ and $\mu = \mu_1 \mathbf{e}$, $Z = \beta_1 \mathbf{e}$ with $|\beta_1| = \lambda_1 > 0$ and $W = \gamma_1 \mathbf{e}$ with $|\gamma_1| = \mu_1 > 0$; then $Z \odot W := \beta_1 \gamma_1 \mathbf{e} \in \mathbb{S}_{\lambda_1 \mu_1 \mathbf{e}} = \mathbb{S}_{\lambda \star \mu}$. Hence, with this new multiplication the set $\mathcal{S}_{\mathbf{e},\mathbb{D}}$ becomes a multiplicative group. The same for $\mathcal{S}_{\mathbf{e}^\dagger,\mathbb{D}}$.

The analogues of $\tilde{\mathbf{S}}_\mathbb{D}$ are the sets

$$\tilde{\mathcal{S}}_{\mathbf{e},\mathbb{D}} := \{\mathbb{S}_\lambda\}_{\lambda \in \mathbb{D}_{\mathbf{e}}^+} \quad \text{and} \quad \tilde{\mathcal{S}}_{\mathbf{e}^\dagger,\mathbb{D}} := \{\mathbb{S}_\lambda\}_{\lambda \in \mathbb{D}_{\mathbf{e}^\dagger}^+} .$$

The multiplication $\odot$ of bicomplex spheres in $\tilde{\mathcal{S}}_{\mathbf{e},\mathbb{D}}$ is defined by the formula

$$\mathbb{S}_\lambda \odot \mathbb{S}_\mu := \mathbb{S}_{\lambda \star \mu}.$$

Similarly for $\tilde{\mathcal{S}}_{\mathbf{e}^\dagger,\mathbb{D}}$. The reader is invited to complete the proofs.

Most of the material covered in this Chapter is original and does not have counterparts in the existing literature. Our detailed study of complex and hyperbolic lines and curves in $\mathbb{BC}$ is of great importance, as mentioned at the end of Chapter 3, e.g. for applications in the mathematics of Special Relativity, [11].

Bicomplex spheres and balls have also been studied in [49, 56] from an Euclidean point of view, and, from a manifold theory point of view, in [7]. We stress once more that our approach to this subject involves the hyperbolic norm $|\cdot|_\mathbf{k}$, a notion that was first developed in [2]. We strongly believe that a comprehensive study of the applications of this norm in the context of Special Relativity is well worth pursuing in the future, a realm of study that is beyond the scope of this book.

Chapter 5

Limits and Continuity

The notion of limit for complex functions is well known and we will not rediscuss it here. Note that the formal proofs of its properties depend strongly on the properties of the modulus of a complex number;

$$|ab| = |a| \cdot |b|, \qquad |a + b| \leq |a| + |b|, \qquad \left|\frac{1}{a}\right| = \frac{1}{|a|} \quad \text{for } a \neq 0. \tag{5.1}$$

In particular, the existence of the limit of a function is equivalent to the existence of the limits of its real and imaginary parts; moreover, the limit of a function exists if and only if the limit of its conjugate function exists.

Our aim in this chapter is to extend the above to bicomplex functions showing that there exist many similarities but differences as well. Indeed, instead of properties (5.1) for the Euclidean modulus one has their analogs of the form

$$|Z \cdot W| \leq \sqrt{2}|Z| \cdot |W|, \qquad |Z + W| \leq |Z| + |W|, \tag{5.2}$$

and it turns out that they allow us to repeat most of the proofs almost literally, although one needs to take into account the fact that $|Z^{-1}|$ is not always equal to $\frac{1}{|Z|}$.

5.1 Bicomplex sequences

Definition 5.1.1. *A sequence of bicomplex numbers $\{Z_n\}_{n\in\mathbb{N}}$ is called* convergent *if there exists $Z_0 \in \mathbb{BC}$ such that for any $\epsilon > 0$ there exists $N \in \mathbb{N}$ such that for any $n \geq N$ there holds:*

$$|Z_n - Z_0| < \epsilon.$$

In this case we say that Z_0 is the limit of the sequence which we write as

$$\lim_{n\to\infty} Z_n = Z_0,$$

and we say that the sequence $\{Z_n\}_{n\in\mathbb{N}}$ converges to Z_0.

Recalling the different forms of writing the bicomplex numbers (1.3)-(1.9), including the idempotent representations (1.24), we have that

$$
\begin{aligned}
|Z|^2 &= |z_1|^2 + |z_2|^2 = |\zeta_1|^2 + |\zeta_2|^2 \\
&= |\mathfrak{z}_1|^2 + |\mathfrak{z}_2|^2 = |\mathfrak{w}_1|^2 + |\mathfrak{w}_2|^2 \\
&= |w_1|^2 + |w_2|^2 = |\omega_1|^2 + |\omega_2|^2 \\
&= x_1^2 + y_1^2 + x_2^2 + y_2^2 \\
&= \frac{1}{2}\left(|a_1|^2 + |a_2|^2\right) = \frac{1}{2}\left(|\alpha_1|^2 + |\alpha_2|^2\right),
\end{aligned}
$$

which gives directly that a sequence $\{Z_n\}_{n\in\mathbb{N}}$ converges to Z_0 if and only if the corresponding coordinate sequences converge to the respective components of Z_0. For instance, writing $Z_n = z_{1n} + \mathbf{j}z_{2n} = a_{1n}\mathbf{e} + a_{2n}\mathbf{e}^\dagger$ we conclude that $\{Z_n\}_{n\in\mathbb{N}}$ converges to $Z_0 = z_{10} + \mathbf{j}z_{20} = a_{10}\mathbf{e} + a_{20}\mathbf{e}^\dagger$ if and only if $\{z_{1n}\}_{n\in\mathbb{N}}$ converges to z_{10} and $\{z_{2n}\}_{n\in\mathbb{N}}$ converges to z_{20}, and if and only if $\{a_{1n}\}_{n\in\mathbb{N}}$ converges to a_{10} and $\{a_{2n}\}_{n\in\mathbb{N}}$ converges to a_{20}. The same happens for the other representations.

Consider now an arbitrary bicomplex convergent sequence $Z_n \to Z_0$; if $Z_n = a_{1n}\mathbf{e} + a_{2n}\mathbf{e}^\dagger$, $Z_0 = a_{10}\mathbf{e} + a_{20}\mathbf{e}^\dagger$ and assuming that Z_0 is not a zero-divisor, then $a_{10} \neq 0$, $a_{20} \neq 0$; since $a_{1n} \to a_{10}$ and $a_{2n} \to a_{20}$ and these are complex sequences, then there exists $n_0 \in \mathbb{N}$ such that $a_{1n} \neq 0$ and $a_{2n} \neq 0$ for all $n \geq n_0$. In other words starting from $n = n_0$ our sequence contains only invertible terms.

For the sum, product and quotient of two bicomplex sequences the usual statements hold, and we comment on the cases of the product and of the quotient. Let $\lim_{n\to\infty} Z_n = Z_0$ and $\lim_{n\to\infty} W_n = W_0$. For any $n \in \mathbb{N}$ there holds:

$$
\begin{aligned}
|Z_n \cdot W_n - Z_0 \cdot W_0| &= |W_n(Z_n - Z_0) + Z_0(W_n - W_0)| \\
&\leq |W_n(Z_n - Z_0)| + |Z_0(W_n - W_0)| \\
&\leq \sqrt{2}\left(|W_n| \cdot |Z_n - Z_0| + |Z_0| \cdot |W_n - W_0|\right),
\end{aligned}
$$

where we have used properties (5.2). Take $M > 0$ such that $|Z_n| < M$ and $|W_n| < M$ for all $n \in \mathbb{N}$. Now given an arbitrary $\epsilon > 0$ there exists $N \in \mathbb{N}$ such that simultaneously

$$
|Z_n - Z_0| < \frac{\epsilon}{2\sqrt{2}M} \quad \text{and} \quad |W_n - Z_0| < \frac{\epsilon}{2\sqrt{2}M}, \quad \forall n \geq N.
$$

We have just proved that the sequence $\{Z_n \cdot W_n\}_{n\in\mathbb{N}}$ converges to $Z_0 \cdot W_0$.

We proceed now to the case of a quotient using here the idempotent representations. Let $\{Z_n\}_{n\in\mathbb{N}}$ and $\{W_n\}_{n\in\mathbb{N}}$ be convergent sequences of bicomplex numbers: $\lim_{n\to\infty} Z_n = Z_0$, $\lim_{n\to\infty} W_n = W_0 \notin \mathfrak{S}_0$; then the quotients $\dfrac{Z_n}{W_n}$ are well defined for sufficiently large n. We write:

$$Z_n = \beta_{1n}\mathbf{e} + \beta_{2n}\mathbf{e}^\dagger, \quad W_n = \gamma_{1n}\mathbf{e} + \gamma_{2n}\mathbf{e}^\dagger, \tag{5.3}$$

$$Z_0 = \beta_{10}\mathbf{e} + \beta_{20}\mathbf{e}^\dagger, \quad W_0 = \gamma_{10}\mathbf{e} + \gamma_{20}\mathbf{e}^\dagger, \tag{5.4}$$

then

$$\frac{Z_n}{W_n} = \frac{\beta_{1n}}{\gamma_{1n}}\mathbf{e} + \frac{\beta_{2n}}{\gamma_{2n}}\mathbf{e}^\dagger.$$

By hypothesis, the following limits exist:

$$\lim_{n\to\infty}\beta_{1n} = \beta_{10}, \quad \lim_{n\to\infty}\beta_{2n} = \beta_{20}, \quad \lim_{n\to\infty}\gamma_{1n} = \gamma_{10}, \quad \lim_{n\to\infty}\gamma_{2n} = \gamma_{20}.$$

What is more, $\gamma_{10} \neq 0$, $\gamma_{20} \neq 0$, hence the sequences $\left\{\dfrac{\beta_{1n}}{\gamma_{1n}}\right\}_{n\in\mathbb{N}}$ and $\left\{\dfrac{\beta_{2n}}{\gamma_{2n}}\right\}_{n\in\mathbb{N}}$ are well defined and converge to $\dfrac{\beta_{10}}{\gamma_{10}}$ and $\dfrac{\beta_{20}}{\gamma_{20}}$ respectively. Thus, $\left\{\dfrac{Z_n}{W_n}\right\}_{n\in\mathbb{N}}$ is convergent with its limit being

$$\frac{\beta_{10}}{\gamma_{10}}\mathbf{e} + \frac{\beta_{20}}{\gamma_{20}}\mathbf{e}^\dagger = \frac{\beta_{10}\mathbf{e} + \beta_{20}\mathbf{e}^\dagger}{\gamma_{10}\mathbf{e} + \gamma_{20}\mathbf{e}^\dagger} = \frac{Z_0}{W_0}.$$

Recalling the quite "good" properties of the hyperbolic modulus, we are in position to introduce another approach to the notion of convergence of bicomplex sequences.

Definition 5.1.2. *A sequence* $\{Z_n\}_{n\in\mathbb{N}}$ *of bicomplex numbers* $\mathbb{D}$-*converges (synonymous: hyperbolically converges; converges with respect to the hyperbolic-valued norm* $|\cdot|_\mathbf{k}$*) to the bicomplex number* Z_0 *if for any strictly positive hyperbolic number* ε *there exists* $N \in \mathbb{N}$ *such that for any* $n \geq N$ *there holds:*

$$|Z_n - Z_0|_k \prec \varepsilon.$$

Using the idempotent representations

$$Z_n = \beta_{1n} \cdot \mathbf{e} + \beta_{2n} \cdot \mathbf{e}^\dagger; \qquad Z_0 = \beta_{10} \cdot \mathbf{e} + \beta_{20} \cdot \mathbf{e}^\dagger;$$

$$\varepsilon = \varepsilon_1 \cdot \mathbf{e} + \varepsilon_2 \cdot \mathbf{e}^\dagger,$$

we obtain that, equivalently,

$$|\beta_{1n} - \beta_{10}| < \varepsilon_1 \quad \text{and} \quad |\beta_{2n} - \beta_{20}| < \varepsilon_2;$$

which means that the sequence $\{Z_n\}_{n\in\mathbb{N}}$ converges to the bicomplex number Z_0 with respect to the hyperbolic-valued norm if and only if it converges to Z_0 with respect to the Euclidean norm. Notice that even though the two norms cannot be compared, as they take values in different rings, one still obtains the same sets of convergent and divergent sequences.

Thus we will usually write "a convergent sequence" without specifying which type of convergence is meant.

5.2   The Euclidean topology on $\mathbb{BC}$

Since we have already endowed $\mathbb{BC}$ with the Euclidean norm, which is associated with the identifications $\mathbb{BC} = \mathbb{R}^4 = \mathbb{C}^2(\mathbf{i}) = \mathbb{C}^2(\mathbf{j})$, we will consider the topological space $(\mathbb{BC}, \tau_{euc})$ where τ_{euc} is the Euclidean topology on $\mathbb{R}^4$: its basis consists of all open balls in $\mathbb{R}^4$. Since for any bicomplex numbers Z and W there holds:

$$|Z + W| \leq |Z| + |W|;$$
$$|Z \cdot W| \leq \sqrt{2}\,|Z| \cdot |W|,$$

then the operations of addition and of multiplication are continuous with respect to τ_{euc}, and we can speak about the respective linearity of the topological space $(\mathbb{BC}, \tau_{euc})$; more exactly, it is a real, a $\mathbb{C}(\mathbf{i})$-complex and a $\mathbb{C}(\mathbf{j})$-complex linear topological space, but also a hyperbolic and a bicomplex linear topological module.

Besides the Euclidean open balls we can consider also open bicomplex balls with non-zero-divisor radius:

$$\left\{ Z \mid |Z|_{\mathbf{k}} \prec \gamma \text{ with } \gamma \in \mathbb{D}^+_{inv} \right\}.$$

Geometrically, such a ball can be seen as a bidisk in $\mathbb{C}^2$, thus all such balls form another basis in the topology τ_{euc}.

In Section 5.1 we introduced two formally different types of convergent sequences. Now we realize they are the same convergence with respect to the Euclidean topology but one of the definitions deals with the Euclidean basis of it and the other deals with the bicomplex balls with non-zero-divisor radius. Thus, in accordance with the problem we are faced, we will use one or another basis of the topology, depending on which one of them is more appropriate.

5.3   Bicomplex functions

Given a set $\Omega \subset \mathbb{BC}$, any mapping $F : \Omega \to \mathbb{BC}$ will be called a bicomplex function of the bicomplex variable $Z \in \Omega$. Since both Z and $F(Z)$ are bicomplex numbers, each of them admits any of the representations (see (1.3)–(1.9) and (1.24)), then F can be interpreted in different ways: as a mapping from $\mathbb{C}^2(\mathbf{i})$ to $\mathbb{C}^2(\mathbf{i})$, from $\mathbb{C}^2(\mathbf{j})$ to $\mathbb{C}^2(\mathbf{j})$, from $\mathbb{C}^2(\mathbf{i})$ generated by the idempotent representation to itself, etc. All these mappings are different but they all coincide when the bicomplex structure is considered.

We illustrate the above with an example. Consider the bicomplex function $F(Z) = Z^2$ and several mappings generated by it. To do this we write:

$$\begin{aligned}
F(Z) = Z^2 &= (z_1 + \mathbf{j}z_2)^2 = (z_1^2 - z_2^2) + \mathbf{j}(2z_1 z_2) \\
&= (\gamma_1 + \mathbf{i}\gamma_2)^2 = (\gamma_1^2 - \gamma_2^2) + \mathbf{i}(2\gamma_1\gamma_2) \\
&= (\beta_1 \mathbf{e} + \beta_2 \mathbf{e}^\dagger)^2 = \beta_1^2 \mathbf{e} + \beta_2^2 \mathbf{e}^\dagger \\
&= (\mathfrak{z}_1 + \mathbf{i}\mathfrak{z}_2)^2 = (\mathfrak{z}_1^2 - \mathfrak{z}_2^2) + \mathbf{i}(2\mathfrak{z}_1\mathfrak{z}_2).
\end{aligned}$$

This generates the following maps:

$$\mathbb{C}^2(\mathbf{i}) \ni (z_1, z_2) \mapsto (z_1^2 - z_2^2, 2z_1z_2) \in \mathbb{C}^2(\mathbf{i});$$
$$\mathbb{C}^2(\mathbf{j}) \ni (\gamma_1, \gamma_2) \mapsto (\gamma_1^2 - \gamma_2^2, 2\gamma_1\gamma_2) \in \mathbb{C}^2(\mathbf{j});$$
$$\mathbb{C}^2(\mathbf{i}) \ni (\beta_1, \beta_2) \mapsto (\beta_1^2, \beta_2^2) \in \mathbb{C}^2(\mathbf{i});$$
$$\mathbb{D}^2 \ni (\mathfrak{z}_1, \mathfrak{z}_2) \mapsto (\mathfrak{z}_1^2 - \mathfrak{z}_2^2, 2\mathfrak{z}_1\mathfrak{z}_2) \in \mathbb{D}^2.$$

Let Z_0 be a point in the closure of Ω. The function F has the limit A at Z_0 if for any $\epsilon > 0$ there exists $\delta > 0$ such that the condition $|Z - Z_0| < \delta$ implies that $|F(Z) - A| < \epsilon$. As usual in metric spaces, it is equivalent to say that for any sequence $\{Z_n\}_{n \in \mathbb{N}} \subset \Omega$ such that $\lim_{n \to \infty} Z_n = Z_0$, the sequence $\{F(Z_n)\}_{n \in \mathbb{N}}$ converges to A. We obtain immediately:

(I) if the limit $\lim_{Z \to Z_0} F(Z)$ exists, it is unique;

(II) if the limit $\lim_{Z \to Z_0} F(Z)$ exists, then the function F is bounded in a Euclidean ball with center in Z_0 and it is $\mathbb{D}$-bounded in a bicomplex ball with a non-zero-divisor radius;

(III) if $\lim_{Z \to Z_0} F(Z) = A \notin \mathfrak{S}_0$, then there exists a ball B with center in Z_0 such that for all $Z \in B$, $F(Z) \notin \mathfrak{S}_0$;

(IV) if $\lim_{Z \to Z_0} F(Z) = A$, $\lim_{Z \to Z_0} G(Z) = B$, then the sum, the product and the quotient (if $B \notin \mathfrak{S}_0$) have limits at Z_0 and the usual formulas hold.

A bicomplex function is *continuous* at a point $Z_0 \in \Omega \subset \mathbb{BC}$, if $\lim_{Z \to Z_0} F(Z)$ exists and

$$\lim_{Z \to Z_0} F(Z) = F(Z_0).$$

Then we say that a bicomplex function $F : \Omega \to \mathbb{BC}$, where $\Omega \subset \mathbb{BC}$, is continuous on Ω if and only if F is continuous at every $Z_0 \in \Omega$.

As in the complex case, it is easy to prove that if two functions are continuous at a point, then their sum and product are also continuous at that point. Moreover, if the second function takes at Z_0 an invertible value, then the quotient is continuous at this point also. Furthermore, the composition of continuous functions is continuous.

Bicomplex sequences and notions of convergence, limits and continuity, have also been studied in works such as [41, 45, 84]. Applications to dynamical systems, e.g. Mandelbrot and Julia sets in the bicomplex setup, have been developed in [29, 59, 60, 62, 91, 92, 93, 102].

Chapter 6

Elementary Bicomplex Functions

Historically, a small collection of real functions was assigned the name of elementary functions: polynomials, rational functions, the exponential, trigonometric functions; together with their inverses: the n-th root, logarithm, inverse trigonometric functions. Later, for their complex extensions the same name has been preserved, although many other functions have emerged which could rightly be called elementary as well.

The aim of this chapter is to show that the structure and the properties of bicomplex numbers allow us to further extend to $\mathbb{BC}$ all those elementary functions in such a way that the extensions keep having amazingly many properties and features of their real and complex antecedents.

6.1 Polynomials of a bicomplex variable

6.1.1 Complex and real polynomials.

A complex polynomial is a function of the form

$$p(z) = \sum_{k=0}^{n} a_k z^k$$

where the a_k are complex numbers and where z is a complex variable. We assume that the leading coefficient $a_n \neq 0$ so that the polynomial is said to have degree n. In particular, a polynomial of degree zero is, by definition, a non-zero constant. When all a_k are real numbers, the polynomial $p(z)$ is called a real polynomial of one complex variable. Observe that $p(z)$ is a real polynomial of a complex variable if and only if

$$\overline{p(\overline{z})} = p(z) \tag{6.1}$$

for all $z \in \mathbb{C}$. From this it follows that, for such a polynomial, if $p(a) = 0$, then $p(\bar{a}) = 0$; therefore either a is real or p has the conjugate pair of zeros a and $\bar{a}$.

Rewriting (6.1) as $p(\bar{z}) = \overline{p(z)}$, one concludes that the range $p(\mathbb{C})$ of the real polynomial p is symmetric with respect to the real axis.

One of the most remarkable properties of complex polynomials is captured by Gauss's Fundamental Theorem of Algebra which states that a complex polynomial of degree n, $n > 0$, has exactly n zeros considering the multiplicities. The usual proofs of this use methods of analysis or topology (it is not a result which follows purely from the algebraic field property of the complex numbers). In particular, we have then that if $p(z)$ and $q(z)$ are polynomials of degree not exceeding n and if the equation

$$p(z) = q(z)$$

is satisfied at $n + 1$ distinct points, then $p = q$.

Notice that real polynomials of a complex variable obey the Fundamental Theorem of Algebra, but the situation with real polynomials of a real variable is totally different: such a polynomial of degree $n > 0$ might have any number of zeros up to n. We will see what happens, in this sense, with bicomplex polynomials.

6.1.2   Bicomplex polynomials

Let

$$p(Z) = \sum_{k=0}^{n} A_k Z^k$$

be a bicomplex polynomial of degree n of a bicomplex variable Z. Let us write $Z = z_1 + \mathbf{j}z_2$ in its $\mathbb{C}(\mathbf{i})$-idempotent representation: $Z = \beta_1 \mathbf{e} + \beta_2 \mathbf{e}^\dagger$ with $\beta_1 := z_1 - \mathbf{i}z_2$ and $\beta_2 := z_1 + \mathbf{j}z_2$. We write also the bicomplex coefficients as $A_k = \gamma_k \mathbf{e} + \delta_k \mathbf{e}^\dagger$, for $k = 0, \ldots, n$. Then $Z^k = \beta_1^k \mathbf{e} + \beta_2^k \mathbf{e}^\dagger$ and we rewrite our polynomial as

$$p(Z) = \sum_{k=0}^{n} \left(\gamma_k \beta_1^k\right) \mathbf{e} + \sum_{k=0}^{n} \left(\delta_k \beta_2^k\right) \mathbf{e}^\dagger =: \phi(\beta_1)\mathbf{e} + \psi(\beta_2)\mathbf{e}^\dagger.$$

If we denote the set of distinct roots of ϕ and ψ by $\mathcal{S}_1$ and $\mathcal{S}_2$, and if we denote by $\mathcal{S}$ the set of distinct roots of the polynomial p, then

$$\mathcal{S} = \mathcal{S}_1 \mathbf{e} + \mathcal{S}_2 \mathbf{e}^\dagger,$$

so that the structure of the null-set of a bicomplex polynomial $p(Z)$ of degree n is fully described by the following three cases:

1. If both polynomials ϕ and ψ are of degree at least one, and if $\mathcal{S}_1 = \{\beta_{1,1}, \ldots, \beta_{1,k}\}$ and $\mathcal{S}_2 = \{\beta_{2,1}, \ldots, \beta_{2,\ell}\}$, then the set of distinct roots of p is given by

$$\mathcal{S} = \left\{ Z_{s,t} = \beta_{1,s}\mathbf{e} + \beta_{2,t}\mathbf{e}^\dagger \mid s = 1, \ldots, k, \ t = 1, \ldots, \ell \right\}.$$

2. If ϕ is identically zero, then $\mathcal{S}_1 = \mathbb{C}$ and $\mathcal{S}_2 = \{\beta_{2,1}, \ldots, \beta_{2,\ell}\}$, with $\ell \leq n$. Hence

$$\mathcal{S} = \{Z_t = \lambda \mathbf{e} + \beta_{2,t} \mathbf{e}^\dagger \mid \lambda \in \mathbb{C}, \, t = 1, \ldots, \ell\}.$$

Similarly, if ψ is identically zero, then $\mathcal{S}_2 = \mathbb{C}$ and $\mathcal{S}_1 = \{\beta_{1,1}, \ldots, \beta_{1,k}\}$, where $k \leq n$. Then

$$\mathcal{S} = \{Z_s = \beta_{1,s} \mathbf{e} + \lambda \mathbf{e}^\dagger \mid \lambda \in \mathbb{C}, \, s = 1, \ldots, k\}.$$

3. If all the coefficients A_k with the exception of $A_0 = \gamma_0 \mathbf{e} + \delta_0 \mathbf{e}^\dagger$ are complex multiples of $\mathbf{e}$ (respectively of $\mathbf{e}^\dagger$), but $\delta_0 \neq 0$ (respectively $\gamma_0 \neq 0$), then p has no roots.

We now discuss a few examples, to give a flavor for computations in $\mathbb{BC}$.

Example 6.1.2.1. First, consider the polynomial

$$p(Z) = \left(\frac{1}{2} + \mathbf{j}\frac{\mathbf{i}}{2}\right) Z^5 + (-(1+4\mathbf{i}) + 2\mathbf{j}(2-\mathbf{i}))Z^4$$

$$+ ((-11 + 6\mathbf{i}) - \mathbf{j}(12 + 11\mathbf{i})) Z^3$$

$$+ \left(\left(\frac{29}{2} + 13\mathbf{i}\right) + \mathbf{j}\left(-13 + \frac{47}{2}\mathbf{i}\right)\right) Z^2$$

$$+ \left(\left(\frac{13}{2} - 17\mathbf{i}\right) + \mathbf{j}\left(17 + \frac{13}{2}\mathbf{i}\right)\right) Z$$

$$- \left(\frac{11}{2} + \mathbf{i}\right) + \mathbf{j}\left(1 - \frac{11}{2}\mathbf{i}\right).$$

The corresponding $\mathbb{C}(\mathbf{i})$-complex polynomials are:

$$\phi(\beta_1) = \beta_1^5 - (3+8\mathbf{i})\beta_1^4 + 2(-11+9\mathbf{i})\beta_1^3 + 2(19+13\mathbf{i})\beta_1^2$$
$$+ (13 - 34\mathbf{i})\beta_1 - (11 + 2\mathbf{i}),$$
$$\psi(\beta_2) = \beta_2^4 - 6\mathbf{i}\beta_2^3 - 9\beta_2^2.$$

Their distinct roots are $\mathcal{S}_1 = \{\mathbf{i}, 1+2\mathbf{i}\}$ and $\mathcal{S}_2 = \{0, 3\mathbf{i}\}$. Then p has the following four roots:

$$\mathcal{S} = \left\{\frac{1}{2}\mathbf{i} - \frac{1}{2}\mathbf{j}, \, 2\mathbf{i} + \mathbf{j}, \, \frac{1+2\mathbf{i}}{2} + \mathbf{j}\frac{-2+\mathbf{i}}{2}, \, \frac{1+5\mathbf{i}}{2} + \mathbf{j}\frac{1+\mathbf{i}}{2}\right\}.$$

Example 6.1.2.2. Consider the polynomial

$$p(Z) = (1 + \mathbf{ji}) Z^2 - (\mathbf{i} - \mathbf{j}).$$

The associated complex polynomials are:

$$\phi(\beta_1) = 2(\beta_1^2 - \mathbf{i}), \qquad \psi(\beta_2) \equiv 0.$$

The null set of p is

$$S = \left\{ \pm \left(\frac{\sqrt{2}}{2} + \mathbf{i}\frac{\sqrt{2}}{2} \right) \mathbf{e} + \lambda \mathbf{e}^\dagger \mid \lambda \in \mathbb{C} \right\}.$$

Example 6.1.2.3. Slightly adjusting the previous example, i.e., taking $\psi(\beta_2) \equiv 2$, we get the polynomial

$$p(Z) = (1 + \mathbf{j}\mathbf{i}) Z^2 + (1 - \mathbf{i}) + \mathbf{j} (1 - \mathbf{i}),$$

which has no roots.

It is also important to note that a bicomplex polynomial may not have a unique factorization into linear polynomials. For example, the polynomial $p(Z) = Z^3 - 1$ has 9 zeroes. Indeed, the associated complex polynomials are

$$\phi(\beta_1) = \beta_1^3 - 1, \qquad \phi(\beta_2) = \beta_2^3 - 1.$$

The set of zeros of ϕ and ψ are, respectively:

$$S_1 = \left\{ \beta_{1,1} = 1, \ \beta_{1,2} = -\frac{1}{2} + \mathbf{i}\frac{\sqrt{3}}{2}, \ \beta_{1,3} = -\frac{1}{2} - \mathbf{i}\frac{\sqrt{3}}{2} \right\}$$

$$S_2 = \left\{ \beta_{2,1} = 1, \ \beta_{2,2} = -\frac{1}{2} + \mathbf{i}\frac{\sqrt{3}}{2}, \ \beta_{2,3} = -\frac{1}{2} - \mathbf{i}\frac{\sqrt{3}}{2} \right\}$$

Then the set of zeroes of p is

$$S = \left\{ Z_{kl} = \beta_{1,k}\mathbf{e} + \beta_{2,\ell}\mathbf{e}^\dagger \mid k, \ell = 1 \ldots 3 \right\},$$

and we have at least two distinct factorizations:

$$Z^3 - 1 = (Z - 1) \left(Z + \frac{1}{2} - \frac{\sqrt{3}}{2}\mathbf{i} \right) \left(Z + \frac{1}{2} + \frac{\sqrt{3}}{2}\mathbf{i} \right)$$

and

$$Z^3 - 1 = (Z - 1) \left(Z + \frac{1}{2} - \mathbf{j}\frac{\sqrt{3}}{2} \right) \left(Z + \frac{1}{2} + \mathbf{j}\frac{\sqrt{3}}{2} \right).$$

It is therefore clear from what we have indicated that bicomplex polynomials do not satisfy the Fundamental Theorem of Algebra in its original form. At the same time, the following results are true and summarize the comments above.

Theorem 6.1.2.4 (Analogue of the Fundamental Theorem of Algebra for bicomplex polynomials). *Consider a bicomplex polynomial* $p(Z) = \sum\limits_{k=0}^{n} A_k Z^k$. *If all the coefficients* A_k *with the exception of the free term* $A_0 = \gamma_0\mathbf{e} + \delta_0\mathbf{e}^\dagger$ *are complex multiples of* $\mathbf{e}$ *(respectively of* $\mathbf{e}^\dagger$*), but* $\delta_0 \neq 0$ *(respectively* $\gamma_0 \neq 0$*), then* p *has no roots. In all other cases,* p *has at least one root.*

Corollary 6.1.2.5. *Assume that a bicomplex polynomial p of degree $n \geq 1$ has at least one root. Then:*

1. *If at least one of the coefficients A_k, for $k = 1, \ldots, n$, is invertible, then p has at most n^2 distinct roots.*

2. *If all coefficients are complex multiples of $\mathbf{e}$ (respectively $\mathbf{e}^\dagger$), then p has infinitely many roots.*

Note that zeros of bicomplex polynomials were originally investigated in [56] and [55].

Formula (6.1) has several analogues in the actual context. Let all the coefficients of the polynomial $p(Z) = \sum_{k=0}^{n} A_k Z^k$ be real numbers, then

$$p(Z) = \overline{p(\overline{Z})} = p^\dagger(Z^\dagger) = p^*(Z^*) \tag{6.2}$$

for all $Z \in \mathbb{BC}$. This implies that if $p(Z_0) = 0$, then $p(\overline{Z}_0) = 0$, $p(Z_0^\dagger) = 0$, $p(Z_0^*) = 0$; now if $Z_0 \in \mathbb{R}$ in this case Z_0 does not have "associated" roots of p. If Z_0 is not real, then the following situations arise:

- $Z_0 \in \mathbb{C}(\mathbf{i}) \setminus \mathbb{R}$, then $\overline{Z}_0 \neq Z_0$ is also a root of p.

- $Z_0 \in \mathbb{C}(\mathbf{j}) \setminus \mathbb{R}$, then $Z_0^\dagger \neq Z_0$ is also a root of p.

- $Z_0 \in \mathbb{D} \setminus \mathbb{R}$, then $\overline{Z}_0 = Z_0^\dagger \neq Z_0$ is also a root of p.

- Z_0 does not belong to any of the above three sets, hence the polynomial p has together with Z_0 all the three conjugates $\overline{Z}_0$, $Z_0^\dagger$, Z_0^* as its roots.

Equations (6.2) can be written as

$$p(\overline{Z}) = \overline{p(Z)}; \quad p(Z^\dagger) = p(Z)^\dagger; \quad p(Z^*) = p(Z)^*$$

meaning that the range $p(\mathbb{BC})$ of the polynomial p with real coefficients possesses all the three symmetries generated by the three conjugations.

Assume now that the coefficients of the polynomial $p(Z) = \sum_{k=0}^{n} A_k Z^k$ are in $\mathbb{C}(\mathbf{i})$, then

$$p(Z) = p(Z^\dagger)^\dagger \tag{6.3}$$

for all Z in $\mathbb{BC}$. This implies that if Z_0 is a root of p, $p(Z_0) = 0$, then $p(Z_0^\dagger) = 0$; if Z_0 is a $\mathbb{C}(\mathbf{i})$-complex number, then Z_0 does not have "associated" roots of p. But if Z_0 is not in $\mathbb{C}(\mathbf{i})$, then $Z_0^\dagger$ is also a root of p. Of course, since (6.3) is equivalent to $p(Z^\dagger) = p(Z)^\dagger$, one may conclude that the range $p(\mathbb{BC})$ of such a polynomial has the symmetry determined by the $\dagger$-conjugation.

As the next step, we assume that the coefficients of the polynomial p are in $\mathbb{C}(\mathbf{j})$, then

$$p(Z) = \overline{p(\overline{Z})} \tag{6.4}$$

for all Z in $\mathbb{BC}$. This implies that if Z_0 is a root of p, $p(Z_0) = 0$, then $p(\overline{Z}_0) = 0$; again if Z_0 is a $\mathbb{C}(\mathbf{j})$-complex number, then Z_0 does not have "associated" roots of p. But if Z_0 is not in $\mathbb{C}(\mathbf{j})$, then $\overline{Z}_0$ is also a root of p. Of course, since (6.4) is equivalent to $\overline{p(Z)} = p(\overline{Z})$, one may conclude that the range $p(\mathbb{BC})$ of such a polynomial has the symmetry determined by the bar-conjugation.

Finally, we consider the case of hyperbolic coefficients of the polynomial p. Here

$$p(Z) = p(Z^*)^* \tag{6.5}$$

for all Z in $\mathbb{BC}$. This implies that if Z_0 is a root of p, $p(Z_0) = 0$, then $p(Z_0^*) = 0$; if Z_0 is a hyperbolic number, then Z_0 does not have "associated" roots of p. But if Z_0 is not in $\mathbb{D}$, then $Z_0^* \neq Z_0$ is also a root of p. Of course, since (6.5) is equivalent to $p(Z)^* = p(Z^*)$, one may conclude that the range $p(\mathbb{BC})$ of such a polynomial has the symmetry determined by the $*$-conjugation.

Definition 6.1.2.6. *We will use the name* bicomplex rational functions *for functions of the form*

$$\frac{p(Z)}{q(Z)}$$

with two bicomplex polynomials p and q.

Such a function is well-defined for those values of Z for which the polynomial $q(Z) = B_m Z^m + B_{m-1} Z^{m-1} + \cdots + B_1 Z + B_0$ takes values in $\mathbb{BC}_{inv} := \mathbb{BC} \setminus \mathfrak{S}_0$. If both polynomials are of degree one: $p(Z) = A_1 Z + A_0$, $q(Z) = B_1 Z + B_0$, then we have a fractional linear transformation. If the coefficients A_1 and B_1 are both invertible, then the fractional linear transform takes the form

$$C_1 + \frac{C_2}{Z + C_3}$$

with bicomplex C_1, C_2, C_3.

6.2 Exponential functions

6.2.1 The real and complex exponential functions

As Ahlfors writes in [1], if one approaches calculus exclusively from the point of view of real numbers, there is no reason to expect any relationship between the exponential function e^x and the trigonometric functions $\cos x$ and $\sin x$. Indeed, these functions seem to be derived from completely different sources and with different purposes in mind. The reader will notice, no doubt, a similarity between the Taylor expansions of these functions, and if willing to use imaginary arguments, the reader will be able to derive Euler's formula $e^{\mathbf{i}x} = \cos x + \mathbf{i}\sin x$ as a formal identity.

The exponential function can be derived in many different ways. If one recalls that the Euler's number e arises as the limit of the sequence $\left(1 + \dfrac{1}{n}\right)^n$, then one is tempted to consider, for any real number x, the sequence $\left(1 + \dfrac{x}{n}\right)^n$ and its limit which brings us to the definition of the real exponential function as

$$exp(x) = e^x := \lim_{n \to \infty} \left(1 + \frac{x}{n}\right)^n. \tag{6.6}$$

This approach has the advantage of using minimal mathematical tools; as a matter of fact, only the properties of sequences and of their limits are necessary, meanwhile the definitions as the sum of a convergent power series or as a solution of a differential equation require much more elaborated techniques.

Defined by (6.6), the real exponential function preserves all the expected properties; for example, if $x \in \mathbb{N}$, then $e^x = \underbrace{e \cdot e \cdots e}_{x \;\; \text{times}}$. It has the property

$$e^{x_1 + x_2} = e^{x_1} \cdot e^{x_2}, \tag{6.7}$$

thus, it is a homomorphism of the additive group of real numbers into the multiplicative group $\mathbb{R} \setminus \{0\}$. Since the real exponential function is monotone, then it realizes an isomorphism of these groups.

It turns out that definition (6.6) extends to the complex numbers in the sense that for any $z \in \mathbb{C}$

$$exp(z) = e^z := \lim_{n \to \infty} \left(1 + \frac{z}{n}\right)^n. \tag{6.8}$$

This implies the famous Euler formula

$$e^z = e^x (\cos y + \mathbf{i} \sin y)$$

which says, in particular that $|e^z| = e^x$ and $\text{Arg}\, e^z = \{y + 2\pi k \,|\, k \in \mathbb{Z}\}$. The analogue of (6.7) is

$$e^{z+w} = e^z e^w.$$

Thus the complex exponential function is a homomorphism (but not an isomorphism) of the additive group of complex numbers into the multiplicative group $\mathbb{C} \setminus \{0\}$.

6.2.2 The bicomplex exponential function

In this section, we introduce the exponential function of a bicomplex variable extending directly the ideas of the previous section.

Theorem 6.2.2.1. *Let Z be any bicomplex number. Then the sequence*

$$Z_n := \left(1 + \frac{Z}{n}\right)^n$$

is convergent, and if Z is written as $Z = z_1 + \mathbf{j}z_2$, then the limit is

$$e^{z_1}\left(\cos(z_2) + \mathbf{j}\sin(z_2)\right).$$

Proof. The computation below proves that the sequence is convergent. Set as before $Z = \beta_1 \mathbf{e} + \beta_2 \mathbf{e}^\dagger$. Then

$$\left(1 + \frac{Z}{n}\right)^n = \left(1 + \frac{\beta_1}{n}\mathbf{e} + \frac{\beta_2}{n}\mathbf{e}^\dagger\right)^n = \left(\mathbf{e} + \mathbf{e}^\dagger + \frac{\beta_1}{n}\mathbf{e} + \frac{\beta_2}{n}\mathbf{e}^\dagger\right)^n$$

$$= \left(\left(1 + \frac{\beta_1}{n}\right)\mathbf{e} + \left(1 + \frac{\beta_2}{n}\right)\mathbf{e}^\dagger\right)^n = \left(1 + \frac{\beta_1}{n}\right)^n\mathbf{e} + \left(1 + \frac{\beta_2}{n}\right)^n\mathbf{e}^\dagger.$$

Relying on the fact that the corresponding sequences of complex numbers $\left(1 + \frac{\beta_1}{n}\right)^n$ and $\left(1 + \frac{\beta_2}{n}\right)^n$ are convergent to the complex exponentials e^{β_1} and e^{β_2}, respectively, we get that the limit of the right-hand side exists. What is more,

$$
\begin{aligned}
\lim_{n\to\infty}\left(1 + \frac{Z}{n}\right)^n &= \lim_{n\to\infty}\left(\left(1 + \frac{\beta_1}{n}\right)^n\mathbf{e} + \left(1 + \frac{\beta_2}{n}\right)^n\mathbf{e}^\dagger\right) = e^{\beta_1}\mathbf{e} + e^{\beta_2}\mathbf{e}^\dagger \\
&= \frac{1}{2}(e^{\beta_1} + e^{\beta_2}) + \mathbf{j}\frac{\mathbf{i}}{2}(e^{\beta_1} - e^{\beta_2}) \\
&= \frac{1}{2}(e^{z_1 - \mathbf{i}z_2} + e^{z_1 + \mathbf{i}z_2}) + \mathbf{j}\frac{\mathbf{i}}{2}(e^{z_1 - \mathbf{i}z_2} - e^{z_1 + \mathbf{i}z_2}) \qquad (6.9) \\
&= e^{z_1}\left(\frac{1}{2}(e^{-\mathbf{i}z_2} + e^{\mathbf{i}z_2}) + \mathbf{j}\frac{\mathbf{i}}{2}(e^{-\mathbf{i}z_2} - e^{\mathbf{i}z_2})\right) \\
&= e^{z_1}\left(\cos(z_2) + \mathbf{j}\sin(z_2)\right).
\end{aligned}
$$

This concludes our proof.　　　　　　　　　　　　　　　　　　　　　　　　　□

Clearly, the theorem justifies the following definition.

Definition 6.2.2.2. *We set*

$$e^Z := \lim_{n\to\infty}\left(1 + \frac{Z}{n}\right)^n.$$

Hence, if Z is written as $Z = z_1 + \mathbf{j}z_2$, then we obtain the bicomplex Euler formula:

$$e^Z = e^{z_1}\left(\cos(z_2) + \mathbf{j}\sin(z_2)\right).$$

The reader may rewrite e^Z in many other forms of writing of bicomplex numbers. For instance, if $Z = \zeta_1 + \mathbf{i}\zeta_2$ with ζ_1 and ζ_2 in $\mathbb{C}(\mathbf{j})$, then

$$e^Z = e^{\zeta_1}\left(\cos(\zeta_2) + \mathbf{i}\sin(\zeta_2)\right),$$

which exemplifies once more the peculiarity of the different writings of bicomplex numbers.

We pass now to the properties of this newly introduced bicomplex exponential function.

- First we note that the bicomplex exponential is an extension to $\mathbb{BC}$ of the complex exponential function: indeed, for $Z = z_1 + \mathbf{j}0 \in \mathbb{C}$, we have that

$$e^Z = e^{z_1}(\cos(0) + \mathbf{j}\sin(0)) = e^{z_1},$$

which is the usual complex exponential function.

In the same way, the restriction of e^Z onto $\mathbb{C}(\mathbf{j})$ coincides, obviously, with the $\mathbb{C}(\mathbf{j})$-complex exponential function.

The situation with the restriction onto $\mathbb{D}$ is more subtle; if $Z = a + b\mathbf{k}$ with reals a and b, then

$$e^Z = e^{a+b\mathbf{k}} = e^{a+b\mathbf{ij}} = e^a\left(\cos(b\mathbf{i}) + \mathbf{j}\sin(b\mathbf{i})\right)$$
$$= e^a\left(\cosh(b) + \mathbf{ji}\sinh(b)\right) = e^a\left(\cosh(b) + \mathbf{k}\sinh(b)\right),$$

where cosh and sinh are the classical hyperbolic cosine and sine functions, thus arriving at the definition of the hyperbolic (in the sense of hyperbolic numbers) exponential function.

- Note that e^{z_1} is the complex modulus of the bicomplex number e^Z and z_2 is the complex argument of the same bicomplex number e^Z. The reader may find it instructive to compare this fact with what happens in the complex case.

- For $Z = \mathbf{0} = 0\mathbf{e} + 0\mathbf{e}^\dagger$, we have: $e^0 = 1\mathbf{e} + 1\mathbf{e}^\dagger = 1$.

- For any bicomplex number Z, the exponential e^Z is invertible. This is because

$$e^Z = e^{z_1 - \mathbf{i}z_2}\mathbf{e} + e^{z_1 + \mathbf{i}z_2}\mathbf{e}^\dagger$$

and the exponential terms $e^{z_1 - \mathbf{i}z_2}$ and $e^{z_1 + \mathbf{i}z_2}$ are complex exponential functions, so they are never zero. The inverse of e^Z is

$$e^{-Z} = e^{-(z_1 - \mathbf{i}z_2)}\mathbf{e} + e^{-(z_1 + \mathbf{i}z_2)}\mathbf{e}^\dagger = e^{-z_1}\left(\cos(z_2) - \mathbf{j}\sin(z_2)\right).$$

Thus, the range of the bicomplex exponential function does not contain either the zero or any zero-divisors.

- For $\mathbf{e} = 1 \cdot \mathbf{e} + 0 \cdot \mathbf{e}^\dagger$, and $\mathbf{e}^\dagger = 0 \cdot \mathbf{e} + 1 \cdot \mathbf{e}^\dagger$, we have:

$$e^{\mathbf{e}} = e \cdot \mathbf{e} + 1 \cdot \mathbf{e}^\dagger = e^{\frac{1}{2}} \left(\cos\left(\frac{\mathbf{i}}{2}\right) + \mathbf{j} \sin\left(\frac{\mathbf{i}}{2}\right) \right)$$

$$= e^{\frac{1}{2}} \left(\cosh\left(\frac{1}{2}\right) + \mathbf{j}\mathbf{i} \sinh\left(\frac{1}{2}\right) \right).$$

Similarly:

$$e^{\mathbf{e}^\dagger} = 1 \cdot \mathbf{e} + e \cdot \mathbf{e}^\dagger = e^{\frac{1}{2}} \left(\cos\left(\frac{\mathbf{i}}{2}\right) - \mathbf{j} \sin\left(\frac{\mathbf{i}}{2}\right) \right)$$

$$= e^{\frac{1}{2}} \left(\cosh\left(\frac{1}{2}\right) - \mathbf{j}\mathbf{i} \sinh\left(\frac{1}{2}\right) \right) = (e^{\mathbf{e}})^\dagger.$$

Notice that both numbers, $e^{\mathbf{e}}$ and $e^{\mathbf{e}^\dagger}$, are hyperbolic numbers. This is because so are the idempotents $\mathbf{e}$ and $\mathbf{e}^\dagger$ and because the restriction of the bicomplex exponential is the hyperbolic exponential.

- Due to the commutativity of the multiplication in $\mathbb{BC}$, we can show that for any Z_1 and Z_2 in $\mathbb{BC}$, the following formula holds:

$$e^{Z_1} e^{Z_2} = e^{Z_1 + Z_2}. \tag{6.10}$$

Indeed, writing $Z_1 = z_{11} + \mathbf{j} z_{12}$ and $Z_2 = z_{21} + \mathbf{j} z_{22}$ we have:

$$e^{Z_1} e^{Z_2} = \left(e^{z_{11}} \left(\cos(z_{12}) + \mathbf{j} \sin(z_{12}) \right) \right) \left(e^{z_{21}} \left(\cos(z_{22}) + \mathbf{j} \sin(z_{22}) \right) \right)$$

$$= e^{z_{11}} e^{z_{21}} \left((\cos(z_{12}) \cos(z_{22}) - \sin(z_{12}) \sin(z_{22}) \right)$$

$$+ \mathbf{j} (\sin(z_{12}) \cos(z_{22}) + \sin(z_{22}) \cos(z_{12})))$$

$$= e^{z_{11} + z_{21}} \left(\cos(z_{12} + z_{22}) + \mathbf{j} \sin(z_{12} + z_{22}) \right) = e^{Z_1 + Z_2}.$$

This equality means that the exponential function is a homomorphism from the additive group of bicomplex numbers into the multiplicative group $\mathbb{BC}_{inv}$ of invertible bicomplex numbers.

- In the case $Z = 0 + \mathbf{j} z_2$, we have:

$$e^Z = e^{\mathbf{j} z_2} = \cos(z_2) + \mathbf{j} \sin(z_2).$$

- The complex formula $e^{\mathbf{i}\pi} + 1 = 0$ remains valid for bicomplex numbers, but it is complemented with its mirror image $e^{\mathbf{j}\pi} + 1 = 0$.

- For any $Z = \beta_1 \mathbf{e} + \beta_2 \mathbf{e}^\dagger \in \mathbb{BC}$ and any invertible bicomplex number $W = \gamma_1 \mathbf{e} + \gamma_2 \mathbf{e}^\dagger$, i.e., $\gamma_1 \gamma_2 \neq 0$, the equation $e^Z = W$ is equivalent to the system $e^{\beta_1} = \gamma_1$ and $e^{\beta_2} = \gamma_2$. Because $\gamma_1 \gamma_2 \neq 0$, it follows that there is always a solution. Of course, this is the first step to talk about the bicomplex logarithm which will be commented on below.

- Recalling that the complex exponential function and the complex trigonometric functions are periodic, we obtain that

$$e^Z = e^{z_1}\left(\cos(z_2) + \mathbf{j}\sin(z_2)\right)$$
$$= e^{z_1 + 2\pi i m}\left(\cos(z_2 + 2\pi n) + \mathbf{j}\sin(z_2 + 2\pi n)\right)$$
$$= e^{Z + 2\pi(m\mathbf{i} + n\mathbf{j})},$$

for integer numbers m and n. Thus the bicomplex exponential function is periodic with bicomplex periods $2\pi(m\mathbf{i} + n\mathbf{j})$. One can prove that these are the only periods.

6.3 Trigonometric and hyperbolic functions of a bicomplex variable

6.3.1 Complex Trigonometric Functions

The complex trigonometric functions $\cos(z)$ and $\sin(z)$ are defined in terms of the complex exponential as

$$\cos(z) := \frac{e^{\mathbf{i}z} + e^{-\mathbf{i}z}}{2}, \quad \sin(z) := \frac{e^{\mathbf{i}z} - e^{-\mathbf{i}z}}{2\mathbf{i}}. \tag{6.11}$$

The hyperbolic functions of a complex variable, $\cosh(z)$ and $\sinh(z)$ are defined in terms of the complex exponential as follows:

$$\cosh(z) := \frac{e^z + e^{-z}}{2}, \quad \sinh(z) := \frac{e^z - e^{-z}}{2}. \tag{6.12}$$

All four of them are extensions of the respective functions of a real variable. The inverse of the complex cosine function is obtained by solving the equation

$$\cos(z) = \frac{e^{\mathbf{i}z} + e^{-\mathbf{i}z}}{2} = w.$$

This is a quadratic equation in $e^{\mathbf{i}z}$ with roots

$$e^{\mathbf{i}z} = w \pm \sqrt{w^2 - 1}.$$

Because of the periodicity of the complex exponential this equation, with unknown z, has a countable family of solutions and the formulas

$$\arccos(w) := z = -\mathbf{i}\log(w \pm \sqrt{w^2 - 1}) = \pm\mathbf{i}\log(w + \sqrt{w^2 - 1}) \tag{6.13}$$

show how to treat precisely the arccosine function of a complex variable. In a similar fashion, the other inverse functions are dealt with.

We will now follow the same process to define bicomplex trigonometric functions.

6.3.2   Bicomplex Trigonometric Functions

Adding and subtracting the formulas $e^{\mathbf{j}z_2} = \cos(z_2) + \mathbf{j}\sin(z_2)$ and $e^{-\mathbf{j}z_2} = \cos(z_2) - \mathbf{j}\sin(z_2)$, for any $z_2 \in \mathbb{C}(\mathbf{i})$, we express the complex cosine and sine via the bicomplex exponential:

$$\cos z_2 = \frac{e^{\mathbf{j}z_2} + e^{-\mathbf{j}z_2}}{2},$$

$$\sin z_2 = \frac{e^{\mathbf{j}z_2} - e^{-\mathbf{j}z_2}}{2\mathbf{j}}.$$

Thus we are in a position to introduce the bicomplex sine and cosine functions which are direct extensions of their complex antecedents.

Definition 6.3.2.1. *Let $Z \in \mathbb{BC}$. We define the bicomplex cosine and sine functions of a bicomplex variable as follows:*

$$\cos Z := \frac{e^{\mathbf{j}Z} + e^{-\mathbf{j}Z}}{2} = \frac{e^{\mathbf{i}Z} + e^{-\mathbf{i}Z}}{2},$$

$$\sin Z := \frac{e^{\mathbf{j}Z} - e^{-\mathbf{j}Z}}{2\mathbf{j}} = \frac{e^{\mathbf{i}Z} - e^{-\mathbf{i}Z}}{2\mathbf{i}}. \tag{6.14}$$

Both are well defined since a direct computation gives that, indeed,

$$\frac{e^{\mathbf{j}Z} + e^{-\mathbf{j}Z}}{2} = \frac{e^{\mathbf{i}Z} + e^{-\mathbf{i}Z}}{2} \quad \text{and} \quad \frac{e^{\mathbf{j}Z} - e^{-\mathbf{j}Z}}{2\mathbf{j}} = \frac{e^{\mathbf{i}Z} - e^{-\mathbf{i}Z}}{2\mathbf{i}}.$$

Note that $\dfrac{e^{\mathbf{k}Z} + e^{-\mathbf{k}Z}}{2}$ does not give the same $\cos Z$ but it gives the hyperbolic cosine of a bicomplex number. See Section 6.3.3 below.

Given $Z = z_1 + \mathbf{j}z_2 = \beta_1\mathbf{e} + \beta_2\mathbf{e}^\dagger \in \mathbb{BC}$, the properties of the bicomplex exponential bring us immediately to the idempotent representation of $\cos Z$ and $\sin Z$:

$$\cos Z = \cos(\beta_1)\mathbf{e} + \cos(\beta_2)\mathbf{e}^\dagger,$$

$$\sin Z = \sin(\beta_1)\mathbf{e} + \sin(\beta)\mathbf{e}^\dagger. \tag{6.15}$$

In terms of the components of the cartesian representation, one gets:

$$\cos Z = \cos(z_1 - \mathbf{i}z_2)\mathbf{e} + \cos(z_1 + \mathbf{i}z_2)\mathbf{e}^\dagger.$$

Since for a complex variable z the following formulas hold:

$$\cosh(z) = \cos(\mathbf{i}z), \qquad \sinh(z) = -\mathbf{i}\sin(\mathbf{i}z),$$

we obtain that

$$\cos Z = \cosh(z_2)\cos(z_1) - \mathbf{j}\sinh(z_2)\sin(z_1).$$

One may write Z in different forms to observe how the above formulas change.

We continue with a description of some basic properties of the bicomplex trigonometric functions.

- Since the complex sine and cosine functions are periodic with principal period 2π, then taking $Z = \beta_1 \mathbf{e} + \beta_2 \mathbf{e}^\dagger$ and setting $Z_{k,\ell} = (\beta_1 + 2k\pi)\mathbf{e} + (\beta_2 + 2\ell\pi)\mathbf{e}^\dagger$ for arbitrary integers k, ℓ we have:

$$\cos(Z_{k,\ell}) = \cos(Z), \qquad \sin(Z_{k,\ell}) = \sin(Z).$$

Thus the real number $(2\pi)\mathbf{e} + (2\pi)\mathbf{e}^\dagger = 2\pi$ remains the principal period of both bicomplex sine and cosine functions.

- From (6.15), the equation $\cos Z = 0$ is equivalent to the equations in complex variables β_1 and β_2:

$$\cos(\beta_1) = 0, \qquad \cos(\beta_2) = 0.$$

The solutions are $\beta_1 = \dfrac{\pi}{2} + k\pi$, and $\beta_2 = \dfrac{\pi}{2} + \ell\pi$, for $k, \ell \in \mathbb{Z}$. Note that β_1 and β_2 are never zero, so the bicomplex solutions Z to $\cos Z = 0$ are always hyperbolic invertible numbers. In the $\{\mathbf{1}, \mathbf{j}\}$ basis, we get the general solution to $\cos Z = 0$ as

$$Z = z_1 + \mathbf{j}z_2 = ((1 + k + \ell) + \mathbf{j}\,\mathbf{i}(k - \ell))\,\frac{\pi}{2}, \tag{6.16}$$

a set of hyperbolic numbers.

- Similarly, the equation $\sin Z = 0$ is equivalent to

$$\sin(\beta_1) = 0, \qquad \sin(\beta_2) = 0.$$

The solutions are $\beta_1 = k\pi$, and $\beta_2 = \ell\pi$, for $k, l \in \mathbb{Z}$. Note that there are non-invertible solutions for $\sin Z = 0$, e.g., for $\beta_1 = 0$, i.e., $k = 0$, and $\beta_2 \neq 0$. In the $\{\mathbf{1}, \mathbf{j}\}$ basis, we get the general solution for $\sin Z = 0$ as

$$Z = z_1 + \mathbf{j}z_2 = (k + \ell + \mathbf{j}\,\mathbf{i}(k - \ell))\,\frac{\pi}{2},$$

again a set of hyperbolic numbers.

- Formulas (6.15) guarantee that the usual trigonometric identities are true, e.g., the sums and differences of angle formulas, the double angle identities, etc. For example:

$$\sin^2 Z + \cos^2 Z = (\sin^2(\beta_1) + \cos^2(\beta_1))\mathbf{e} + (\sin^2(\beta_2) + \cos^2(\beta_2))\mathbf{e}^\dagger = 1.$$

- Taking the argument equal to the idempotents $\mathbf{e}$ and $\mathbf{e}^\dagger$ gives again funny formulas. Indeed, if $Z = \mathbf{e}$, i.e., $\beta_1 = 1$ and $\beta_2 = 0$, then

$$\cos\mathbf{e} = \cos(1)\mathbf{e} + \mathbf{e}^\dagger = \left(\frac{\cos(1)+1}{2}\right) - \mathbf{j}\left(\mathbf{i}\frac{\cos(1)-1}{2}\right),$$

$$\sin\mathbf{e} = \sin(1)\mathbf{e} = \frac{\sin(1)}{2} - \mathbf{j}\left(\mathbf{i}\frac{\sin(1)}{2}\right).$$

Similarly, if $Z = \mathbf{e}^\dagger$, i.e., $\beta_1 = 0$ and $\beta_2 = 1$, then

$$\cos\mathbf{e}^\dagger = 1\mathbf{e} + \cos(1)\mathbf{e}^\dagger = \left(\frac{\cos(1)+1}{2}\right) + \mathbf{j}\left(\mathbf{i}\frac{\cos(1)-1}{2}\right),$$

$$\sin\mathbf{e}^\dagger = \sin(1)\mathbf{e}^\dagger = \frac{\sin(1)}{2} + \mathbf{j}\left(\mathbf{i}\frac{\sin(1)}{2}\right).$$

Notice that all four are hyperbolic numbers.

As in the complex case, the other bicomplex trigonometric functions are defined in terms of the bicomplex sine and cosine functions. For example, the tangent function is the following.

Definition 6.3.2.2. *Let Z be in $\mathbb{BC}$. We define the* bicomplex *tangent function of a bicomplex variable:*

$$\tan Z := \frac{\sin Z}{\cos Z} \tag{6.17}$$

whenever $\cos Z$ is invertible, i.e., both complex numbers β_1 and β_2 are not equal to $\dfrac{\pi}{2}$ plus integer multiples of π.

A direct computation yields:

$$\tan Z = -\mathbf{j}\frac{e^{\mathbf{j}Z} - e^{-\mathbf{j}Z}}{e^{\mathbf{j}Z} + e^{-\mathbf{j}Z}} = \tan(\beta_1)\mathbf{e} + \tan(\beta_2)\mathbf{e}^\dagger.$$

In a similar fashion we have

Definition 6.3.2.3. *Let Z be in $\mathbb{BC}$. We define the* bicomplex *cotangent function of a bicomplex variable:*

$$\cot Z := \frac{\cos Z}{\sin Z} \tag{6.18}$$

whenever $\sin Z$ is invertible, i.e., both complex numbers β_1 and β_2 are not integer multiples of π.

A direct computation yields:

$$\cot Z = \mathbf{j}\frac{e^{\mathbf{j}Z} + e^{-\mathbf{j}Z}}{e^{\mathbf{j}Z} - e^{-\mathbf{j}Z}} = \cot(\beta_1)\mathbf{e} + \cot(\beta_2)\mathbf{e}^\dagger.$$

We have described the fundamentals of the theory of trigonometric bicomplex functions. Of course, now the theory can be continued in many directions.

We mention now a curious fact. The real tangent function takes any real value. It is easy to show that the complex tangent does not contain $\pm\mathbf{i}$ in its range. For the bicomplex tangent the excluded values are $\{\pm\mathbf{i},\ \pm\mathbf{j}\}$.

6.3.3 Hyperbolic functions of a bicomplex variable

We want to extend directly the notion of a hyperbolic function of a complex variable onto a bicomplex variable. A direct way is clear.

Definition 6.3.3.1. *Let Z in $\mathbb{BC}$. Then we define the bicomplex hyperbolic sine and cosine functions as*

$$\cosh Z := \frac{e^Z + e^{-Z}}{2} = \frac{e^{\mathbf{k}Z} + e^{-\mathbf{k}Z}}{2},$$
$$\sinh Z := \frac{e^Z - e^{-Z}}{2} = \frac{e^{\mathbf{k}Z} - e^{-\mathbf{k}Z}}{2\mathbf{k}}. \tag{6.19}$$

Again, both are well defined since a direct computation gives now that

$$\frac{e^Z + e^{-Z}}{2} = \frac{e^{\mathbf{k}Z} + e^{-\mathbf{k}Z}}{2} \qquad \text{and} \qquad \frac{e^Z - e^{-Z}}{2} = \frac{e^{\mathbf{k}Z} - e^{-\mathbf{k}Z}}{2\mathbf{k}}.$$

As above, in the idempotent representation $Z = \beta_1\mathbf{e} + \beta_2\mathbf{e}^\dagger$ we get that

$$\cosh Z = \cosh(\beta_1)\mathbf{e} + \cosh(\beta_2)\mathbf{e}^\dagger,$$
$$\sinh Z = \sinh(\beta_1)\mathbf{e} + \sinh(\beta_2)\mathbf{e}^\dagger. \tag{6.20}$$

As in the previous section, these formulas would yield the usual properties analogous to hyperbolic complex functions. For example,

$$\cosh^2 Z - \sinh^2 Z = \left(\cosh^2(\beta_1) - \sinh^2(\beta_1)\right)\mathbf{e} + \left(\cosh^2(\beta_2) - \sinh^2(\beta_2)\right)\mathbf{e}^\dagger = 1.$$

The addition formulas are:

$$\cosh(Z_1 + Z_2) = \cosh Z_1 \cosh Z_2 + \sinh Z_1 \sinh Z_2,$$
$$\sinh(Z_1 + Z_2) = \sinh Z_1 \cosh Z_2 + \cosh Z_1 \sinh Z_2.$$

Moreover, for $Z = \mathbf{j}z_2$, we have:

$$\cosh(\mathbf{j}z_2) = \cos(z_2), \qquad \sinh(\mathbf{j}z_2) = \mathbf{j}\sin(z_2).$$

Then for $Z = z_1 + \mathbf{j}z_2$ we have:

$$\sinh Z = \sinh(z_1 + \mathbf{j}z_2) = \sinh(z_1)\cos(z_2) + \mathbf{j}\cosh(z_1)\sin(z_2).$$

A similar formula holds for $\cosh Z$.

6.4 Bicomplex radicals

In this and the next section we begin the study of inverse functions in $\mathbb{BC}$. We start by looking at the equation $Z^n = W$, where the bicomplex numbers Z and W are written as $Z = z_1 + \mathbf{j}z_2 = \beta_1\mathbf{e} + \beta_2\mathbf{e}^\dagger$, and $W = w_1 + \mathbf{j}w_2 = \gamma_1\mathbf{e} + \gamma_2\mathbf{e}^\dagger$. This equation is equivalent to the following two complex equations in variables β_1 and β_2:

$$\beta_1^n = \gamma_1, \qquad \beta_2^n = \gamma_2.$$

If W is invertible, i.e., $\gamma_1\gamma_2 \neq 0$, then each complex equation has n distinct complex solutions, and the equations are independent of each other. Denote these solutions by $\gamma_{1,k} \in \sqrt[n]{\gamma_1}$ and $\gamma_{2,\ell} \in \sqrt[n]{\gamma_2}$, respectively, where the symbol $\sqrt[n]{\gamma_1}$ denotes the set of all solutions of the corresponding equation; the same for $\sqrt[n]{\gamma_2}$. Therefore the bicomplex equation $Z^n = W$ has n^2 solutions given by the bicomplex numbers

$$Z_{k\ell} = \gamma_{1,k}\,\mathbf{e} + \gamma_{2,\ell}\,\mathbf{e}^\dagger = \frac{\gamma_{1,k} + \gamma_{2,\ell}}{2} + \mathbf{j}\frac{\gamma_{2,\ell} - \gamma_{1,k}}{2\mathbf{i}}$$

for all $k,\ell = 1\ldots n$. We define the n-th root of W to be the set of all of these solutions, $\sqrt[n]{W} := \{Z_{k\ell}\}$.

Note that if we start with formula (3.5) for the bicomplex number $W = w_1 + \mathbf{j}w_2$, i.e.,

$$W = |W|_{\mathbf{i}}(\cos\theta + \mathbf{j}\sin\theta)$$

where $|W|_{\mathbf{i}} = \sqrt{w_1^2 + w_2^2}$ is the complex modulus of W, and θ is the complex argument of W, then the solutions $Z_{k\ell}$ of the equation $Z^n = W$ have complex moduli $\sqrt[n]{|W|_{\mathbf{i}}}$, which is a set of n complex numbers, and arguments $\dfrac{\theta + 2\ell\pi}{n}$, for $\ell = 1,\ldots,n$. In conclusion, we find again that there are n^2 bicomplex $n-th$ roots, and more precisely

$$\sqrt[n]{W} = \left\{ \sqrt[n]{|W|_{\mathbf{i}}}\left(\cos\frac{\theta + 2\ell\pi}{n} + \mathbf{j}\sin\frac{\theta + 2\ell\pi}{n}\right) : \ell \in \{0,1,\ldots,n-1\}\right\}.$$

If $W = \gamma_1\mathbf{e} + \gamma_2\mathbf{e}^\dagger$ is a zero-divisor, then exactly one of the complex numbers γ_1 or γ_2 is zero, so the bicomplex equation $Z^n = W$ has exactly n solutions, all of them zero divisors. Obviously if $W = 0$ there is only one solution, $Z = 0$, to the equation $Z^n = 0$.

6.5 The bicomplex logarithm

6.5.1 The real and complex logarithmic functions.

Since the real exponential function is monotone, then the real logarithmic function being its inverse does not cause much problem. The situation with its complex extension is much more sophisticated. The complex logarithm, that is, the notion of the logarithm of a complex number, is introduced from the study of the equation

$e^z = w$. Because the periodicity of the complex exponential function, this equation has, for $w \neq 0$, a countable family of solutions of the form

$$\ln |w| + \mathbf{i}(\arg w + 2\pi m)$$

where $\ln |w|$ is the real logarithm of the positive number $|w|$, m is any integer. Hence, for any non-zero complex number there are infinitely many complex numbers which can be equally called its logarithm.

We will use the following notations and names:

$$\ln(w) = \ln w := \ln |w| + \mathbf{i} \arg w$$

is called the principal value of the logarithm of w;

$$\ln_m(w) = \ln_m w := \ln |w| + \mathbf{i} (\arg w + 2\pi m),$$

for a fixed $m \in \mathbb{Z}$, is called the m-th branch of logarithm of w;

$$\mathrm{Ln}(w) := \big\{ \ln_m w \,\big|\, m \in \mathbb{Z} \big\}$$

is called the complex logarithm of w. So, $\mathrm{Ln}\, w$ is a set and, thus, it does not generate a univalued function but rather a multivalued function, which explains the name for $\ln_m w$: for each m it determines a function, which can be called a logarithmic function.

Note an important property of logarithm:

$$\mathrm{Ln}\,(w_1 w_2) = \mathrm{Ln}\,(w_1) + \mathrm{Ln}\,(w_2), \tag{6.21}$$

where the plus in the right-hand side among the two sets means that the resulting set is obtained by adding all the elements of the first item and all the elements of the second item. Observe that the various branches of the logarithmic function do not possess this property.

6.5.2 The logarithm of a bicomplex number

In this section we define the notion of the logarithm of a bicomplex number. Take a bicomplex number Z written as $Z = z_1 + \mathbf{j} z_2$, and an invertible bicomplex number $W = w_1 + \mathbf{j} w_2$. We study the solutions to the bicomplex equation $e^Z = W$. Recall that

$$W = |W|_{\mathbf{i}}(\cos \theta + \mathbf{j} \sin \theta),$$

where $|W|_{\mathbf{i}}$ is the complex modulus and θ is the complex argument of W: $|W|_{\mathbf{i}} := \sqrt{w_1^2 + w_2^2}$, $\theta \in \mathrm{Arg}_{\mathbf{i}} W$. From the equation

$$e^{z_1} (\cos z_2 + \mathbf{j} \sin z_2) = |W|_{\mathbf{i}} \cdot (\cos(\mathrm{Arg}_{\mathbf{i}} W) + \mathbf{j} \sin(\mathrm{Arg}_{\mathbf{i}} W))$$

it follows that

$$z_1 \in \mathrm{Ln}\, |W|_{\mathbf{i}},$$

the complex logarithm of the complex number $|W|_\mathbf{i}$, and that

$$z_2 \in \mathrm{Arg}_\mathbf{i} W = \{ \arg_\mathbf{i} W + 2\pi m \mid m \in \mathbb{Z} \} .$$

In analogy with the pattern of complex logarithms we introduce the following definitions for an invertible bicomplex number W: the number

$$\ln(W) = \ln W := \ln\big||W|_\mathbf{i}\big| + \mathbf{i}\arg|W|_\mathbf{i} + \mathbf{j}\arg_\mathbf{i} W$$

is called the principal value of the (bicomplex) logarithm of W; then the number

$$\ln_{m,n}(W) := \ln_m |W|_\mathbf{i} + \mathbf{j} \left(\arg_\mathbf{i}(W) + 2\pi n \right),$$

with two arbitrary integers m and n, is called the (m,n)-th branch of the bicomplex logarithm; and finally the set

$$\mathrm{Ln}(W) := \big\{ \ln_{m,n}(W) \,\big|\, m,\, n \in \mathbb{Z} \big\}$$

is called the bicomplex logarithm of W. Again, $\mathrm{Ln}(W)$ is a set, not a (univalued) function.

If the idempotent representation $W = \gamma_1 \mathbf{e} + \gamma_2 \mathbf{e}^\dagger$ is used, then it turns out that

$$\mathrm{Ln}(W) = \mathrm{Ln}(\gamma_1)\mathbf{e} + \mathrm{Ln}(\gamma_2)\mathbf{e}^\dagger.$$

This is because if $Z = \beta_1 \mathbf{e} + \beta_2 \mathbf{e}^\dagger$, then the equation $e^Z = W$ is equivalent to the two complex equations

$$e^{\beta_1} = \gamma_1, \qquad e^{\beta_2} = \gamma_2$$

which have as solutions the complex logarithms $\mathrm{Ln}(\gamma_1)$ and $\mathrm{Ln}(\gamma_2)$ respectively;

$$\mathrm{Ln}(W) = \frac{1}{2}\mathrm{Ln}(\gamma_1\gamma_2) + \frac{1}{2}\mathbf{ji}\,\mathrm{Ln}\frac{\gamma_1}{\gamma_2} = \mathrm{Ln}\sqrt{\gamma_1\gamma_2} + \mathbf{ji}\,\mathrm{Ln}\sqrt{\frac{\gamma_1}{\gamma_2}}$$
$$= \mathrm{Ln}|W|_\mathbf{i} + \mathbf{j}\,\mathrm{Arg}_\mathbf{i}(W),$$

where we used the fact that

$$\theta = \mathbf{i}\ln\sqrt{\frac{\gamma_1}{\gamma_2}} . \tag{6.22}$$

We state below some properties of the bicomplex logarithm.

- The bicomplex logarithm is not defined for zero-divisors, as the bicomplex exponential $W = e^Z$ is always invertible.

- If $Z = z_1 + \mathbf{j}z_2$ is an invertible bicomplex number and if $m, n \in \mathbb{Z}$, then

$$e^{\ln_{m,n}(Z)} = e^{\ln_m |Z|_\mathbf{i} + \mathbf{j}\arg_\mathbf{i}(Z) + 2n\pi\mathbf{j}} = e^{\ln_m |Z|_\mathbf{i}} e^{\mathbf{j}\arg_\mathbf{i}(Z)}$$
$$= |Z|_\mathbf{i}(\cos(\arg_\mathbf{i}(Z)) + \mathbf{j}\sin(\arg_\mathbf{i}(Z))) = Z.$$

- For $Z = 1 = 1 + \mathbf{j}0$, we have:

$$\ln_{m,n}(1) = 0 + 2m\pi\mathbf{i} + 2n\pi\mathbf{j}$$

for all $m, n \in \mathbb{Z}$.

- For Z_1 and Z_2 two invertible bicomplex numbers, the following formula holds:

$$\mathrm{Ln}(Z_1 Z_2) = \mathrm{Ln}(Z_1) + \mathrm{Ln}(Z_2). \tag{6.23}$$

6.6 On bicomplex inverse trigonometric functions

The inverses of the bicomplex trigonometric functions are defined in complete analogy with the complex case, as we have already properly defined the notions of bicomplex exponential, logarithm, and square root.

For example, the inverse of the bicomplex cosine function is obtained by solving the equation

$$\cos(Z) = \frac{e^{\mathbf{j}Z} + e^{-\mathbf{j}Z}}{2} = W.$$

This is a quadratic equation in $e^{\mathbf{j}Z}$ with roots

$$e^{\mathbf{j}Z} = W \pm \sqrt{W^2 - 1}.$$

Therefore, the set $\mathrm{Arccos}(W)$ should be defined as

$$\mathrm{Arccos}(W) := \left\{ -\mathbf{j}\ln_{m,n}\left(W \pm \sqrt{W^2 - 1}\right) \mid m, n \in \mathbb{Z} \right\}$$
$$= \pm \left\{ \mathbf{j}\ln_{m,n}\left(W + \sqrt{W^2 - 1}\right) \mid m, n \in \mathbb{Z} \right\}.$$

It is clear how to introduce the principal value, the (m, n)-th branch, etc., as well as how to deal with other inverse functions. We leave this to the interested reader, who is recommended also to consider what happens if we solve the equation

$$\cos(Z) = \frac{e^{\mathbf{i}Z} + e^{-\mathbf{i}Z}}{2} = W.$$

6.7 The exponential representations of bicomplex numbers

We will show here that the trigonometric representations of bicomplex numbers can be written with the help of the bicomplex exponential function. Indeed, recall that an invertible bicomplex number Z has a $\mathbb{C}(\mathbf{i})$-trigonometric representation and a $\mathbb{C}(\mathbf{j})$-one:

$$Z = |Z|_{\mathbf{i}} \left(\cos\theta + \mathbf{j}\sin\theta \right)$$

where θ is a $\mathbb{C}(\mathbf{i})$-argument of Z, i.e., any element in $\mathrm{Arg}_{\mathbb{C}(\mathbf{i})}(Z)$, see (3.7); and

$$Z = |Z|_{\mathbf{j}}\,(\cos\psi + \mathbf{i}\sin\psi)$$

where ψ is a $\mathbb{C}(\mathbf{j})$-argument of Z, i.e., any element in $\mathrm{Arg}_{\mathbb{C}(\mathbf{j})}(Z)$, see (3.13). But

$$\cos\theta + \mathbf{j}\sin\theta = e^{\mathbf{j}\theta} \qquad \text{and} \qquad \cos\psi + \mathbf{i}\sin\psi = e^{\mathbf{i}\psi},$$

hence

$$Z = |Z|_{\mathbf{i}}\cdot e^{\mathbf{j}\theta} = |Z|_{\mathbf{i}}\cdot e^{\mathbf{j}\,\mathrm{arg}_{\mathbb{C}(\mathbf{i}),m}(Z)} \tag{6.24}$$

and

$$Z = |Z|_{\mathbf{j}}\cdot e^{\mathbf{i}\psi} = |Z|_{\mathbf{j}}\cdot e^{\mathbf{i}\,\mathrm{arg}_{\mathbb{C}(\mathbf{j}),n}(Z)} \tag{6.25}$$

for any integers m and n. We will refer, sometimes, to (6.24) and (6.25) as the (complex) exponential representations of Z. Note also that the properties of the exponential function imply that

$$Z^{\dagger} = |Z|_{\mathbf{i}}\cdot e^{-\mathbf{j}\theta}$$

and

$$\overline{Z} = |Z|_{\mathbf{j}}\cdot e^{-\mathbf{i}\psi}.$$

The trigonometric representation of a bicomplex number in hyperbolic terms is allowed both for invertible numbers and for zero-divisors. Beginning with an invertible Z we should recall that

$$Z = |Z|_{\mathbf{k}}\,(\cos\Psi_Z + \mathbf{i}\sin\Psi_Z) = |Z|_{\mathbf{k}}\left(e^{\mathbf{i}\nu_1}\cdot\mathbf{e} + e^{\mathbf{i}\nu_2}\cdot\mathbf{e}^{\dagger}\right)$$

with $\Psi_Z = \nu_1\mathbf{e} + \nu_2\mathbf{e}^{\dagger} \in \mathbb{D}^{+}_{inv}$ being any element in $\mathrm{Arg}_{\mathbb{D}}Z$, i.e., a hyperbolic argument of Z. Appealing again to the properties of the exponential function we see that

$$\cos\Psi_Z + \mathbf{i}\sin\Psi_Z = e^{\mathbf{i}\nu_1}\cdot\mathbf{e} + e^{\mathbf{i}\nu_2}\cdot\mathbf{e}^{\dagger} = e^{\mathbf{i}\Psi_Z} = e^{\mathbf{i}\,\mathrm{arg}_{m,n;\mathbb{D}}(Z)}$$

for any integers m and n. Thus, the equality

$$Z = |Z|_{\mathbf{k}}\cdot e^{\mathbf{i}\,\mathrm{arg}_{m,n;\mathbb{D}}(Z)} \tag{6.26}$$

is the exponential representation of an invertible bicomplex number in hyperbolic terms.

If Z is a zero-divisor, then $Z = |Z|_{\mathbf{k}}\cdot e^{\mathbf{i}\nu_1}\mathbf{e}$ or $Z = |Z|_{\mathbf{k}}\cdot e^{\mathbf{i}\nu_2}\mathbf{e}^{\dagger}$ with real numbers ν_1 and ν_2; the corresponding hyperbolic arguments are $\mathrm{arg}_{\mathbb{D}}(Z) = \nu_1\mathbf{e}$ and $\mathrm{arg}_{\mathbb{D}}(Z) = \nu_2\mathbf{e}^{\dagger}$ respectively. But $e^{\mathbf{i}\nu_1}\mathbf{e}$ is equal to $e^{\mathbf{i}\nu_1}\mathbf{e} + \mathbf{e}^{\dagger}$, not to $e^{\mathbf{i}\nu_1}\mathbf{e}$, hence the analogue of the exponential representation in hyperbolic terms for this case is

$$Z = |Z|_{\mathbf{k}}\left(e^{\mathbf{i}\,\mathrm{arg}_{m,\mathbb{D}}(Z)} - \mathbf{e}^{\dagger}\right). \tag{6.27}$$

Similarly we get for $Z = |Z|_{\mathbf{k}} \cdot e^{\mathbf{i}\nu_2} \mathbf{e}^\dagger$

$$Z = |Z|_{\mathbf{k}} \left(e^{\mathbf{i}\arg_{n,\mathbb{D}}(Z)} - \mathbf{e} \right). \qquad (6.28)$$

So although the formulas (6.27)–(6.28) have a different structure than (6.24), (6.25) and (6.26), we believe that the name "exponential representation" is applicable also.

Our study mimics the approach done in usual complex analysis (e.g. [1]) and develops the most important properties of the elementary functions in the bicomplex setup. These are basically the main functions of the bicomplex analysis, on which we "test" the notions of derivability, differentiability, holomorphy, and integration in the following Chapters.

Aspects of the theory of bicomplex elementary functions have been also studied in [45, 55, 57, 99].

Chapter 7

Bicomplex Derivability and Differentiability

7.1 Different kinds of partial derivatives

In this chapter we will study bicomplex-valued functions of a bicomplex variable, and we will examine the notions of derivability and of holomorphy for such
functions (see [46]). We will consider $\mathbb{BC}$ as a topological space endowed with
the Euclidean topology of $\mathbb{R}^4$ and Ω an open set. Let F be a bicomplex function
$F : \Omega \to \mathbb{BC}$ of a bicomplex variable

$$Z = x_1 + \mathbf{i}y_1 + \mathbf{j}x_2 + \mathbf{k}y_2 .$$

Due to the various ways in which bicomplex numbers can be written, the function
F inherits analogous representations, specifically:

$$\begin{aligned} F &= f_1 + \mathbf{j}f_2 = \rho_1 + \mathbf{i}\rho_2 = g_1 + \mathbf{k}g_2 = \gamma_1 + \mathbf{k}\gamma_2 = \mathfrak{f}_1 + \mathbf{i}\mathfrak{f}_2 = \mathfrak{g}_1 + \mathbf{j}\mathfrak{g}_2 \\ &= f_{11} + \mathbf{i}f_{22} + \mathbf{j}f_{21} + \mathbf{k}f_{22} , \end{aligned} \tag{7.1}$$

where f_1, f_2, g_1, g_2 are $\mathbb{C}(\mathbf{i})$-valued functions, $\rho_1, \rho_2, \gamma_1, \gamma_2$ are $\mathbb{C}(\mathbf{j})$-valued functions, $\mathfrak{f}_1, \mathfrak{f}_2, \mathfrak{g}_1, \mathfrak{g}_2$ are hyperbolic-valued functions, and $f_{k\ell}$ are real-valued functions, all of a bicomplex variable Z. We will, at different times in this section, use
all of these representations.

Take now a point $Z_0 \in \Omega$ and let $H = h_{11} + \mathbf{i}h_{12} + \mathbf{j}h_{21} + \mathbf{k}h_{22}$ be the
increment. The partial derivatives of F with respect to the variables x_1, y_1, x_2, y_2
are defined as usual (when they exist) and we give their formulas in order to fix
the notation:

$$\frac{\partial F}{\partial x_1}(Z_0) := \lim_{h_{11} \to 0} \frac{F(Z_0 + h_{11}) - F(Z_0)}{h_{11}},$$

$$\frac{\partial F}{\partial y_1}(Z_0) := \lim_{h_{12} \to 0} \frac{F(Z_0 + ih_{12}) - F(Z_0)}{h_{12}},$$

$$\frac{\partial F}{\partial x_2}(Z_0) := \lim_{h_{21} \to 0} \frac{F(Z_0 + jh_{21}) - F(Z_0)}{h_{21}},$$

$$\frac{\partial F}{\partial y_2}(Z_0) := \lim_{h_{22} \to 0} \frac{F(Z_0 + kh_{22}) - F(Z_0)}{h_{22}}. \qquad (7.2)$$

Recalling formulas (1.3) and (1.4) we can write the increments as

$$H = h_1 + jh_2 := (h_{11} + ih_{12}) + j(h_{21} + ih_{22})$$
$$= \kappa_1 + i\kappa_2 := (h_{11} + jh_{21}) + i(h_{12} + jh_{22}).$$

This leads us to

Definition 7.1.1. *The complex partial derivatives of the bicomplex function F are defined as the following limits (if they exist):*

$$F'_{z_1}(Z_0) := \lim_{h_1 \to 0} \frac{F(Z_0 + h_1) - F(Z_0)}{h_1}, \qquad (7.3)$$

$$F'_{z_2}(Z_0) := \lim_{h_2 \to 0} \frac{F(Z_0 + jh_2) - F(Z_0)}{h_2}, \qquad (7.4)$$

$$F'_{\zeta_1}(Z_0) := \lim_{\kappa_1 \to 0} \frac{F(Z_0 + \kappa_1) - F(Z_0)}{\kappa_1}, \qquad (7.5)$$

$$F'_{\zeta_2}(Z_0) := \lim_{\kappa_2 \to 0} \frac{F(Z_0 + i\kappa_2) - F(Z_0)}{\kappa_2}. \qquad (7.6)$$

Similarly we introduce the hyperbolic partial derivatives: if we recall formula (1.5) and write the increment as

$$H = \mathfrak{h}_1 + i\mathfrak{h}_2 := (h_{11} + kh_{22}) + i(h_{12} + k(-h_{21})),$$

then the hyperbolic partial derivatives are

$$F'_{\mathfrak{z}_1}(Z_0) := \lim_{\mathfrak{h}_1 \notin \mathfrak{S}_0(\mathbb{D}),\, \mathfrak{h}_1 \to 0} \frac{F(Z_0 + \mathfrak{h}_1) - F(Z_0)}{\mathfrak{h}_1},$$

$$F'_{\mathfrak{z}_2}(Z_0) := \lim_{\mathfrak{h}_2 \notin \mathfrak{S}_0(\mathbb{D}),\, \mathfrak{h}_2 \to 0} \frac{F(Z_0 + i\mathfrak{h}_2) - F(Z_0)}{\mathfrak{h}_2}, \qquad (7.7)$$

where with the symbol $\mathfrak{S}_0(\mathbb{D}) := \mathfrak{S}_0 \cap \mathbb{D}$ we indicate the set of the hyperbolic zero-divisors together with $0 \in \mathbb{D}$.

Example 7.1.2. From the point of view of the classic theory of functions of two complex variables, one may think, because of the "symmetry" of the imaginary units $\mathbf{i}$ and $\mathbf{j}$, and thus the analogy between the corresponding complex variables, that there is no reason to define all the complex partial derivatives (7.3)–(7.6). The following simple example illustrates that inside the bicomplex realm the differences are relevant. Consider to this purpose the function $F : \mathbb{BC} \to \mathbb{BC}$, $F(Z) = Z^{\dagger}$. It is immediate to see that

$$F'_{z_1}(Z_0) = 1, \qquad F'_{z_2}(Z_0) = -\mathbf{j}$$

and that neither $F'_{\zeta_1}(Z_0)$ nor $F'_{\zeta_2}(Z_0)$ exist for any $Z_0 \in \mathbb{BC}$.

Remark 7.1.3. *Formulas (1.6)–(1.8) suggest the introduction of six more partial derivatives: four complex and two hyperbolic. But since there exists a direct relation between the corresponding variables, namely,*

$$w_1 = z_1, \quad w_2 = -\mathbf{i}z_2, \quad \omega_1 = \zeta_1, \quad \omega_2 = -\mathbf{j}\zeta_2, \quad \mathfrak{w}_1 = \mathfrak{z}_1, \quad \mathfrak{w}_2 = -\mathbf{k}\mathfrak{z}_2,$$

it is easy to see that we do not get anything essentially new. The situation with partial derivatives with respect to variables arising from the idempotent representations of bicomplex numbers is on the other hand much more interesting, and will be discussed in detail later.

7.2 The bicomplex derivative and the bicomplex derivability

We continue now with the definition of the derivative of a bicomplex function $F : \Omega \subset \mathbb{BC} \to \mathbb{BC}$ of one bicomplex variable Z as follows:

Definition 7.2.1. *The derivative $F'(Z_0)$ of the function F at a point $Z_0 \in \Omega$ is the limit, if it exists,*

$$F'(Z_0) := \lim_{Z \to Z_0} \frac{F(Z) - F(Z_0)}{Z - Z_0} = \lim_{\mathfrak{S}_0 \not\ni H \to 0} \frac{F(Z_0 + H) - F(Z_0)}{H}, \qquad (7.8)$$

for Z in the domain of F such that $H = Z - Z_0$ is an invertible bicomplex number. In this case, the function F is called derivable *at Z_0.*

Corollary 7.2.2. *The function F is derivable at Z_0 if and only if there exists a function α_{F,Z_0} such that*

$$\lim_{\mathfrak{S}_0 \not\ni H \to 0} \alpha_{F,Z_0}(H) = 0$$

and

$$F(Z_0 + H) - F(Z_0) = F'(Z_0) \cdot H + \alpha_{F,Z_0}(H)H \quad \text{for all } H \notin \mathfrak{S}_0. \qquad (7.9)$$

Remark 7.2.3. *It is necessary to make a comment here. Traditionally, see e.g. [103], p. 138 and p. 432, if h is either a real or a complex increment, the symbol $\mathfrak{o}(h)$*

is used to indicate any expression of the form $\alpha(h)|h|$ with $\lim\limits_{h \to 0} \alpha(h) = 0$. Since, both in the real and in the complex case, the expression $\dfrac{|h|}{h}$ remains bounded when $h \to 0$, it is clear that one could replace $\alpha(h)|h|$ by $\alpha(h)h$ in the expression of $\mathbf{o}$. However, the situation is quite different in the bicomplex case. Here, the expression $\dfrac{|H|}{H}$ is not bounded when $H \to 0$, and therefore we need to carefully distinguish the two expressions. In accordance with the usual notation, we will always use $\mathbf{o}(H)$ to denote a function of the form $\alpha(H)|H|$, and therefore the expression in the previous corollary $\alpha_{F,Z_0}(H)H$ is not, in general, $\mathbf{o}(H)$. This distinction is at the basis of the notions of weak and strong Stoltz conditions for bicomplex functions, which are used by G.B.Price in [56].

Remark 7.2.4. *A bicomplex function F derivable at Z_0 enjoys the following property:*

$$\lim_{H \notin \mathfrak{S}_0,\, H \to 0} (F(Z_0 + H) - F(Z_0)) = 0. \tag{7.10}$$

In other words, a function F, which is derivable at a point Z_0, enjoys some sort of "weakened" continuity at that point, in the sense that $F(Z)$ converges to $F(Z_0)$ as long as Z converges to Z_0 in such a way that $Z - Z_0$ is invertible. We will see later on that this restriction can be removed under reasonable assumptions.

It turns out that the arithmetic operations on derivable functions follow the usual rules of their real and complex antecedents.

Theorem 7.2.5. (Derivability and arithmetic operations) *Let F and G be two bicomplex functions defined on $\Omega \subset \mathbb{BC}$ and derivable at $Z_0 \in \Omega$. Then:*

1. *The sum and difference of F and G are derivable at Z_0 and*

$$(F \pm G)'(Z_0) = F'(Z_0) \pm G'(Z_0).$$

2. *For any bicomplex number C the function $C \cdot F$ is derivable at Z_0 and*

$$(C \cdot F)'(Z_0) = C \cdot F'(Z_0).$$

3. *The product of F and G is derivable at Z_0 and*

$$(F \cdot G)'(Z_0) = F'(Z_0) \cdot G(Z_0) + F(Z_0) \cdot G'(Z_0).$$

4. *If G is continuous at Z_0 and $G(Z_0) \notin \mathfrak{S}_0$, then the quotient $\dfrac{F}{G}$ is derivable at Z_0 and*

$$\left(\frac{F}{G}\right)'(Z_0) = \frac{F'(Z_0) \cdot G(Z_0) - F(Z_0) \cdot G'(Z_0)}{(G(Z_0))^2}.$$

Proof. From Corollary 7.2.2, there exist two bicomplex functions α_{F,Z_0} and α_{G,Z_0} such that

$$\lim_{\mathfrak{S}_0 \not\ni H \to 0} \alpha_{F,Z_0}(H) = \lim_{\mathfrak{S}_0 \not\ni H \to 0} \alpha_{G,Z_0}(H) = 0$$

and

$$F(Z_0 + H) - F(Z_0) = F'(Z_0) \cdot H + \alpha_{F,Z_0}(H)H \,, \tag{7.11}$$
$$G(Z_0 + H) - G(Z_0) = G'(Z_0) \cdot H + \alpha_{G,Z_0}(H)H \,, \tag{7.12}$$

for all $H \notin \mathfrak{S}_0$. The first statement about the sum or difference $F \pm G$ is obtained easily by adding (or subtracting) formulas (7.11) and (7.12), and noting that the functions $\alpha_{F \pm G, Z_0} := \alpha_{F,Z_0} \pm \alpha_{G,Z_0}$ respect the properties from Corollary 7.2.2 (which is an "if and only if" statement). Similarly, if we multiply (7.11) by the number C, we obtain statement (2). Note that C needs not necessarily be an invertible bicomplex number. For example, if $C = c\mathbf{e}$, where $c \in \mathbb{C}(\mathbf{i})$, then from (7.11) (multiplied by C) we obtain that

$$c\mathbf{e} \cdot F'(Z_0) = (c F'(Z_0))\mathbf{e} \in \mathbb{BC}_\mathbf{e}$$

is the derivative of $c\mathbf{e} \cdot F$ at Z_0, which is a zero-divisor. The non-invertibility of C does not affect any arguments, as H is always invertible.

If we multiply the terms $F(Z_0 + H)$ and $G(Z_0 + H)$ obtained from the left-hand side of formulas (7.11) and (7.12) we obtain:

$$\begin{aligned}
F(Z_0 + H) \cdot G(Z_0 + H) = {} & F(Z_0)G(Z_0) + (F'(Z_0) \cdot G(Z_0) + F(Z_0) \cdot G'(Z_0)) \cdot H \\
& + (F(Z_0)\alpha_{G,Z_0}(H) + G(Z_0)\alpha_{F,Z_0}(H)) \cdot H \\
& + (F'(Z_0)G'(Z_0) + F'(Z_0)\alpha_{G,Z_0}(H) + G'(Z_0)\alpha_{F,Z_0}(H) \\
& + \alpha_{F,Z_0}(H)\alpha_{G,Z_0}(H)) \cdot H^2.
\end{aligned}$$

Now, writing

$$\begin{aligned}
\alpha_{FG,Z_0}(H) := {} & F(Z_0)\alpha_{G,Z_0}(H) + G(Z_0)\alpha_{F,Z_0}(H) \\
& + (F'(Z_0)G'(Z_0) + F'(Z_0)\alpha_{G,Z_0}(H)+ \\
& + G'(Z_0)\alpha_{F,Z_0}(H) + \alpha_{F,Z_0}(H)\alpha_{G,Z_0}(H)) \cdot H,
\end{aligned}$$

we obtain:

$$\lim_{\mathfrak{S}_0 \not\ni H \to 0} \alpha_{FG,Z_0}(H) = 0$$

and

$$\begin{aligned}
F(Z_0 + H) \cdot G(Z_0 + H) - {} & F(Z_0)G(Z_0) \\
& = (F'(Z_0) \cdot G(Z_0) + F(Z_0) \cdot G'(Z_0)) \cdot H + \alpha_{FG,Z_0}(H)H \,,
\end{aligned}$$

which proves that the product $F \cdot G$ is derivable at Z_0 and the usual product formula holds.

We assume now that $G(Z_0) \notin \mathfrak{S}_0$ and that G is continuous at Z_0, which implies that $G(Z) \notin \mathfrak{S}_0$ in a neighborhood of Z_0. All the computations below are performed in this neighborhood. Then we write:

$$
\begin{aligned}
\frac{F(Z_0 + H)}{G(Z_0 + H)} - \frac{F(Z_0)}{G(Z_0)} &= \frac{F(Z_0 + H)G(Z_0) - G(Z_0 + H)F(Z_0)}{G(Z_0 + H)G(Z_0)} \\
&= \frac{(F(Z_0 + H) - F(Z_0))\, G(Z_0) - (G(Z_0 + H) - G(Z_0))\, F(Z_0)}{G(Z_0 + H)G(Z_0)} \\
&= \frac{(F'(Z_0) \cdot G(Z_0) - G'(Z_0) \cdot F(Z_0)) \cdot H + (\alpha_{F,Z_0}(H)G(Z_0) + \alpha_{G,Z_0}(H)F(Z_0)) \cdot H}{G(Z_0 + H)G(Z_0)} .
\end{aligned}
$$

$$(7.13)$$

Now note that if we write

$$
\frac{1}{G(Z_0 + H)} - \frac{1}{G(Z_0)} = \left(-\frac{G'(Z_0) + \alpha_{G,Z_0}(H)}{G(Z_0 + H)G(Z_0)} \right) \cdot H =: \alpha_1(H) ,
$$

then the function $\alpha_1(H)$ has the property

$$
\lim_{\mathfrak{S}_0 \not\ni H \to 0} \alpha_1(H) = 0 .
$$

Thus using the formula:

$$
\frac{1}{G(Z_0 + H)} = \frac{1}{G(Z_0)} + \alpha_1(H) ,
$$

rewriting (7.13) leads to the expression:

$$
\frac{F(Z_0 + H)}{G(Z_0 + H)} - \frac{F(Z_0)}{G(Z_0)} = \frac{F'(Z_0) \cdot G(Z_0) - G'(Z_0) \cdot F(Z_0)}{(G(Z_0))^2} \cdot H + \alpha_{\frac{F}{G},Z_0}(H) \cdot H ,
$$

where $\alpha_{\frac{F}{G},Z_0}(H)$ is the bicomplex function containing all the remaining terms and which enjoys the property

$$
\lim_{\mathfrak{S}_0 \not\ni H \to 0} \alpha_{\frac{F}{G},Z_0}(H) = 0 .
$$

Using again Corollary 7.2.2, the proof of the derivability of the quotient and of the corresponding formula is finished. $\square$

Theorem 7.2.6. *Let F and G be two bicomplex functions, F is defined on an open set Ω and G is defined on $F(\Omega) \subset \mathbb{BC}$. Assume that there is a point $Z_0 \in \Omega$ such that $W_0 = F(Z_0)$ is an interior point of $F(\Omega)$ and such that*

$$
F(Z) - F(Z_0) \notin \mathfrak{S}_0, \quad \text{implies } Z - Z_0 \notin \mathfrak{S}_0, \ Z \in \Omega .
$$

Assume now that F is derivable at Z_0 and G is derivable at W_0. Then the composition $G \circ F$ is derivable at Z_0 and

$$
(G \circ F)'(Z_0) = G'(F(Z_0)) \cdot F'(Z_0) .
$$

Moreover, if $F'(Z_0) \notin \mathfrak{S}_0$ and if F is a bijective function around Z_0, then its inverse F^{-1} is derivable at $W_0 = F(Z_0)$ and

$$(F^{-1})'(W_0) \cdot F'(Z_0) = 1 \,.$$

Proof. Via Corollary 7.2.2, there exist two bicomplex functions α_{F,Z_0} and α_{G,W_0} such that

$$\lim_{\mathfrak{S}_0 \not\ni H \to 0} \alpha_{F,Z_0}(H) = \lim_{\mathfrak{S}_0 \not\ni K \to 0} \alpha_{G,W_0}(K) = 0$$

and

$$F(Z_0 + H) - F(Z_0) = F'(Z_0) \cdot H + \alpha_{F,Z_0}(H)H \,, \tag{7.14}$$
$$G(W_0 + K) - G(W_0) = G'(W_0) \cdot K + \alpha_{G,W_0}(K)K \,, \tag{7.15}$$

for all $H, K \notin \mathfrak{S}_0$.

Denote by $\mathcal{V}_{W_0}$ a neighborhood of W_0 contained in $F(\Omega)$ which exists by hypothesis. Let then $\mathcal{V}_{W_0}^*$ be the set obtained from $\mathcal{V}_{W_0}$ by eliminating the points W for which $W - W_0$ are zero-divisors. Let $\mathcal{U}_{Z_0}^*$ be the set of all $Z \in \mathcal{U}_{Z_0} :=$ $F^{-1}(\mathcal{V}_{W_0})$ such that $Z - Z_0$ is not a zero-divisor.

Now we note that formula (7.15) is true for those invertible K such that $W_K := W_0 + K \in \mathcal{V}_{W_0}$; such K exist by taking $W_K \in \mathcal{V}_{W_0}^*$. Then by the properties of F, for any such K there exists a $Z_K \in \mathcal{U}_{Z_0}^*$ such that $W_K = W_0 + K = F(Z_K)$. Now we write:

$$Z_K = Z_0 + H := Z_0 + (Z_K - Z_0) \,,$$

where H is invertible by the choice of Z_K. Therefore we replace $W_0 + K = F(Z_K) = F(Z_0 + H)$, $W_0 = F(Z_0)$ and $K = F(Z_0 + H) - F(Z_0)$ in formula (7.15), and we obtain:

$$\begin{aligned}G(F(Z_0 + H)) - G(F(Z_0)) &= G'(F(Z_0)) \cdot (F(Z_0 + H) - F(Z_0)) \\ &\quad + \widetilde{\alpha}_{G,F(Z_0)}(H) \cdot (F(Z_0 + H) - F(Z_0)) \,,\end{aligned}$$

where $\widetilde{\alpha}_{G,F(Z_0)}(H) := \alpha_{G,F(Z_0)}(F(Z_0 + H) - F(Z_0))$ is a function having the property from Corollary 7.2.2, as when $K \to 0$ so does $H \to 0$, in both cases along invertible bicomplex numbers. Now we use formula (7.14) and we get:

$$\begin{aligned}G(F(Z_0 + H)) - G(F(Z_0)) &= G'(F(Z_0)) \cdot (F'(Z_0) \cdot H + \alpha_{F,Z_0}(H)H) \\ &\quad + \widetilde{\alpha}_{G,F(Z_0)}(H) \cdot (F'(Z_0) \cdot H + \alpha_{F,Z_0}(H)H) \\ &=: G'(F(Z_0)) \cdot F'(Z_0) \cdot H + \alpha_{G \circ F}(H)H \,,\end{aligned}$$

where the function $\alpha_{G \circ F}(H)$ has obviously the desired property from Corollary 7.2.2.

The last statement of the theorem can be proved quite analogously to the cases of real and complex functions. If one assumes beforehand the existence of the derivative of the inverse function at W_o, then the corresponding formula becomes a particular case of the chain rule above:

$$1 = (F^{-1} \circ F)'(Z_0) = (F^{-1})'(F(Z_0)) \cdot F'(Z_0) \,,$$

where we used that the derivative of the identity function is one which is obvious.

□

Theorem 7.2.7 (The derivability of bicomplex elementary functions). *All bicomplex elementary functions introduced in Chapter 6 are derivable at any point $Z = Z_0$ where they are defined. In more detail:*

1. *Any constant function is derivable with the derivative zero.*

2. *Bicomplex polynomials are derivable for any $Z \in \mathbb{BC}$. In particular, we have:*

$$(Z^n)' = nZ^{n-1}.$$

3. *The derivative of the bicomplex exponential function e^Z is e^Z.*

4. $(\sin Z)' = \cos Z$ *and* $(\cos Z)' = -\sin Z$.

5. $(\tan Z)' = \dfrac{1}{(\cos Z)^2}$ *for any Z such that $\cos Z$ is an invertible bicomplex number, i.e., using the idempotent representation $Z = \beta_1 \mathbf{e} + \beta_2 \mathbf{e}^\dagger$, the complex numbers β_1 and β_2 are not equal to $\dfrac{\pi}{2}$ plus integer multiples of π.*

   Similarly, $(\cot Z)' = \dfrac{-1}{(\sin Z)^2}$ *for any Z such that $\sin Z$ is an invertible bicomplex number, i.e., whenever β_1 and β_2 are not integer multiples of π.*

6. $(\cosh Z)' = \sinh Z$ *and* $(\sinh Z)' = \cosh Z$.

7. *The (m,n)-branch logarithmic function $F(Z) = \ln_{m,n}(Z)$, which is defined for all invertible bicomplex numbers Z, is derivable for all such Z, and*

$$(\ln_{m,n})'(Z) = \frac{1}{Z}.$$

Proof. Item (1) is obviously true.

Note that the formula

$$A^n - B^n = (A - B)(A^{n-1} + A^{n-2}B + \cdots + B^{n-1})$$

holds for any $A, B \in \mathbb{BC}$, due to the algebraic properties of addition and multiplication of bicomplex numbers. Applying this for $A = Z + H$ and $B = Z$, we obtain:

$$\lim_{\mathfrak{S}_0 \not\ni H \to 0} \frac{(Z+H)^n - Z^n}{H}$$
$$= \lim_{\mathfrak{S}_0 \not\ni H \to 0} \frac{H\left((Z+H)^{n-1} + (Z+H)^{n-2}Z + \cdots + Z^{n-1}\right)}{H}$$
$$= \lim_{\mathfrak{S}_0 \not\ni H \to 0} \left((Z+H)^{n-1} + (Z+H)^{n-2}Z + \cdots + Z^{n-1}\right) = nZ^{n-1}.$$

It follows that any bicomplex polynomial $p(Z) = \sum_{k=0}^{n} A_k Z^k$ is derivable at any $Z \in \mathbb{BC}$, and its derivative is a polynomial of degree one less, given by

$$p'(Z) = \sum_{k=1}^{n} k A_k Z^{k-1} .$$

Let us consider the bicomplex exponential function e^Z. For an arbitrary $Z \in \mathbb{BC}$ and an invertible bicomplex number H, we have:

$$\frac{e^{Z+H} - e^Z}{H} = e^Z \cdot \frac{e^H - 1}{H} .$$

We show now that

$$\lim_{\mathfrak{S}_0 \not\ni H \to 0} \frac{e^H - 1}{H} = 1 .$$

For this, let us write $H = h\mathbf{e} + k\mathbf{e}^\dagger$ in the idempotent representation, where $h, k \in \mathbb{C}(\mathbf{i})$ are such that $hk \neq 0$. Then $e^H = e^h \mathbf{e} + e^k \mathbf{e}^\dagger$, thus

$$\frac{e^H - 1}{H} = \frac{e^h \mathbf{e} + e^k \mathbf{e}^\dagger}{h\mathbf{e} + k\mathbf{e}^\dagger} = \frac{e^h - 1}{h} \mathbf{e} + \frac{e^k - 1}{k} \mathbf{e}^\dagger .$$

Therefore

$$\lim_{\mathfrak{S}_0 \not\ni H \to 0} \frac{e^H - 1}{H} = \lim_{h \to 0} \left(\frac{e^h - 1}{h} \right) \mathbf{e} + \lim_{k \to 0} \left(\frac{e^k - 1}{k} \right) \mathbf{e}^\dagger = \mathbf{e} + \mathbf{e}^\dagger = 1 .$$

It follows that

$$\lim_{\mathfrak{S}_0 \not\ni H \to 0} \frac{e^{Z+H} - e^Z}{H} = e^Z \cdot \lim_{\mathfrak{S}_0 \not\ni H \to 0} \frac{e^H - 1}{H} = e^Z ,$$

so $(e^Z)' = e^Z$, for all $Z \in \mathbb{BC}$.

Recall that the bicomplex trigonometric functions are defined by:

$$\cos Z = \frac{e^{\mathbf{j}Z} + e^{-\mathbf{j}Z}}{2} , \qquad \sin Z = \frac{e^{\mathbf{j}Z} - e^{-\mathbf{j}Z}}{2\mathbf{j}} .$$

Since the exponential function is derivable for all $Z \in \mathbb{BC}$, it follows that $\cos Z$ is derivable for all Z and

$$(\cos Z)' = \frac{\mathbf{j}e^{\mathbf{j}Z} - \mathbf{j}e^{-\mathbf{j}Z}}{2} = -\frac{e^{\mathbf{j}Z} - e^{-\mathbf{j}Z}}{2\mathbf{j}} = \sin Z .$$

Similarly the function $\sin Z$ is derivable for all Z and $(\sin Z)' = \cos Z$.

The bicomplex tangent function is defined as the quotient $\dfrac{\sin Z}{\cos Z}$, whenever $\cos Z$ is invertible. Using the quotient rule, we obtain:

$$(\tan Z)' = \frac{\cos^2 Z + \sin^2 Z}{\cos^2 Z} = \frac{1}{\cos^2 Z}.$$

A similar argument applies for the function $\cot Z$.

The hyperbolic functions of a bicomplex variable are defined as

$$\cosh Z = \frac{e^Z + e^{-Z}}{2}, \qquad \sin Z = \frac{e^Z - e^{-Z}}{2}.$$

Again, using that e^Z is derivable for any Z, it follows that $\cosh Z$ and $\sinh Z$ are derivable for any Z and

$$(\cosh Z)' = \frac{e^Z - e^{-Z}}{2} = \sinh Z, \qquad (\sinh Z)' = \frac{e^Z + e^{-Z}}{2} = \cosh Z.$$

Recall that the (m,n)-branch bicomplex logarithmic function $\ln_{m,n}$, defined for all invertible bicomplex numbers, is the inverse function of e^Z (at least on a subset of its domain of definition). Then we have:

$$1 = \left(e^{\ln_{m,n}(Z)}\right)' = e^{\ln_{m,n}(Z)} \cdot (\ln_{m,n})'(Z) = Z \cdot (\ln_{m,n})'(Z).$$

Since Z runs over all invertible bicomplex numbers, we obtain

$$(\ln_{m,n})'(Z) = \frac{1}{Z},$$

which concludes our proof. $\qquad\square$

7.3 Partial derivatives of bicomplex derivable functions

Since bicomplex numbers admit many representations, in this section we investigate what the existence of the derivative implies for different types of partial derivatives.

Theorem 7.3.1. *Consider a bicomplex function $F : \Omega \subset \mathbb{BC} \to \mathbb{BC}$ derivable at $Z_0 \in \Omega$. Then the following statements are true:*

1. *The real partial derivatives $\dfrac{\partial F}{\partial x_\ell}(Z_0)$ and $\dfrac{\partial F}{\partial y_\ell}(Z_0)$ exist, for $\ell = 1, 2$.*

2. *The real partial derivatives satisfy the identities:*

$$F'(Z_0) = \frac{\partial F}{\partial x_1}(Z_0) = -\mathbf{i}\frac{\partial F}{\partial y_1}(Z_0) = -\mathbf{j}\frac{\partial F}{\partial x_2}(Z_0) = \mathbf{k}\frac{\partial F}{\partial y_2}(Z_0). \qquad (7.16)$$

Proof. Because F is derivable at Z_0, limit (7.8) exists no matter how H converges to zero , as long as H is invertible. Consider first H to be real, i.e., of the form $H = x_1 = x_1 + \mathbf{i}0 + \mathbf{j}0 + \mathbf{k}0 \to 0$, $x_1 \neq 0$, which is always invertible. Then, since there exists $F'(Z_0)$, the following limit also exists:

$$\lim_{x_1 \to 0} \frac{F(Z_0 + x_1) - F(Z_0)}{x_1},$$

and it coincides with $\dfrac{\partial F}{\partial x_1}(Z_0)$.

The rest of the proof follows by considering the three specific forms of the increment H along the units $\mathbf{i}, \mathbf{j}, \mathbf{k}$, i.e., $H = \mathbf{i}y_1$, with $y_1 \neq 0$; $H = \mathbf{j}x_2$, with $x_2 \neq 0$; $H = \mathbf{k}y_2$, with $y_2 \neq 0$, and noting that all are invertible bicomplex numbers. $\qquad \square$

Let us write the bicomplex function as $F = f_{11} + \mathbf{i}f_{12} + \mathbf{j}f_{21} + \mathbf{k}f_{22}$ in terms of its real components, which are all real functions of a bicomplex variable. An immediate consequence of the theorem above is

Corollary 7.3.2. *If F is derivable at Z_0, then the the real Jacobi matrix of F at Z_0 is of the form*

$$J_{Z_0}[F] := \begin{pmatrix} a & -b & -c & d \\ b & a & -d & -c \\ c & -d & a & -b \\ d & c & b & a \end{pmatrix}, \tag{7.17}$$

where

$$a := \frac{\partial f_{11}}{\partial x_1} = \frac{\partial f_{12}}{\partial y_1} = \frac{\partial f_{21}}{\partial x_2} = \frac{\partial f_{22}}{\partial y_2},$$

$$b := -\frac{\partial f_{11}}{\partial y_1} = \frac{\partial f_{12}}{\partial x_1} = -\frac{\partial f_{21}}{\partial y_2} = \frac{\partial f_{22}}{\partial x_2},$$

$$c := -\frac{\partial f_{11}}{\partial x_2} = -\frac{\partial f_{12}}{\partial y_2} = \frac{\partial f_{21}}{\partial x_1} = \frac{\partial f_{22}}{\partial y_1},$$

$$d := \frac{\partial f_{11}}{\partial y_2} = -\frac{\partial f_{12}}{\partial x_2} = -\frac{\partial f_{21}}{\partial y_1} = \frac{\partial f_{22}}{\partial x_1}, \tag{7.18}$$

and where all the partial derivatives are evaluated at the point Z_0.

Proof. The proof relies on a direct computation using equalities (7.16) written in terms of $f_{k\ell}$. $\qquad \square$

Remark 7.3.3. *The reader should notice that the special form of the real Jacobi matrix above encodes several Cauchy-Riemann type conditions on (certain pairs of) the real functions $f_{k\ell}$, a fact which we will exploit in detail below.*

Remark 7.3.4. *Every 4×4 matrix with real entries determines a linear transfor-mation on $\mathbb{R}^4$. Of course, not all of them remain linear when $\mathbb{R}^4$ is seen as $\mathbb{BC}$, i.e., not all of them are $\mathbb{BC}$-linear. Those matrices which are $\mathbb{BC}$-linear are of the form (7.17). Taking into account that the entries of (7.17) are the values of the partial derivatives at Z_0, one may conclude that (7.17) determines a $\mathbb{BC}$-linear operator acting on the tangential $\mathbb{BC}$-linear module at the point Z_0. Moreover, in the matrix (7.17) there are hidden the two linearities with respect to $\mathbb{C}(\mathbf{i})$ and $\mathbb{C}(\mathbf{j})$, in the following sense: fixing the identifications*

$$(x_1 + \mathbf{i}y_1, x_2 + \mathbf{i}y_2) = (z_1, z_2) \longleftrightarrow (x_1, y_1, x_2, y_2)$$

and

$$(x_1 + \mathbf{j}x_2, y_1 + \mathbf{j}y_2) = (\zeta_1, \zeta_2) \longleftrightarrow (x_1, y_1, x_2, y_2)$$

we have two different identifications $\mathbb{C}^2(\mathbf{i}) \leftrightarrow \mathbb{R}^4$ and $\mathbb{C}^2(\mathbf{j}) \leftrightarrow \mathbb{R}^4$. It follows that a 4×4 matrix with real entries determines not only a real transformation but a $\mathbb{C}(\mathbf{i})$-linear one, if and only if it is of the form

$$\begin{pmatrix} l & -m & u & -v \\ m & l & v & u \\ t & -s & g & -h \\ s & t & h & g \end{pmatrix}.$$

Similarly, a 4×4 matrix with real entries represents a $\mathbb{C}(\mathbf{j})$-linear transformation if and only if it is of the form

$$\begin{pmatrix} A & B & -E & -F \\ C & D & -G & -H \\ E & F & A & B \\ G & H & C & D \end{pmatrix}.$$

Thus the structure of the matrix (7.17) includes both complex structures.

Remark 7.3.5. *An easy computation shows that the real Jacobian, i.e., the deter-minant of the matrix (7.17), is given by*

$$\det(J_{Z_0}) = \left((b+c)^2 + (a-d)^2\right) \cdot \left((b-c)^2 + (a+d)^2\right) \in \mathbb{R}_+ .$$

Moreover, a direct computation yields that the expression above is nothing but

$$\left| |F'(Z_0)|_{\mathbf{i}}^2 \right|^2 ,$$

i.e., the square of the usual modulus of the square of the complex modulus of $F'(Z_0)$.

A simple computation shows that $\det(J_{Z_0}) = 0$ if and only if $F'(Z_0) \in \mathfrak{S}_0$, which is equivalent to

$$b = -c \ \text{ and } \ a = d, \qquad or \qquad b = c \ \text{ and } \ a = -d . \tag{7.19}$$

Note that these relations are between the real partial derivatives of the real components of F at Z_0. Recall that the derivative of F at Z_0 is given by

$$F'(Z_0) = a + b\mathbf{i} + c\mathbf{j} + d\mathbf{k}.$$

Indeed, if $b = -c$ and $a = d$, then

$$F'(Z_0) = a(1 + \mathbf{k}) + b(\mathbf{i} - \mathbf{j}) = a(1 + \mathbf{k}) + \mathbf{i}\,b(1 + \mathbf{k}) = 2(a + \mathbf{i}\,b) \cdot \mathbf{e},$$

so $F'(Z_0) \in \mathbb{BC}_{\mathbf{e}} \subset \mathfrak{S}_0$ is a zero-divisor or zero. Similarly, if $b = c$ and $a = -d$, then

$$F'(Z_0) = a(1 - \mathbf{k}) + b(\mathbf{i} + \mathbf{j}) = a(1 - \mathbf{k}) + \mathbf{i}\,b(1 - \mathbf{k}) = 2(a + \mathbf{i}\,b) \cdot \mathbf{e}^{\dagger},$$

so $F'(Z_0) \in \mathbb{BC}_{\mathbf{e}^{\dagger}} \subset \mathfrak{S}_0$. Conversely, we note that $F'(Z_0) = 0$ if and only if $a = b = c = d = 0$, and $F'(Z_0) \in \mathfrak{S}_0$ if and only if relations (7.19) hold.

For example, if all four of the real partial derivatives of any real function $f_{k\ell}$ are zero at Z_0, it follows that $F'(Z_0) = 0$. In particular, if one of $f_{k\ell}$ is constant around Z_0, then $F'(Z_0) = 0$.

As a next step, we investigate the consequence of the existence of $F'(Z_0)$ in terms of the complex variables $z_1, z_2 \in \mathbb{C}(\mathbf{i})$, where we write the bicomplex variable as $Z = z_1 + \mathbf{j}z_2$. We prove the following

Theorem 7.3.6. *Consider a bicomplex function $F = f_1 + \mathbf{j}f_2$ derivable at Z_0. Then we have:*

1. *The $\mathbb{C}(\mathbf{i})$-complex partial derivatives $F'_{z_\ell}(Z_0)$ exist, for $\ell = 1, 2$.*

2. *The complex partial derivatives above verify the identity:*

$$F'(Z_0) = F'_{z_1}(Z_0) = -\mathbf{j}F'_{z_2}(Z_0), \tag{7.20}$$

which is equivalent to the $\mathbb{C}(\mathbf{i})$-complex Cauchy-Riemann system for F (at Z_0), also called the generalized Cauchy-Riemann system in [68] and [69]:

$$f'_{1,z_1}(Z_0) = f'_{2,z_2}(Z_0), \qquad f'_{1,z_2}(Z_0) = -f'_{2,z_1}(Z_0). \tag{7.21}$$

Proof. As in the proof of Theorem 7.3.1, the limit of the difference quotient must be the same regardless of the path on which H approaches zero, as long as H is invertible. Let us choose $H = h_1 = h_1 + \mathbf{j}0 \to 0$, and note that if $Z_0 = z_{01} + \mathbf{j}z_{02}$, then $Z_0 + H = (z_{01} + h_1) + \mathbf{j}z_{02}$. Then, since $F'(Z_0)$ exists, the following limit also exists:

$$\lim_{h_1 \to 0} \frac{F(Z_0 + h_1) - F(Z_0)}{h_1} = F'_{z_1}(Z_0).$$

Similarly, taking $H = \mathbf{j}h_2 \to 0$ we get:

$$F'(Z_0) = \lim_{h_2 \to 0} \frac{F(Z_0 + \mathbf{j}h_2) - F(Z_0)}{\mathbf{j}h_2} = -\mathbf{j}F'_{z_2}(Z_0).$$

In conclusion, the complex partial derivatives of F with respect to z_1 and z_2 exist (at Z_0), and they verify the equality

$$F'(Z_0) = F'_{z_1}(Z_0) = -\mathbf{j}F'_{z_2}(Z_0)\,. \tag{7.22}$$

If we write $F = f_1 + \mathbf{j}f_2$, the complex partial derivatives of F at Z_0 are therefore given by

$$F'_{z_1}(Z_0) = f'_{1,z_1}(Z_0) + \mathbf{j}f'_{2,z_1}(Z_0), \qquad F'_{z_2}(Z_0) = f'_{1,z_2}(Z_0) + \mathbf{j}f'_{2,z_2}(Z_0)\,.$$

Now it is immediate to see that equality (7.22) is equivalent to the $\mathbb{C}(\mathbf{i})$-*complex* Cauchy-Riemann conditions for F at Z_0. $\qquad\qquad\qquad\qquad\qquad\qquad\qquad\qquad\square$

Notice that the symbol f'_{1,z_1} (and the other similarly defined symbols) denotes in our situation the "authentic" complex partial derivative at a point, not the formal operation

$$\frac{\partial f}{\partial z_1} := \frac{1}{2}\left(\frac{\partial f}{\partial x_1} - \mathbf{i}\frac{\partial f}{\partial y_1}\right)$$

defined on $\mathcal{C}^1$-functions. At the same time, these results indicate that bicomplex functions which are derivable in a domain are related to holomorphic mappings of two complex variables. We will come back in more detail to this issue at the end of this section.

Corollary 7.3.7. *Let $F = f_1 + \mathbf{j}f_2$ be a bicomplex function derivable at Z_0, then the real components of the functions $f_1 = f_{11} + \mathbf{i}f_{12}$ and $f_2 = f_{21} + \mathbf{i}f_{22}$ verify the usual real Cauchy-Riemann system (at Z_0) associated with each complex variable $z_1 = x_1 + \mathbf{i}y_1$ and $z_2 = x_2 + \mathbf{i}y_2$; in complex notation this is equivalent to*

$$\frac{\partial F}{\partial \bar{z}_1}(Z_0) = \frac{\partial F}{\partial \bar{z}_2}(Z_0) = 0\,, \tag{7.23}$$

i.e.,

$$\frac{\partial f_1}{\partial \bar{z}_1}(Z_0) = \frac{\partial f_1}{\partial \bar{z}_2}(Z_0) = \frac{\partial f_2}{\partial \bar{z}_1}(Z_0) = \frac{\partial f_2}{\partial \bar{z}_2}(Z_0) = 0\,, \tag{7.24}$$

where the symbols $\dfrac{\partial}{\partial \bar{z}_1}$ and $\dfrac{\partial}{\partial \bar{z}_2}$ are the commonly used formal operations on $\mathbb{C}(\mathbf{i})$-valued functions of z_1 and z_2, having partial derivatives at Z_0, namely,

$$\frac{\partial f}{\partial \bar{z}_1} := \frac{1}{2}\left(\frac{\partial f}{\partial x_1} + \mathbf{i}\frac{\partial f}{\partial y_1}\right)$$

and

$$\frac{\partial f}{\partial \bar{z}_2} := \frac{1}{2}\left(\frac{\partial f}{\partial x_2} + \mathbf{i}\frac{\partial f}{\partial y_2}\right)\,.$$

Proof. Let us employ equalities (7.16) involving the real partial derivatives of $F = f_1 + \mathbf{j}f_2$ at Z_0. The second equality in (7.16) is

$$\frac{\partial F}{\partial x_1}(Z_0) = -\mathbf{i}\frac{\partial F}{\partial y_1}(Z_0), \tag{7.25}$$

which is equivalent to (for simplicity we eliminate the explicit reference to Z_0)

$$\frac{\partial f_{11}}{\partial x_1} + \mathbf{i}\frac{\partial f_{12}}{\partial x_1} + \mathbf{j}\frac{\partial f_{21}}{\partial x_1} + \mathbf{ij}\frac{\partial f_{22}}{\partial x_1} = -\mathbf{i}\left(\frac{\partial f_{11}}{\partial y_1} + \mathbf{i}\frac{\partial f_{12}}{\partial y_1} + \mathbf{j}\frac{\partial f_{21}}{\partial y_1} + \mathbf{ij}\frac{\partial f_{22}}{\partial y_1}\right).$$

Because the functions $f_{k\ell}$ are real, this is equivalent to the system

$$\frac{\partial f_{11}}{\partial x_1} = \frac{\partial f_{12}}{\partial y_1}, \qquad \frac{\partial f_{11}}{\partial y_1} = -\frac{\partial f_{12}}{\partial x_1},$$

$$\frac{\partial f_{21}}{\partial x_1} = \frac{\partial f_{22}}{\partial y_1}, \qquad \frac{\partial f_{21}}{\partial y_1} = -\frac{\partial f_{22}}{\partial x_1}. \tag{7.26}$$

These are the real Cauchy-Riemann conditions for the complex functions $f_1 = f_{11} + \mathbf{i}f_{12}$ and $f_2 = f_{21} + \mathbf{i}f_{22}$, with respect to the complex variable $z_1 = x_1 + \mathbf{i}y_1$. In complex notation, if we write equality (7.25) in terms of the complex differential operator $\frac{1}{2}\left(\frac{\partial}{\partial \bar{z}_1} = \frac{\partial}{\partial x_1} + \mathbf{i}\frac{\partial}{\partial y_1}\right)$, it becomes equivalent to the first part of (7.23).

We repeat this reasoning for the equality

$$-\mathbf{j}\frac{\partial F}{\partial x_2}(Z_0) = \mathbf{ij}\frac{\partial F}{\partial y_2}(Z_0), \tag{7.27}$$

which, after division by $-\mathbf{j}$, is equivalent to the second equality in (7.23). This leads to the conclusion that f_1 and f_2 verify the real Cauchy-Riemann system with respect to the variable z_2 at Z_0. □

Of course, this corollary shows the relation between bicomplex derivability and classical holomorphy in two complex variables.

Let us now express the bicomplex variable as $Z = \zeta_1 + \mathbf{i}\,\zeta_2$, and the bicomplex function as $F = \rho_1 + \mathbf{i}\rho_2$, where $\rho_1 = f_{11} + \mathbf{j}f_{21}$ and $\rho_2 = f_{12} + \mathbf{j}f_{22}$ are $\mathbb{C}(\mathbf{j})$-valued functions. We prove the following

Theorem 7.3.8. *Consider a bicomplex function F derivable at Z_0. Then*

1. *The $\mathbb{C}(\mathbf{j})$-complex partial derivatives $F'_{\zeta_\ell}(Z_0)$ exist, for $\ell = 1, 2$.*

2. *The complex partial derivatives above verify the equality:*

$$F'(Z_0) = F'_{\zeta_1}(Z_0) = -\mathbf{i}F'_{\zeta_2}(Z_0), \tag{7.28}$$

which is equivalent to the $\mathbb{C}(\mathbf{j})$-complex Cauchy-Riemann system (at Z_0):

$$\rho'_{1,\zeta_1}(Z_0) = \rho'_{2,\zeta_2}(Z_0), \qquad \rho'_{1,\zeta_2}(Z_0) = -\rho'_{2,\zeta_1}(Z_0). \tag{7.29}$$

Proof. In the definition of $F'(Z_0)$ we consider the limit of the difference quotient as $H \to 0$ first through invertible values of the form $H = \kappa_1 + i0$, and then through values of the form $H = 0 + i\kappa_2$, where $\kappa_1, \kappa_2 \in \mathbb{C}(\mathbf{j})$. A computation as in Theorem 7.3.6 leads to the existence of the complex partial derivatives of F at Z_0 with respect to the variables $\zeta_1, \zeta_2 \in \mathbb{C}(\mathbf{j})$, and one can show that they verify the equation:

$$F'(Z_0) = F'_{\zeta_1}(Z_0) = -iF'_{\zeta_2}(Z_0),\qquad(7.30)$$

where

$$F'_{\zeta_1}(Z_0) = \rho'_{1,\zeta_1}(Z_0) + i\rho'_{2,\zeta_1}(Z_0),\qquad F'_{\zeta_2}(Z_0) = \rho'_{1,\zeta_2}(Z_0) + i\rho'_{2,\zeta_2}(Z_0).$$

Now it is immediate to see that the equality (7.30) is equivalent to the system (7.29). $\qquad\square$

Corollary 7.3.9. *If $F = \rho_1 + i\rho_2$ is bicomplex derivable at Z_0, then the real components of the functions $\rho_1 = f_{11} + \mathbf{j}f_{21}$ and $\rho_2 = f_{12} + \mathbf{j}f_{22}$ verify the usual real Cauchy-Riemann system (at Z_0) associated to each complex variable $\zeta_1 = x_1 + \mathbf{j}x_2$ and $\zeta_2 = y_1 + \mathbf{j}y_2$; in complex notation this is equivalent to*

$$\frac{\partial F}{\partial \zeta_1^*}(Z_0) = \frac{\partial F}{\partial \zeta_2^*}(Z_0) = 0,\qquad(7.31)$$

i.e.,

$$\frac{\partial \rho_1}{\partial \zeta_1^*}(Z_0) = \frac{\partial \rho_1}{\partial \zeta_2^*}(Z_0) = \frac{\partial \rho_2}{\partial \zeta_1^*}(Z_0) = \frac{\partial \rho_2}{\partial \zeta_2^*}(Z_0) = 0.\qquad(7.32)$$

Here

$$\frac{\partial}{\partial \zeta_1^*} := \frac{1}{2}\left(\frac{\partial}{\partial x_1} + \mathbf{j}\frac{\partial}{\partial x_2}\right),\qquad \frac{\partial}{\partial \zeta_2^*} := \frac{1}{2}\left(\frac{\partial}{\partial y_1} + \mathbf{j}\frac{\partial}{\partial y_2}\right).$$

Proof. The following equalities from (7.16) among the real partial derivatives of F

$$\frac{\partial F}{\partial x_1}(Z_0) = -\mathbf{j}\frac{\partial F}{\partial x_2}(Z_0),\qquad -i\frac{\partial F}{\partial y_1}(Z_0) = i\mathbf{j}\frac{\partial F}{\partial y_2}(Z_0)\qquad(7.33)$$

are equivalent to (7.31). $\qquad\square$

Remark 7.3.10. *We showed in Example 7.1.2 that it is possible for a bicomplex function to have partial derivatives with respect to z_1, z_2, and not to have partial derivatives with respect to ζ_1, ζ_2. Now we see that if the function is bicomplex derivable, then such a situation is not possible: a bicomplex derivable function has complex partial derivatives with respect to the $\mathbb{C}(\mathbf{i})$-complex variables z_1 and z_2 as well as with respect to the $\mathbb{C}(\mathbf{j})$-variables ζ_1 and ζ_2. Moreover, formulas (7.20) and (7.28) show that such derivatives are related by*

$$F'_{z_1}(Z_0) = F'_{\zeta_1}(Z_0) = -\mathbf{j}F'_{z_2}(Z_0) = -iF'_{\zeta_2}(Z_0).$$

We express now the bicomplex variable as $Z = \mathfrak{z}_1 + \mathbf{i}\mathfrak{z}_2$, where $\mathfrak{z}_1 = x_1 + \mathbf{k}y_2$ and $\mathfrak{z}_2 = y_1 + \mathbf{k}(-x_2)$ are hyperbolic numbers, and the function as $F = \mathfrak{f}_1 + \mathbf{i}\mathfrak{f}_2$, where $\mathfrak{f}_1 = f_{11} + \mathbf{k}f_{22}$ and $\mathfrak{f}_2 = f_{12} + \mathbf{k}(-f_{21})$. We prove the following

Theorem 7.3.11. *Consider a bicomplex function $F = \mathfrak{f}_1 + \mathbf{i}\mathfrak{f}_2$ derivable at a point Z_0. Then*

1. *The hyperbolic partial derivatives $F'_{\mathfrak{z}_\ell}$ exist for $\ell = 1, 2$.*

2. *The partial derivatives above verify the equality*

$$F'(Z_0) = F'_{\mathfrak{z}_1}(Z_0) = -\mathbf{i}F'_{\mathfrak{z}_2}(Z_0),$$

which is equivalent to the following Cauchy-Riemann type system for the hyperbolic components of a bicomplex derivable function:

$$\mathfrak{f}'_{1,\mathfrak{z}_1}(Z_0) = \mathfrak{f}'_{2,\mathfrak{z}_2}(Z_0), \qquad \mathfrak{f}'_{1,\mathfrak{z}_2}(Z_0) = -\mathfrak{f}'_{2,\mathfrak{z}_1}(Z_0). \tag{7.34}$$

Proof. The existence of the hyperbolic partial derivatives of $F = F(\mathfrak{z}_1, \mathfrak{z}_2)$ is obtained in a similar fashion as above, letting $H \to 0$ through invertible hyperbolic values $H = \mathfrak{h}_1 + \mathbf{i}0$ and then on paths of the form $H = \mathbf{i}\mathfrak{h}_2$, where $\mathfrak{h}_1$ and $\mathfrak{h}_2$ are invertible hyperbolic numbers. We obtain also:

$$F'(Z_0) = F'_{\mathfrak{z}_1}(Z_0) = -\mathbf{i}F'_{\mathfrak{z}_2}(Z_0),$$

which is equivalent to system (7.34). $\qquad\square$

Corollary 7.3.12. *If $F = \mathfrak{f}_1 + \mathbf{i}\mathfrak{f}_2$ is derivable at Z_0, then the real components of the hyperbolic functions $\mathfrak{f}_1 = f_{11} + \mathbf{k}f_{22}$ and $\mathfrak{f}_2 = f_{12} + \mathbf{k}(-f_{21})$ verify the Cauchy-Riemann type systems with respect to both variables $\mathfrak{z}_1, \mathfrak{z}_1 \in \mathbb{D}$; in hyperbolic terms this is equivalent to*

$$\frac{\partial F}{\partial \mathfrak{z}_1^\diamond}(Z_0) = \frac{\partial F}{\partial \mathfrak{z}_2^\diamond}(Z_0) = 0, \tag{7.35}$$

i.e.,

$$\frac{\partial \mathfrak{f}_1}{\partial \mathfrak{z}_1^\diamond}(Z_0) = \frac{\partial \mathfrak{f}_1}{\partial \mathfrak{z}_2^\diamond}(Z_0) = \frac{\partial \mathfrak{f}_2}{\partial \mathfrak{z}_1^\diamond}(Z_0) = \frac{\partial \mathfrak{f}_2}{\partial \mathfrak{z}_2^\diamond}(Z_0) = 0, \tag{7.36}$$

where

$$\frac{\partial}{\partial \mathfrak{z}_1^\diamond} = \frac{1}{2}\left(\frac{\partial}{\partial x_1} - \mathbf{k}\frac{\partial}{\partial y_2}\right), \qquad \frac{\partial}{\partial \mathfrak{z}_2^\diamond} = \frac{1}{2}\left(\frac{\partial}{\partial y_1} + \mathbf{k}\frac{\partial}{\partial x_2}\right). \tag{7.37}$$

Proof. We use once again the following equalities from (7.16):

$$\frac{\partial F}{\partial x_1}(Z_0) = \mathbf{k}\frac{\partial F}{\partial y_2}(Z_0), \qquad \frac{\partial F}{\partial y_1}(Z_0) = -\mathbf{k}\frac{\partial F}{\partial x_2}(Z_0), \tag{7.38}$$

where $\mathbf{k} = \mathbf{ij}$. Recalling formulas (7.37) we obtain equality (7.35). $\qquad\square$

Remark 7.3.13. *Appealing again to formulas (1.6)–(1.8) which deal with the complex variables $w_1, w_2, \omega_1, \omega_2$ and the hyperbolic variables $\mathfrak{w}_1, \mathfrak{w}_2$, one may wonder: what about the Cauchy-Riemann conditions with respect to the corresponding partial derivatives? Remark 7.1.3 explains how they can be obtained directly from the previous statements. We omit the details.*

Definition 7.3.14. *Let F be a bicomplex function defined on a non-empty open set $\Omega \subset \mathbb{BC}$. If F has bicomplex derivative at each point of Ω, we will say that F is a bicomplex holomorphic, or $\mathbb{BC}$-holomorphic, function.*

Thus for a $\mathbb{BC}$-holomorphic function F all the conclusions made in this section hold in the whole domain. Theorem 7.3.6 says that F is holomorphic with respect to z_1 for any z_2 fixed and F is holomorphic with respect to z_2 for any z_1 fixed. Thus, see for instance [39, pages 4-5], F is holomorphic in the classical sense of two complex variables. This implies immediately many quite useful properties of F, in particular, it is of class $\mathcal{C}^\infty(\Omega)$, and the reader may compare this with Remark 7.2.4 where we were able, working with just one point, not with a domain, to state a weakened continuity at the point.

7.4 Interplay between real differentiability and derivability of bicomplex functions

7.4.1 Real differentiability in complex and hyperbolic terms.

We begin now assuming that Ω is an open set in $\mathbb{BC}$ and $F : \Omega \subset \mathbb{BC} \to \mathbb{BC}$ is a bicomplex function of class $\mathcal{C}^1(\Omega)$ with respect to the canonical coordinates x_1, y_1, x_2, y_2. We are going to work with $\mathbb{BC}$-holomorphic functions and we want to determine the place that $\mathbb{BC}$-holomorphic functions occupy among the bicomplex $\mathcal{C}^1$-functions. The condition $F \in \mathcal{C}^1(\Omega, \mathbb{BC})$ ensures that F is real differentiable for any $Z \in \Omega$, i.e., that

$$F(Z + H) - F(Z) = \frac{\partial F}{\partial x_1}(Z)h_{11} + \frac{\partial F}{\partial y_1}(Z)h_{12}$$
$$+ \frac{\partial F}{\partial x_2}(Z)h_{21} + \frac{\partial F}{\partial y_2}(Z)h_{22} + \mathfrak{o}(H), \qquad (7.39)$$

where $Z = x_1 + \mathbf{i}y_1 + \mathbf{j}x_2 + \mathbf{k}y_2$, $H = h_{11} + \mathbf{i}h_{12} + \mathbf{j}h_{21} + \mathbf{k}h_{22}$. As a matter of fact, this formula does not depend on how the function F and the variable Z are written, and therefore it will be quite helpful to analyze the structure of such functions.

First of all, let us write (7.39) in terms of the $\mathbb{C}(\mathbf{i})$-complex variables $h_1 := h_{11} + \mathbf{i}h_{12}$ and $h_2 := h_{21} + \mathbf{i}h_{22}$, so that

$$h_{11} = \frac{h_1 + \overline{h}_1}{2}, \qquad h_{12} = \frac{h_1 - \overline{h}_1}{2\mathbf{i}},$$

$$h_{21} = \frac{h_2 + \overline{h}_2}{2}, \qquad h_{22} = \frac{h_2 - \overline{h}_2}{2\mathbf{i}}.$$

Using this in (7.39) and grouping adequately, we obtain:

$$F(Z+H) - F(Z) = \frac{1}{2}\left(\frac{\partial F}{\partial x_1}(Z) - \mathbf{i}\frac{\partial F}{\partial y_1}(Z)\right) \cdot h_1$$

$$+ \frac{1}{2}\left(\frac{\partial F}{\partial x_1}(Z) + \mathbf{i}\frac{\partial F}{\partial y_1}(Z)\right) \cdot \overline{h}_1 + \frac{1}{2}\left(\frac{\partial F}{\partial x_2}(Z) - \mathbf{i}\frac{\partial F}{\partial y_2}(Z)\right) \cdot h_2$$

$$+ \frac{1}{2}\left(\frac{\partial F}{\partial x_2}(Z) + \mathbf{i}\frac{\partial F}{\partial y_2}(Z)\right) \cdot \overline{h}_2 + \mathfrak{o}(H).$$

If we employ the $\mathbb{C}(\mathbf{i})$-complex variables $z_1 = x_1 + \mathbf{i}y_1$ and $z_2 = x_2 + \mathbf{i}y_2$, and the usual complex differential operators

$$\frac{\partial}{\partial z_1} := \frac{1}{2}\left(\frac{\partial}{\partial x_1} - \mathbf{i}\frac{\partial}{\partial y_1}\right), \qquad \frac{\partial}{\partial \overline{z}_1} := \frac{1}{2}\left(\frac{\partial}{\partial x_1} + \mathbf{i}\frac{\partial}{\partial y_1}\right),$$

$$\frac{\partial}{\partial z_2} := \frac{1}{2}\left(\frac{\partial}{\partial x_2} - \mathbf{i}\frac{\partial}{\partial y_2}\right), \qquad \frac{\partial}{\partial \overline{z}_2} := \frac{1}{2}\left(\frac{\partial}{\partial x_2} + \mathbf{i}\frac{\partial}{\partial y_2}\right), \qquad (7.40)$$

we obtain the formula

$$F(Z+H) - F(Z) = \frac{\partial F}{\partial z_1}(Z) \cdot h_1 + \frac{\partial F}{\partial \overline{z}_1}(Z) \cdot \overline{h}_1$$

$$+ \frac{\partial F}{\partial z_2}(Z) \cdot h_2 + \frac{\partial F}{\partial \overline{z}_2}(Z) \cdot \overline{h}_2 + \mathfrak{o}(H). \qquad (7.41)$$

We emphasize that (7.41) does not express a new notion; indeed, this is simply the condition of real differentiability for a $\mathcal{C}^1$-bicomplex function, although it is now expressed in $\mathbb{C}(\mathbf{i})$-complex terms.

Note that in Section 7 we introduced the symbols $F'_{z_1}(Z)$ and $F'_{z_2}(Z)$ instead of the symbols $\frac{\partial F}{\partial z_1}(Z)$ and $\frac{\partial F}{\partial z_2}(Z)$ because the former are complex partial derivatives, defined, as usual, as limits of suitable difference quotients, meanwhile the latter indicates well known operators acting on $\mathcal{C}^1$-functions. The relationship between these two notions is clarified by the following definition and theorem.

Definition 7.4.1. *A bicomplex $\mathcal{C}^1$-function F is called $\mathbb{C}(\mathbf{i})$-complex differentiable if*

$$F(Z+H) - F(Z) = F'_{z_1}(Z) \cdot h_1 + F'_{z_2}(Z) \cdot h_2 + \mathfrak{o}(H).$$

Theorem 7.4.2. *A $\mathcal{C}^1$-bicomplex function F is $\mathbb{C}(\mathbf{i})$-complex differentiable if and only if both its components f_1, f_2 are holomorphic functions in the sense of two complex variables.*

Proof. The partial derivative $F'_{z_1}(Z)$ exists in Ω if and only if the operator $\dfrac{\partial}{\partial \bar{z}_1}$ (which one can think of as a dual to the operator $\dfrac{\partial}{\partial z_1}$) annihilates the function F; that is, if and only if F is holomorphic as a function of z_1; this can be proved by taking $h_2 = 0$ and $h_1 \neq 0$ in (7.41). What remains is

$$F(Z + H) - F(Z) = F(Z + h_1) - F(Z) = \frac{\partial F}{\partial z_1}(Z) \cdot h_1 + \frac{\partial F}{\partial \bar{z}_1}(Z) \cdot \bar{h}_1 + \mathfrak{o}(H),$$

and since $\dfrac{\bar{h}_1}{h_1}$ has no limit when $h_1 \to 0$, then we conclude that $F'_{z_1}(Z)$ exists if and only if $\dfrac{\partial F}{\partial \bar{z}_1}(Z) = 0$. In this case, $F'_{z_1}(Z) = \dfrac{\partial F}{\partial z_1}(Z)$.

Similarly, the partial derivative $F'_{z_2}(Z)$ exists in Ω if and only if the operator $\dfrac{\partial}{\partial \bar{z}_2}$ annihilates the function F; this is because we can take now $h_1 = 0, h_2 \neq 0$ in (7.41). Therefore $F'_{z_2}(Z) = \dfrac{\partial F}{\partial z_2}(Z)$.

Finally, we can assume both conditions

$$\frac{\partial F}{\partial \bar{z}_1}(Z) = \frac{\partial F}{\partial \bar{z}_2}(Z) = 0$$

to be fulfilled in Ω, with (7.41) becoming

$$F(Z + H) - F(Z) = F'_{z_1}(Z) \cdot h_1 + F'_{z_2}(Z) \cdot h_2 + \mathfrak{o}(H). \tag{7.42}$$

This concludes the proof.   □

Note that for an arbitrary $\mathbb{C}(\mathbf{i})$-complex differentiable bicomplex function, in general, there is no relation between its complex partial derivatives.

Very similar calculations can be made if we write the bicomplex number as $Z = \zeta_1 + \mathbf{i}\zeta_2$ and the $\mathbb{C}(\mathbf{j})$-complex increments as $\kappa_1 := h_{11} + \mathbf{j}h_{21}$ and $\kappa_2 := h_{12} + \mathbf{j}h_{22}$. If one follows the steps indicated above, one eventually obtains the analogue of (7.41):

$$F(Z + H) - F(Z) = \frac{\partial F}{\partial \zeta_1}(Z)\kappa_1 + \frac{\partial F}{\partial \bar{\zeta}_1}(Z)\bar{\kappa}_1$$

$$+ \frac{\partial F}{\partial \zeta_2}(Z)\kappa_2 + \frac{\partial F}{\partial \bar{\zeta}_2}(Z)\bar{\kappa}_2 + \mathfrak{o}(H), \tag{7.43}$$

where the expressions $\dfrac{\partial}{\partial \zeta_1}$ (and similar) are the usual complex differential operators in $\mathbb{C}(\mathbf{j})$. Again, this simply represents the real differentiability of a C^1-bicomplex function in $\mathbb{C}(\mathbf{j})$-complex terms.

The same analysis as above applies to equation (7.43). In particular, the definition of the $\mathbb{C}(\mathbf{j})$ complex differentiability of a bicomplex function is

$$F(Z+H) - F(Z) = F'_{\zeta_1}(Z) \cdot \kappa_1 + F'_{\zeta_2}(Z) \cdot \kappa_2 + \mathrm{o}(H).$$

Notice that Remark 7.1.3 explains why the other complex variables $w_1, w_2, \omega_1, \omega_2$ do not present any interest in the analysis of the increment of the function.

Our last step consists in writing $Z = \mathfrak{z}_1 + \mathbf{i}\mathfrak{z}_2$, where $\mathfrak{z}_1 := x_1 + \mathbf{k}y_2$ and $\mathfrak{z}_2 := y_1 + \mathbf{k}(-x_2)$ are hyperbolic variables. The hyperbolic increments are $\mathfrak{h}_1 := h_{11} + \mathbf{k}h_{22}$ and $\mathfrak{h}_2 := h_{12} + \mathbf{k}(-h_{21})$. Then, using the formulas

$$h_{11} = \frac{\mathfrak{h}_1 + \mathfrak{h}_1^\diamond}{2}, \qquad h_{22} = \frac{\mathfrak{h}_1 - \mathfrak{h}_1^\diamond}{2\mathbf{k}},$$

$$h_{12} = \frac{\mathfrak{h}_2 + \mathfrak{h}_2^\diamond}{2}, \qquad h_{21} = -\frac{\mathfrak{h}_2 - \mathfrak{h}_2^\diamond}{2\mathbf{k}},$$

we regroup the right-hand side of (7.39) as follows:

$$
\begin{aligned}
F&(Z+H) - F(Z)\\
&= \frac{\partial F}{\partial x_1}(Z) \cdot \frac{\mathfrak{h}_1 + \mathfrak{h}_1^\diamond}{2} + \frac{\partial F}{\partial y_1}(Z) \cdot \frac{\mathfrak{h}_2 + \mathfrak{h}_2^\diamond}{2}\\
&\quad - \frac{\partial F}{\partial x_2}(Z) \cdot \frac{\mathfrak{h}_2 - \mathfrak{h}_2^\diamond}{2\mathbf{k}} + \frac{\partial F}{\partial y_2}(Z) \cdot \frac{\mathfrak{h}_1 - \mathfrak{h}_1^\diamond}{2\mathbf{k}} + \mathrm{o}(H)\\
&= \frac{1}{2}\left(\frac{\partial F}{\partial x_1}(Z) + \mathbf{k}\frac{\partial F}{\partial y_2}(Z) \right)\mathfrak{h}_1 + \frac{1}{2}\left(\frac{\partial F}{\partial x_1}(Z) - \mathbf{k}\frac{\partial F}{\partial y_2}(Z) \right)\mathfrak{h}_1^\diamond\\
&\quad + \frac{1}{2}\left(\frac{\partial F}{\partial y_1}(Z) - \mathbf{k}\frac{\partial F}{\partial x_2}(Z) \right)\mathfrak{h}_2 + \frac{1}{2}\left(\frac{\partial F}{\partial y_1}(Z) + \mathbf{k}\frac{\partial F}{\partial x_2}(Z) \right)\mathfrak{h}_2^\diamond\\
&\quad + \mathrm{o}(H).
\end{aligned}
$$

The "hyperbolic" differential operators appearing above are consistent with the ones in the context of hyperbolic analysis. For the hyperbolic variable $\mathfrak{z} = x + \mathbf{k}y$, the formal hyperbolic partial derivatives are given by the formulas:

$$\frac{\partial}{\partial \mathfrak{z}} = \frac{1}{2}\left(\frac{\partial}{\partial x} + \mathbf{k}\frac{\partial}{\partial y} \right), \qquad \frac{\partial}{\partial \mathfrak{z}^\diamond} = \frac{1}{2}\left(\frac{\partial}{\partial x} - \mathbf{k}\frac{\partial}{\partial y} \right),$$

where $\mathfrak{z}^\diamond = x - \mathbf{k}y$ is the hyperbolic conjugate of $\mathfrak{z}$. Therefore, in terms of the hyperbolic variables $\mathfrak{z}_1 = x_1 + \mathbf{k}y_2$ and $\mathfrak{z}_2 = y_1 + \mathbf{k}(-x_2)$ and the corresponding hyperbolic differential operators, we obtain:

$$
\begin{aligned}
F&(Z+H) - F(Z)\\
&= \frac{\partial F}{\partial \mathfrak{z}_1}(Z)\mathfrak{h}_1 + \frac{\partial F}{\partial \mathfrak{z}_1^\diamond}(Z)\mathfrak{h}_1^\diamond + \frac{\partial F}{\partial \mathfrak{z}_2}(Z)\mathfrak{h}_2 + \frac{\partial F}{\partial \mathfrak{z}_2^\diamond}(Z)\mathfrak{h}_2^\diamond + \mathrm{o}(H),
\end{aligned}
\qquad (7.44)
$$

which is a hyperbolic reformulation of the real differentiability of a $\mathcal{C}^1$-bicomplex function.

Again, we can apply to the equation (7.44) a similar reasoning as made above. In particular, the definition of the hyperbolic differentiability of a bicomplex function is

$$F(Z+H) - F(Z) = \frac{\partial F}{\partial \mathfrak{z}_1}(Z)\mathfrak{h}_1 + \frac{\partial F}{\partial \mathfrak{z}_2}(Z)\mathfrak{h}_2 + \mathfrak{o}(H).$$

We are not aware of any other work that studies what could be called hyperbolic holomorphic mappings from $\mathbb{D}^2$ to $\mathbb{D}^2$, that is, pairs of holomorphic functions of two hyperbolic variables.

7.4.2 Real differentiability in bicomplex terms

Formula (7.39) as well as any of the formulas (7.41), (7.43) and (7.44) expresses the real differentiability of a bicomplex function, although written in different languages: the first of them is in real language, the next two in complex ($\mathbb{C}(\mathbf{i})$- and $\mathbb{C}(\mathbf{j})$-) language and the last is given in the hyperbolic one. Now we are going to see what the bicomplex language will give.

Let us first consider the bicomplex increment $H = h_1 + \mathbf{j}h_2$, for which we have:

$$h_1 = \frac{H + H^\dagger}{2}, \qquad h_2 = \frac{H - H^\dagger}{2\mathbf{j}},$$

$$\overline{h}_1 = \frac{\overline{H} + H^*}{2}, \qquad \overline{h}_2 = \frac{\overline{H} - H^*}{2\mathbf{j}}. \tag{7.45}$$

In this setup, formula (7.41) becomes:

$$F(Z+H) - F(Z) = \frac{1}{2}\left(\frac{\partial F}{\partial z_1} - \mathbf{j}\frac{\partial F}{\partial z_2}\right)(Z) \cdot H + \frac{1}{2}\left(\frac{\partial F}{\partial z_1} + \mathbf{j}\frac{\partial F}{\partial z_2}\right)(Z) \cdot H^\dagger$$

$$+ \frac{1}{2}\left(\frac{\partial F}{\partial \overline{z}_1} - \mathbf{j}\frac{\partial F}{\partial \overline{z}_2}\right)(Z) \cdot \overline{H} + \frac{1}{2}\left(\frac{\partial F}{\partial \overline{z}_1} - \mathbf{j}\frac{\partial F}{\partial \overline{z}_2}\right)(Z) \cdot H^*$$

$$+ \mathfrak{o}(H). \tag{7.46}$$

We introduce the following bicomplex differential operators:

$$\frac{\partial}{\partial Z} := \frac{1}{2}\left(\frac{\partial}{\partial z_1} - \mathbf{j}\frac{\partial}{\partial z_2}\right), \qquad \frac{\partial}{\partial Z^\dagger} := \frac{1}{2}\left(\frac{\partial}{\partial z_1} + \mathbf{j}\frac{\partial}{\partial z_2}\right),$$

$$\frac{\partial}{\partial \overline{Z}} := \frac{1}{2}\left(\frac{\partial}{\partial \overline{z}_1} - \mathbf{j}\frac{\partial}{\partial \overline{z}_2}\right), \qquad \frac{\partial}{\partial Z^*} := \frac{1}{2}\left(\frac{\partial}{\partial \overline{z}_1} + \mathbf{j}\frac{\partial}{\partial \overline{z}_2}\right). \tag{7.47}$$

We obtain the following intrinsic expression of the real differentiability of the bicomplex function F in terms of bicomplex differential operators and variables:

$$F(Z+H) - F(Z)$$

$$= \frac{\partial F}{\partial Z}(Z)H + \frac{\partial F}{\partial Z^\dagger}(Z)H^\dagger + \frac{\partial F}{\partial \overline{Z}}(Z)\overline{H} + \frac{\partial F}{\partial Z^*}(Z)H^* + \mathfrak{o}(H). \tag{7.48}$$

While real differentiability uniquely defines the coefficients in (7.48), one may think that if we had begun with another writing of Z and H, then formula (7.48) would be different, that is, other operators would have appeared in it. Direct computations however confirm that no matter in which form we write the functions and the variables (recall that the operators in (7.47) are given in terms of $\mathbb{C}(\mathbf{i})$-complex differential operators) the operators $\dfrac{\partial}{\partial Z}, \dfrac{\partial}{\partial Z^\dagger}, \dfrac{\partial}{\partial \overline{Z}}$, and $\dfrac{\partial}{\partial Z^*}$ in the right-hand side of (7.48) are uniquely defined. But this requires us to clarify what is meant by this "uniqueness". These are the same operators but only when acting on bicomplex functions, without taking into account any concrete intrinsic substructure in $\mathbb{BC}$. For instance, if the functions are considered $\mathbb{C}^2(\mathbf{i})$-valued or $\mathbb{C}^2(\mathbf{j})$-valued, then the operators are of course different; this is just because they act on objects of different nature. For this reason, one should be careful when working with bicomplex functions and operators, having in mind all the time what is exactly the structure of $\mathbb{BC}$ which is of interest for a specific goal.

We write here each operator in the various possible ways:

$$
\begin{aligned}
\frac{\partial}{\partial Z} &= \frac{1}{2}\left(\frac{\partial}{\partial z_1} - \mathbf{j}\frac{\partial}{\partial z_2}\right) = \frac{1}{2}\left(\frac{\partial}{\partial \zeta_1} - \mathbf{i}\frac{\partial}{\partial \zeta_2}\right) \\
&= \frac{1}{2}\left(\frac{\partial}{\partial \mathfrak{z}_1} - \mathbf{i}\frac{\partial}{\partial \mathfrak{z}_2}\right) = \frac{1}{2}\left(\frac{\partial}{\partial w_1} + \mathbf{k}\frac{\partial}{\partial w_2}\right) \\
&= \frac{1}{2}\left(\frac{\partial}{\partial \omega_1} + \mathbf{k}\frac{\partial}{\partial \omega_2}\right) = \frac{1}{2}\left(\frac{\partial}{\partial \mathfrak{w}_1} + \mathbf{j}\frac{\partial}{\partial \mathfrak{w}_2}\right) \\
&= \frac{1}{4}\left(\frac{\partial}{\partial x_1} - \mathbf{i}\frac{\partial}{\partial y_1} - \mathbf{j}\frac{\partial}{\partial x_2} + \mathbf{k}\frac{\partial}{\partial y_2}\right),
\end{aligned} \tag{7.49}
$$

$$
\begin{aligned}
\frac{\partial}{\partial Z^\dagger} &= \frac{1}{2}\left(\frac{\partial}{\partial z_1} + \mathbf{j}\frac{\partial}{\partial z_2}\right) = \frac{1}{2}\left(\frac{\partial}{\partial \zeta_1^*} - \mathbf{i}\frac{\partial}{\partial \zeta_2^*}\right) \\
&= \frac{1}{2}\left(\frac{\partial}{\partial \mathfrak{z}_1^\diamond} - \mathbf{i}\frac{\partial}{\partial \mathfrak{z}_2^\diamond}\right) = \frac{1}{2}\left(\frac{\partial}{\partial w_1} - \mathbf{k}\frac{\partial}{\partial w_2}\right) \\
&= \frac{1}{2}\left(\frac{\partial}{\partial \omega_1^*} - \mathbf{k}\frac{\partial}{\partial \omega_2^*}\right) = \frac{1}{2}\left(\frac{\partial}{\partial \mathfrak{w}_1^\diamond} + \mathbf{j}\frac{\partial}{\partial \mathfrak{w}_2^\diamond}\right) \\
&= \frac{1}{4}\left(\frac{\partial}{\partial x_1} - \mathbf{i}\frac{\partial}{\partial y_1} + \mathbf{j}\frac{\partial}{\partial x_2} - \mathbf{k}\frac{\partial}{\partial y_2}\right),
\end{aligned} \tag{7.50}
$$

$$
\begin{aligned}
\frac{\partial}{\partial \overline{Z}} &= \frac{1}{2}\left(\frac{\partial}{\partial \overline{z}_1} - \mathbf{j}\frac{\partial}{\partial \overline{z}_2}\right) = \frac{1}{2}\left(\frac{\partial}{\partial \overline{\zeta}_1} + \mathbf{i}\frac{\partial}{\partial \overline{\zeta}_2}\right) \\
&= \frac{1}{2}\left(\frac{\partial}{\partial \mathfrak{z}_1^\diamond} + \mathbf{i}\frac{\partial}{\partial \mathfrak{z}_2^\diamond}\right) = \frac{1}{2}\left(\frac{\partial}{\partial \overline{w}_1} - \mathbf{k}\frac{\partial}{\partial \overline{w}_2}\right) \\
&= \frac{1}{2}\left(\frac{\partial}{\partial \omega_1} - \mathbf{k}\frac{\partial}{\partial \omega_2}\right) = \frac{1}{2}\left(\frac{\partial}{\partial \mathfrak{w}_1^\diamond} - \mathbf{j}\frac{\partial}{\partial \mathfrak{w}_2^\diamond}\right)
\end{aligned}
$$

$$= \frac{1}{4}\left(\frac{\partial}{\partial x_1} + \mathbf{i}\frac{\partial}{\partial y_1} - \mathbf{j}\frac{\partial}{\partial x_2} - \mathbf{k}\frac{\partial}{\partial y_2}\right),\tag{7.51}$$

and

$$\begin{aligned}
\frac{\partial}{\partial Z^*} &= \frac{1}{2}\left(\frac{\partial}{\partial \bar{z}_1} + \mathbf{j}\frac{\partial}{\partial \bar{z}_2}\right) = \frac{1}{2}\left(\frac{\partial}{\partial \zeta_1^*} + \mathbf{i}\frac{\partial}{\partial \zeta_2^*}\right) \\
&= \frac{1}{2}\left(\frac{\partial}{\partial \mathfrak{z}_1} + \mathbf{i}\frac{\partial}{\partial \mathfrak{z}_2}\right) = \frac{1}{2}\left(\frac{\partial}{\partial \overline{w}_1} + \mathbf{k}\frac{\partial}{\partial \overline{w}_2}\right) \\
&= \frac{1}{2}\left(\frac{\partial}{\partial w_1^*} + \mathbf{k}\frac{\partial}{\partial w_2^*}\right) = \frac{1}{2}\left(\frac{\partial}{\partial \mathfrak{w}_1} - \mathbf{j}\frac{\partial}{\partial \mathfrak{w}_2}\right) \\
&= \frac{1}{4}\left(\frac{\partial}{\partial x_1} + \mathbf{i}\frac{\partial}{\partial y_1} + \mathbf{j}\frac{\partial}{\partial x_2} + \mathbf{k}\frac{\partial}{\partial y_2}\right).
\end{aligned}\tag{7.52}$$

Again, take a bicomplex function F and apply to it, say, the operator $\dfrac{\partial}{\partial Z^*}$; the resulting function $\dfrac{\partial F}{\partial Z^*}$ is a bicomplex function. But if we omit the bicomplex structure and consider F to be a mapping from $\mathbb{C}^2(\mathbf{i})$ to $\mathbb{C}^2(\mathbf{i})$, then such an F does understand already what the action of the operator

$$\frac{1}{2}\left(\frac{\partial}{\partial \bar{z}_1} + \mathbf{j}\frac{\partial}{\partial \bar{z}_2}\right) = \frac{1}{2}\begin{pmatrix} \dfrac{\partial}{\partial \bar{z}_1} & -\dfrac{\partial}{\partial \bar{z}_2} \\[2mm] \dfrac{\partial}{\partial \bar{z}_2} & \dfrac{\partial}{\partial \bar{z}_1} \end{pmatrix}$$

means, but it does not understand the action of the operator $\dfrac{1}{2}\left(\dfrac{\partial}{\partial \zeta_1^*} + \mathbf{i}\dfrac{\partial}{\partial \zeta_2^*}\right)$.

As a consequence of the previous discussion, we obtain the following result:

Theorem 7.4.3. *Let* $F \in \mathcal{C}^1(\Omega, \mathbb{BC})$: *if* F *is* $\mathbb{BC}$*-holomorphic, then*

$$\frac{\partial F}{\partial Z^\dagger}(Z) = \frac{\partial F}{\partial \overline{Z}}(Z) = \frac{\partial F}{\partial Z^*}(Z) = 0.\tag{7.53}$$

holds on Ω.

Proof. Since F is $\mathbb{BC}$-holomorphic, formula (7.9) holds for all $H \notin \mathfrak{S}_0$. But F is a $\mathcal{C}^1$-function, hence (7.48) holds as well for any $H \neq 0$, thus both formulas hold for non-zero-divisors. Then (7.53) follows directly by recalling that both (7.9) and (7.48) are unique representations for a given function F, and by comparing them.   □

Remark 7.4.4. *As we will show later the converse of this result is true as well, but we need some additional steps before we can prove it.*

In order to have more consistency with the previous reasonings of this section and in analogy with the cases of functions of real or complex variables, we introduce the following definition.

Definition 7.4.5. *A bicomplex function $F \in \mathcal{C}^1(\Omega, \mathbb{BC})$ is called* bicomplex- $(\mathbb{BC}\text{-})$ differentiable *on* Ω *if*

$$F(Z + H) - F(Z) = A_Z \cdot H + \alpha(H)H \qquad (7.54)$$

with $\alpha(H) \to 0$ when $H \to 0$ and A_Z a bicomplex constant.

Note that in this definition H is allowed to be a zero-divisor but taking H in (7.54) to be any non-zero-divisor, we see from (7.9) that $\mathbb{BC}$-differentiability implies $\mathbb{BC}$-derivability. The reciprocal statement is more delicate and will not be treated immediately.

It turns out that Theorem 7.4.3 has many deep and far-reaching consequences which we will discuss in the next section.

7.5 Bicomplex holomorphy versus holomorphy in two (complex or hyperbolic) variables

Take a $\mathbb{BC}$-holomorphic function F, which we write as $F = f_1 + \mathbf{j}f_2$ on a domain $\Omega \subset \mathbb{BC}$ with the independent variable written as $Z = z_1 + \mathbf{j}z_2$. By Theorem 7.4.3, and Remark 7.4.4, this is equivalent to saying that F verifies in Ω the system

$$\frac{\partial F}{\partial Z^\dagger} = \frac{\partial F}{\partial \overline{Z}} = \frac{\partial F}{\partial Z^*} = 0 .$$

For the operators involved we use the appropriate representations from the table in the previous section. Using such a representation for the operators $\dfrac{\partial}{\partial \overline{Z}}$ and $\dfrac{\partial}{\partial Z^*}$ leads to the system

$$\frac{\partial F}{\partial \overline{z}_1} - \mathbf{j}\frac{\partial F}{\partial \overline{z}_2} = 0, \qquad \frac{\partial F}{\partial \overline{z}_1} + \mathbf{j}\frac{\partial F}{\partial \overline{z}_2} = 0,$$

implying that $\dfrac{\partial F}{\partial \overline{z}_1} = 0 = \dfrac{\partial F}{\partial \overline{z}_2}$. The latter is equivalent to the holomorphy, in the sense of complex functions of two $\mathbb{C}(\mathbf{i})$-complex variables, of the components f_1, f_2 of the function F. Thus F can be seen as a holomorphic mapping from $\Omega \subset \mathbb{C}^2(\mathbf{i}) \to \mathbb{C}^2(\mathbf{i})$. But we still have more information. Since $\dfrac{\partial F}{\partial Z^\dagger} = 0$, then $F = f_1 + \mathbf{j}f_2$ verifies

$$\begin{aligned}
\frac{\partial F}{\partial Z^\dagger}(Z) &= \frac{1}{2}\left(\frac{\partial}{\partial z_1} + \mathbf{j}\frac{\partial}{\partial z_2}\right)(f_1 + \mathbf{j}f_2)(Z)\\
&= \frac{1}{2}\left(\left(\frac{\partial f_1}{\partial z_1} - \frac{\partial f_2}{\partial z_2}(Z)\right) + \mathbf{j}\left(\frac{\partial f_2}{\partial z_1}(Z) + \frac{\partial f_1}{\partial z_2}(Z)\right)\right) = 0.
\end{aligned}$$

Thus

$$\frac{\partial f_1}{\partial z_1} = \frac{\partial f_2}{\partial z_2}, \qquad \frac{\partial f_1}{\partial z_2} = -\frac{\partial f_2}{\partial z_1}, \tag{7.55}$$

that is, the complex partial derivatives of the holomorphic functions f_1, f_2 are not independent; they are tied by the Cauchy-Riemann type conditions (7.55). We reformulate the reasoning as

Proposition 7.5.1. *A function $F = f_1 + \mathbf{j} f_2 : \Omega \subset \mathbb{BC} \to \mathbb{BC}$ is $\mathbb{BC}$-holomorphic if and only if, seen as a mapping from $\Omega \subset \mathbb{C}^2(\mathbf{i}) \to \mathbb{C}^2(\mathbf{i})$, it is a holomorphic mapping with its components related by the Cauchy-Riemann type conditions* (7.55).

In other words, the theory of bicomplex holomorphic functions can be seen as a theory of a proper subset of holomorphic mappings in two complex variables. Each equation in Theorem 7.4.3 plays a different role: two of them together guarantee the holomorphy of the $\mathbb{C}(\mathbf{i})$-complex components and the third one provides the relation between them.

It is worth noting that the operator $\dfrac{\partial}{\partial Z^\dagger}$ arises in the works of J. Ryan [69] on complex Clifford analysis as a Cauchy-Riemann operator which is defined directly on holomorphic mappings with values in a complex Clifford algebra.

Next, take a bicomplex holomorphic function F in the form $F = g_1 + \mathbf{i} g_2$, where g_1 and g_2 take values in $\mathbb{C}(\mathbf{j})$ and we write now $Z = \zeta_1 + \mathbf{i} \zeta_2$, then the corresponding differential operators are:

$$\frac{\partial}{\partial Z^\dagger} = \frac{1}{2}\left(\frac{\partial}{\partial \zeta_1^*} - \mathbf{i}\frac{\partial}{\partial \zeta_2^*} \right),$$

$$\frac{\partial}{\partial \overline{Z}} = \frac{1}{2}\left(\frac{\partial}{\partial \zeta_1} + \mathbf{i}\frac{\partial}{\partial \zeta_2} \right),$$

$$\frac{\partial}{\partial Z^*} = \frac{1}{2}\left(\frac{\partial}{\partial \zeta_1^*} + \mathbf{i}\frac{\partial}{\partial \zeta_2^*} \right).$$

Using again Theorem 7.4.3 and Remark 7.4.4 we have that to be a bicomplex holomorphic function means for F that

$$\frac{\partial F}{\partial Z^\dagger} = \frac{1}{2}\left(\frac{\partial F}{\partial \zeta_1^*} - \mathbf{i}\frac{\partial F}{\partial \zeta_2^*} \right) = 0,$$

$$\frac{\partial F}{\partial \overline{Z}} = \frac{1}{2}\left(\frac{\partial F}{\partial \zeta_1} + \mathbf{i}\frac{\partial F}{\partial \zeta_2} \right) = 0,$$

$$\frac{\partial F}{\partial Z^*} = \frac{1}{2}\left(\frac{\partial F}{\partial \zeta_1^*} + \mathbf{i}\frac{\partial F}{\partial \zeta_2^*} \right) = 0.$$

The first and the third equations together give, again, that the components g_1 and g_2 of F are holomorphic functions of the $\mathbb{C}(\mathbf{j})$-complex variables ζ_1 and ζ_2, while the second equation gives:

$$\frac{\partial g_1}{\partial \zeta_1} = \frac{\partial g_2}{\partial \zeta_2}, \qquad \frac{\partial g_1}{\partial \zeta_2} = -\frac{\partial g_2}{\partial \zeta_1}. \tag{7.56}$$

Proposition 7.5.2. *A function $F = g_1 + \mathbf{i}g_2 : \Omega \subset \mathbb{BC} \to \mathbb{BC}$ is $\mathbb{BC}$-holomorphic if and only if, seen as a mapping from $\Omega \subset \mathbb{C}^2(\mathbf{j}) \to \mathbb{C}^2(\mathbf{j})$, it is a holomorphic mapping with its components related by the Cauchy-Riemann type conditions (7.56).*

Finally, take a $\mathbb{BC}$-holomorphic function $F : \Omega \subset \mathbb{BC} \to \mathbb{BC}$ in the form $F = \mathfrak{u}_1 + \mathbf{i}\mathfrak{u}_2$, where $\mathfrak{u}_1, \mathfrak{u}_2$ take values in $\mathbb{D}$ and write $Z = \mathfrak{z}_1 + \mathbf{i}\mathfrak{z}_2$, then the corresponding operators are:

$$\frac{\partial}{\partial Z^\dagger} = \frac{1}{2}\left(\frac{\partial}{\partial \mathfrak{z}_1^\diamond} - \mathbf{i}\frac{\partial}{\partial \mathfrak{z}_2^\diamond}\right),$$

$$\frac{\partial}{\partial \overline{Z}} = \frac{1}{2}\left(\frac{\partial}{\partial \mathfrak{z}_1^\diamond} + \mathbf{i}\frac{\partial}{\partial \mathfrak{z}_2^\diamond}\right),$$

$$\frac{\partial}{\partial Z^*} = \frac{1}{2}\left(\frac{\partial}{\partial \mathfrak{z}_1} + \mathbf{i}\frac{\partial}{\partial \mathfrak{z}_2}\right).$$

Using again Theorem 7.4.3 and Remark 7.4.4 we have that to be a $\mathbb{BC}$-holomorphic function means for F that

$$\frac{\partial F}{\partial Z^\dagger} = \frac{1}{2}\left(\frac{\partial F}{\partial \mathfrak{z}_1^\diamond} - \mathbf{i}\frac{\partial F}{\partial \mathfrak{z}_2^\diamond}\right) = 0,$$

$$\frac{\partial F}{\partial \overline{Z}} = \frac{1}{2}\left(\frac{\partial F}{\partial \mathfrak{z}_1^\diamond} + \mathbf{i}\frac{\partial F}{\partial \mathfrak{z}_2^\diamond}\right) = 0,$$

$$\frac{\partial F}{\partial Z^*} = \frac{1}{2}\left(\frac{\partial F}{\partial \mathfrak{z}_1} + \mathbf{i}\frac{\partial F}{\partial \mathfrak{z}_2}\right) = 0.$$

The first and second equations together imply that the components $\mathfrak{u}_1$ and $\mathfrak{u}_2$ of F are holomorphic functions of hyperbolic variables $\mathfrak{z}_1$ and $\mathfrak{z}_2$, meanwhile the last equation gives the Cauchy-Riemann type conditions:

$$\frac{\partial \mathfrak{u}_1}{\partial \mathfrak{z}_1} = \frac{\partial \mathfrak{u}_2}{\partial \mathfrak{z}_2} \qquad \frac{\partial \mathfrak{u}_1}{\partial \mathfrak{z}_2} = -\frac{\partial \mathfrak{u}_2}{\partial \mathfrak{z}_1}. \qquad (7.57)$$

They look exactly as their antecedent in one complex variable, but this is a totally different thing: we deal here with $\mathbb{D}$-valued functions of two hyperbolic variables and with hyperbolic partial derivatives.

Proposition 7.5.3. *A function $F = \mathfrak{u}_1 + \mathbf{i}\mathfrak{u}_2 : \Omega \subset \mathbb{BC} \to \mathbb{BC}$ is $\mathbb{BC}$-holomorphic if and only if, seen as a mapping from $\Omega \subset \mathbb{D}^2 \to \mathbb{D}^2$, it is a holomorphic mapping with its components related by the Cauchy-Riemann type conditions (7.57).*

7.6   Bicomplex holomorphy: the idempotent representation

Take a bicomplex function $F : \Omega \subset \mathbb{BC} \to \mathbb{BC}$ on a domain Ω. We write all the bicomplex numbers involved in $\mathbb{C}(\mathbf{i})$-idempotent form, for instance,

$$Z = \beta_1 \mathbf{e} + \beta_2 \mathbf{e}^{\dagger} = (\ell_1 + i m_1) \mathbf{e} + (\ell_2 + i m_2) \mathbf{e}^{\dagger},$$

$$F(Z) = G_1(Z) \mathbf{e} + G_2(Z) \mathbf{e}^{\dagger},$$

$$H = \eta_1 \mathbf{e} + \eta_2 \mathbf{e}^{\dagger} = (u_1 + i v_1) \mathbf{e} + (u_2 + i v_2) \mathbf{e}^{\dagger}.$$

Let us introduce the sets

$$\Omega_1 := \{ \beta_1 \,|\, \beta_1 \mathbf{e} + \beta_2 \mathbf{e}^{\dagger} \in \Omega \} \subset \mathbb{C}(\mathbf{i}) \tag{7.58}$$

and

$$\Omega_2 := \{ \beta_2 \,|\, \beta_1 \mathbf{e} + \beta_2 \mathbf{e}^{\dagger} \in \Omega \} \subset \mathbb{C}(\mathbf{i}). \tag{7.59}$$

It is easy to prove that Ω_1 and Ω_2 are domains in $\mathbb{C}(\mathbf{i})$, see book [56].

We assume that $F \in \mathcal{C}^1(\Omega)$ where the real partial derivatives are taken with respect to the "idempotent real variables": ℓ_1, m_1, ℓ_2, m_2; how this is related with the cartesian variables x_1, y_1, x_2, y_2 will be discussed later. The condition $F \in \mathcal{C}^1(\Omega)$ ensures the real differentiability of F in Ω:

$$F(Z + H) - F(Z) = \frac{\partial F}{\partial \ell_1}(Z) \cdot u_1 + \frac{\partial F}{\partial m_1}(Z) \cdot v_1$$

$$+ \frac{\partial F}{\partial \ell_2}(Z) \cdot u_2 + \frac{\partial F}{\partial m_2}(Z) \cdot v_2 + \mathbf{o}(H) \tag{7.60}$$

for $H \to 0$. We are going to follow Section 7.4 so we omit many details. First of all, let us translate formula (7.60) in $\mathbb{C}(\mathbf{i})$-complex language: since

$$u_1 = \frac{1}{2}(\eta_1 + \bar{\eta}_1); \qquad u_2 = \frac{1}{2}(\eta_2 + \bar{\eta}_2);$$

$$v_1 = \frac{\mathbf{i}}{2}(\bar{\eta}_1 - \eta_1); \qquad v_2 = \frac{\mathbf{i}}{2}(\bar{\eta}_2 - \eta_2),$$

then

$$F(Z + H) - F(Z) = \frac{\partial F}{\partial \ell_1}(Z) \cdot \frac{1}{2}(\eta_1 + \bar{\eta}_1) + \frac{\partial F}{\partial m_1}(Z) \cdot \frac{\mathbf{i}}{2}(\bar{\eta}_1 - \eta_1)$$

$$+ \frac{\partial F}{\partial \ell_2}(Z) \cdot \frac{1}{2}(\eta_2 + \bar{\eta}_2) + \frac{\partial F}{\partial m_2}(Z) \cdot \frac{\mathbf{i}}{2}(\bar{\eta}_2 - \eta_2) + \mathbf{o}(H)$$

$$= \eta_1 \frac{1}{2}\left(\frac{\partial F}{\partial \ell_1}(Z) - \mathbf{i}\frac{\partial F}{\partial m_1}(Z) \right) + \bar{\eta}_1 \frac{1}{2}\left(\frac{\partial F}{\partial \ell_1}(Z) + \mathbf{i}\frac{\partial F}{\partial m_1}(Z) \right)$$

$$+ \eta_2 \frac{1}{2}\left(\frac{\partial F}{\partial \ell_2}(Z) - \mathbf{i}\frac{\partial F}{\partial m_2}(Z) \right) + \bar{\eta}_2 \frac{1}{2}\left(\frac{\partial F}{\partial \ell_2}(Z) + \mathbf{i}\frac{\partial F}{\partial m_2}(Z) \right) + \mathbf{o}(H)$$

$$=: \eta_1 \frac{\partial F}{\partial \beta_1}(Z) + \bar{\eta}_1 \frac{\partial F}{\partial \bar{\beta}_1}(Z) + \eta_2 \frac{\partial F}{\partial \beta_2}(Z) + \bar{\eta}_2 \frac{\partial F}{\partial \bar{\beta}_2}(Z) + \mathbf{o}(H).$$

Remark 7.6.1. *Note that the above calculations show that, as is well known, the bicomplex function F of class $\mathcal{C}^1$, seen as a mapping from $\mathbb{C}^2(\mathbf{i}) \to \mathbb{C}^2(\mathbf{i})$, is holomorphic with respect to β_q $(q = 1, 2)$ if and only if $\dfrac{\partial F}{\partial \overline{\beta}_q}(Z) = 0$ in Ω. Note that if F is a bicomplex function, and we express it in cartesian coordinates, it turns out that F is $\mathbb{BC}$-holomorphic if and only if its components are holomorphic as functions of two complex variables and satisfy a Cauchy-Riemann type relation between them. As we will show later, this is definitely not the case when we express F in the idempotent representation. In this case, $\mathbb{BC}$-holomorphy will be equivalent to the requirement that each component is a holomorphic function of a single complex variable and there are no relations between the components.*

We emphasize that now we are considering the identification between $\mathbb{BC}$ and $\mathbb{C}^2(\mathbf{i})$,

$$Z = \beta_1 \mathbf{e} + \beta_2 \mathbf{e}^\dagger \longleftrightarrow (\beta_1, \beta_2) \in \mathbb{C}^2(\mathbf{i}),$$

where however the basis in $\mathbb{C}^2(\mathbf{i})$ is not the canonical basis $\{1, \mathbf{j}\}$, but rather the idempotent basis $\{\mathbf{e}, \mathbf{e}^\dagger\}$.

For the next step recall the formulas

$$H = \eta_1 \mathbf{e} + \eta_2 \mathbf{e}^\dagger, \qquad H^\dagger = \eta_2 \mathbf{e} + \eta_1 \mathbf{e}^\dagger,$$
$$\overline{H} = \overline{\eta}_2 \mathbf{e} + \overline{\eta}_1 \mathbf{e}^\dagger, \qquad H^* = \overline{\eta}_1 \mathbf{e} + \overline{\eta}_2 \mathbf{e}^\dagger,$$

which imply that

$$\eta_1 = H\mathbf{e} + H^\dagger \mathbf{e}^\dagger, \qquad \overline{\eta}_1 = H^* \mathbf{e} + \overline{H}\mathbf{e}^\dagger,$$
$$\eta_2 = H^\dagger \mathbf{e} + H\mathbf{e}^\dagger, \qquad \overline{\eta}_2 = \overline{H}\mathbf{e} + H^* \mathbf{e}^\dagger.$$

The condition of real differentiability after substitutions becomes:

$$F(Z + H) - F(Z)$$
$$= H \left(\frac{\partial F}{\partial \beta_1}(Z)\mathbf{e} + \frac{\partial F}{\partial \beta_2}(Z)\mathbf{e}^\dagger \right) + H^\dagger \left(\frac{\partial F}{\partial \beta_2}(Z)\mathbf{e} + \frac{\partial F}{\partial \beta_1}(Z)\mathbf{e}^\dagger \right) \qquad (7.61)$$
$$+ \overline{H} \left(\frac{\partial F}{\partial \overline{\beta}_2}(Z)\mathbf{e} + \frac{\partial F}{\partial \overline{\beta}_1}(Z)\mathbf{e}^\dagger \right) + H^* \left(\frac{\partial F}{\partial \overline{\beta}_1}(Z)\mathbf{e} + \frac{\partial F}{\partial \overline{\beta}_2}(Z)\mathbf{e}^\dagger \right) + \mathfrak{o}(H).$$

Note that the expressions in the parentheses are not, yet, the idempotent forms of anything, since the coefficients of $\mathbf{e}$ and $\mathbf{e}^\dagger$ are bicomplex numbers, not $\mathbb{C}(\mathbf{i})$-complex numbers. Thus we are required to make one more step. Using the formula $F = G_1 \mathbf{e} + G_2 \mathbf{e}^\dagger$ we arrive at

$$F(Z + H) - F(Z)$$
$$= H \left(\frac{\partial G_1}{\partial \beta_1}(Z)\mathbf{e} + \frac{\partial G_2}{\partial \beta_2}(Z)\mathbf{e}^\dagger \right) + H^\dagger \left(\frac{\partial G_1}{\partial \beta_2}(Z)\mathbf{e} + \frac{\partial G_2}{\partial \beta_1}(Z)\mathbf{e}^\dagger \right)$$
$$+ \overline{H} \left(\frac{\partial G_1}{\partial \overline{\beta}_2}(Z)\mathbf{e} + \frac{\partial G_2}{\partial \overline{\beta}_1}(Z)\mathbf{e}^\dagger \right) + H^* \left(\frac{\partial G_1}{\partial \overline{\beta}_1}(Z)\mathbf{e} + \frac{\partial G_2}{\partial \overline{\beta}_2}(Z)\mathbf{e}^\dagger \right) + \mathfrak{o}(H).$$
$$(7.62)$$

This formula is valid for any F in $\mathcal{C}^1(\Omega)$, so let us analyze how $\mathbb{BC}$-holomorphic functions are singled out among those of class $\mathcal{C}^1$.

Theorem 7.6.2. *The $\mathcal{C}^1$-function F is $\mathbb{BC}$-holomorphic if and only if the three bicomplex coefficients of $H^\dagger, \overline{H}$ and H^* in (7.62) are all zero for any Z in Ω.*

Proof. The *if* direction follows as in Theorem 7.4.3. Specifically, since F is $\mathbb{BC}$-holomorphic, formula (7.9) holds for all $H \notin \mathfrak{S}_0$. But F is a $\mathcal{C}^1$-function, hence (7.62) holds as well for any $H \neq 0$, thus both formulas hold for non-zero-divisors. Then the result follows directly by recalling that both (7.9) and (7.62) are unique representations for a given function F, and by comparing them.

In order to prove the *only if*, it is helpful to write explicitly the meaning of the vanishing of these coefficients, namely:

$$\frac{\partial G_1}{\partial \beta_2}(Z)\mathbf{e} + \frac{\partial G_2}{\partial \beta_1}(Z)\mathbf{e}^\dagger = 0\,,$$

$$\frac{\partial G_1}{\partial \overline{\beta}_2}(Z)\mathbf{e} + \frac{\partial G_2}{\partial \overline{\beta}_1}(Z)\mathbf{e}^\dagger = 0\,,$$

$$\frac{\partial G_1}{\partial \overline{\beta}_1}(Z)\mathbf{e} + \frac{\partial G_2}{\partial \overline{\beta}_2}(Z)\mathbf{e}^\dagger = 0\,. \tag{7.63}$$

Now note that the second and the third equations, because of the independence of $\mathbf{e}$ and $\mathbf{e}^\dagger$, impose that G_1 and G_2 are $\mathbb{C}(\mathbf{i})$-valued holomorphic functions of the complex variables β_1, β_2 and thus they have authentic complex partial derivatives. What is more, the first equation in (7.63) says that one of the partial derivatives of each G_1 and G_2 is identically zero: $\dfrac{\partial G_1}{\partial \beta_2}(Z) = 0$, $\dfrac{\partial G_2}{\partial \beta_1}(Z) = 0$ for any $Z \in \Omega$. Hence, using (7.58) and (7.59), G_1 is a holomorphic function of the single variable $\beta_1 \in \Omega_1$ and G_2 is a holomorphic function of the single variable $\beta_2 \in \Omega_2$. We now want to show that these equations imply that F is $\mathbb{BC}$-holomorphic. But in fact, because of these equations, we have that for any invertible H there holds:

$$\frac{F(Z+H) - F(Z)}{H} = \frac{G_1(\beta_1 + \eta_1) - G_1(\beta_1)}{\eta_1}\mathbf{e} + \frac{G_2(\beta_2 + \eta_2) - G_2(\beta_2)}{\eta_2}\mathbf{e}^\dagger\,,$$

where Z is an arbitrary point in Ω.

Now, by the properties of G_1 and G_2 we deduce that the right-hand side has, for $\mathfrak{S}_0 \not\ni H \to 0$, the limit $G_1'(\beta_1)\mathbf{e} + G_2'(\beta_2)\mathbf{e}^\dagger$, which concludes the proof: the limit in the left-hand side exists also for any $Z \in \Omega$ with $\mathfrak{S}_0 \not\ni H \to 0$ and it coincides with the derivative $F'(Z)$ making F $\mathbb{BC}$-holomorphic in Ω. $\qquad\square$

As a matter of fact, the proof allows us to make a more precise characterization of $\mathcal{C}^1$-functions which are $\mathbb{BC}$-holomorphic.

Theorem 7.6.3. *A bicomplex function $F = G_1\mathbf{e} + G_2\mathbf{e}^\dagger : \Omega \subset \mathbb{BC} \to \mathbb{BC}$ of class $\mathcal{C}^1$ is $\mathbb{BC}$-holomorphic if and only if the following two conditions hold:*

(I) *The component G_1, seen as a $\mathbb{C}(\mathbf{i})$-valued function of two complex variables (β_1, β_2) is holomorphic; moreover, it does not depend on the variable β_2 and thus G_1 is a holomorphic function of the variable β_1.*

(II) *The component G_2, seen as a $\mathbb{C}(\mathbf{i})$-valued function of two complex variables (β_1, β_2) is holomorphic; moreover, it does not depend on the variable β_1 and thus G_2 is a holomorphic function of the variable β_2.*

Remark 7.6.4. *The functions G_1 and G_2 are independent in the sense that there are no Cauchy-Riemann type conditions relating them.*

We are in a position now to prove that the converse to Theorem 7.4.3 is true as well.

Theorem 7.6.5. *Given $F \in \mathcal{C}^1(\Omega, \mathbb{BC})$, then condition (7.53) implies that F is $\mathbb{BC}$-holomorphic.*

Proof. If (7.53) holds, then a direct computation shows that all the three formulas in (7.63) are true, and by Theorem 7.6.2 F is $\mathbb{BC}$-holomorphic. $\square$

The direct computation mentioned above is quite useful and instructive and we will perform it later.

Corollary 7.6.6. *Let F be a $\mathbb{BC}$-holomorphic function in Ω, then F is of the form $F(Z) = G_1(\beta_1)\mathbf{e} + G_2(\beta_2)\mathbf{e}^\dagger$ with $Z = \beta_1\mathbf{e} + \beta_2\mathbf{e}^\dagger \in \Omega$ and its derivative is given by*

$$F'(Z) = G_1'(\beta_1)\mathbf{e} + G_2'(\beta_2)\mathbf{e}^\dagger .$$

Taking into account the relations between β_1, β_2 and the cartesian components z_1, z_2, we have also that

$$F'(z_1 + \mathbf{j}z_2) = G_1'(z_1 - \mathbf{i}z_2)\mathbf{e} + G_2'(z_1 + \mathbf{i}z_2)\mathbf{e}^\dagger;$$
$$F'(Z) = G_1'(Z\mathbf{e} + Z^\dagger\mathbf{e}^\dagger)\mathbf{e} + G_2'(Z^\dagger\mathbf{e} + Z\mathbf{e}^\dagger)\mathbf{e}^\dagger .$$

This implies that a $\mathbb{BC}$-holomorphic function has derivatives of any order and

$$F^{(n)}(Z) = G_1^{(n)}(\beta_1)\,\mathbf{e} + G_2^{(n)}(\beta_2)\,\mathbf{e}^\dagger$$
$$= G_1^{(n)}(Z\mathbf{e} + Z^\dagger\mathbf{e}^\dagger)\,\mathbf{e} + G_2^{(n)}(Z^\dagger\mathbf{e} + Z\mathbf{e}^\dagger)\,\mathbf{e}^\dagger.$$

Remark 7.6.7. *Although formula (7.62) is quite similar to formula (7.48) its consequences for the function F are paradoxically different: while formula (7.48) has allowed us to conclude that the cartesian components f_1, f_2 are holomorphic functions of two complex variables which are not independent, formula (7.62) explains to us that the idempotent components G_1, G_2 are usual holomorphic functions of one complex variable which are, besides, independent.*

Remark 7.6.8. *We have proved that if F is $\mathbb{BC}$-holomorphic, then for any $Z = \beta_1\mathbf{e} + \beta_2\mathbf{e}^\dagger \in \Omega$ it is of the form*

$$F(Z) = G_1(\beta_1)\mathbf{e} + G_2(\beta_2)\mathbf{e}^\dagger\,.$$

But the right-hand side of the latter is well-defined on the wider set $\widetilde{\Omega} := \Omega_1 \cdot \mathbf{e} + \Omega_2 \cdot \mathbf{e}^\dagger \supset \Omega$ (in general, this inclusion is proper), with the notations as in (7.58) and (7.59). Moreover, by Theorem 7.6.3 the function $\widetilde{F}$ defined by

$$\widetilde{F}(Z) := G_1(\beta_1)\mathbf{e} + G_2(\beta_2)\mathbf{e}^\dagger, \qquad Z \in \widetilde{\Omega}\,,$$

is $\mathbb{BC}$-holomorphic in $\widetilde{\Omega}$. Since $\widetilde{F}\big|_\Omega \equiv F$ we see that, unlike what happens in the complex case, not every domain in $\mathbb{BC}$ is a domain of $\mathbb{BC}$-holomorphy: every function which is $\mathbb{BC}$-holomorphic in a domain Ω extends $\mathbb{BC}$-holomorphically up to the minimal set of the form $X_1 \cdot \mathbf{e} + X_2 \cdot \mathbf{e}^\dagger$ containing Ω. One can compare this with [81].

Remark 7.6.9. *We recall that if $\Omega \subset \mathbb{C}(\mathbf{i})$ is an open set bounded by a simple closed curve, then there exists a holomorphic function f on Ω with the following property: if $\widetilde{\Omega}$ is any open set which strictly contains Ω, then there is no holomorphic function $\widetilde{f}$ on $\widetilde{\Omega}$ such that $\widetilde{f}$ restricted to Ω equals f. Usually Ω is called a* domain of holomorphy *for f.*

Consider now a bicomplex holomorphic function F on a domain Ω. Suppose that Ω_1 and Ω_2 are domains of holomorphy for G_1 and G_2, respectively, i.e., G_1 and G_2 cannot be holomorphically extended to any bigger open set in $\mathbb{C}(\mathbf{i})$. Then the bicomplex function $F = G_1\mathbf{e} + G_2\mathbf{e}^\dagger$, defined on $\widetilde{\Omega} := \Omega_1 \cdot \mathbf{e} + \Omega_2 \cdot \mathbf{e}^\dagger$ cannot be $\mathbb{BC}$-holomorphically extended to any open set containing $\widetilde{\Omega}$. Combining this fact with the comments of the previous Remark 7.6.8, we can say that $\widetilde{\Omega}$ is a *domain of bicomplex holomorphy* for F.

Remark 7.6.10. *The same analysis can be done for the idempotent representation with $\mathbb{C}(\mathbf{j})$ coefficients.*

Remark 7.6.11. *Recall that at the beginning of this chapter we worked with the cartesian representation of bicomplex numbers and we investigated many properties of derivable bicomplex functions, in particular, such functions proved to have complex partial derivatives with respect to z_1 and z_2. This approach fails immediately when one tries to apply it to the case of the idempotent representation: this is because the definition of the derivative excludes precisely the values of H which are necessary for the complex partial derivatives with respect to β_1, β_2. But in the proof of Theorem 7.4.3 we have shown, as a matter of fact, that such partial derivatives of a $\mathbb{BC}$-holomorphic functions do exist and, moreover, $\dfrac{\partial F}{\partial \beta_1}(Z) = G_1'(\beta_1) \cdot \mathbf{e}$ and $\dfrac{\partial F}{\partial \beta_2}(Z) = G_2'(\beta_2) \cdot \mathbf{e}^\dagger$.*

Remark 7.6.12. *Using the results of this section, we can give the proofs of the statements of Theorem 7.2.7 in terms of the idempotent writing of the bicomplex elementary functions. For example, to prove that the bicomplex exponential function $F(Z) = e^Z$ is $\mathbb{BC}$-holomorphic, we write:*

$$F(Z) = e^Z = e^{\beta_1}\mathbf{e} + e^{\beta_2}\mathbf{e}^\dagger = G_1(\beta_1)\mathbf{e} + G_2(\beta_2)\mathbf{e}^\dagger,$$

for all $Z = \beta_1\mathbf{e} + \beta_2\mathbf{e}^\dagger \in \mathbb{BC}$. But G_1 is a complex holomorphic function depending on β_1, similar for G_2, thus Theorem 7.6.3 guarantees that F is a bicomplex holomorphic function. Besides,

$$(e^Z)' = (e^{\beta_1})'\mathbf{e} + (e^{\beta_2})'\mathbf{e}^\dagger = e^{\beta_1}\mathbf{e} + e^{\beta_2}\mathbf{e}^\dagger = e^Z.$$

7.7 Cartesian versus idempotent representations in $\mathbb{BC}$-holomorphy

In the previous sections we worked with the real linear space $\mathbb{R}^4$ which was endowed sometimes with the standard coordinates $\vec{x} := (x_1, y_1, x_2, y_2)$ and thus we denote it by $\mathbb{R}^4_{\vec{x}}$, and sometimes endowed with the idempotent coordinates $\vec{\ell} := (\ell_1, m_1, \ell_2, m_2)$ thus denoted by $\mathbb{R}^4_{\vec{\ell}}$. The relation between both is given by

$$\ell_1 = x_1 + y_2, \quad m_1 = y_1 - x_2, \quad \ell_2 = x_1 - y_2, \quad m_2 = y_1 + x_2, \tag{7.64}$$

or, equivalently, by

$$(\ell_1, m_1, \ell_2, m_2)^t := \begin{pmatrix} 1 & 0 & 0 & 1 \\ 0 & 1 & -1 & 0 \\ 1 & 0 & 0 & -1 \\ 0 & 1 & 1 & 0 \end{pmatrix} \cdot (x_1, y_1, x_2, y_2)^t. \tag{7.65}$$

The 4×4 matrix from the right-hand side of (7.65) has determinant equal 4, so it is invertible, hence

$$(x_1, y_1, x_2, y_2)^t := \frac{1}{2}\begin{pmatrix} 1 & 0 & 1 & 0 \\ 0 & 1 & 0 & 1 \\ 0 & -1 & 0 & 1 \\ 1 & 0 & -1 & 0 \end{pmatrix} \cdot (\ell_1, m_1, \ell_2, m_2)^t, \tag{7.66}$$

that is,

$$x_1 = \frac{\ell_1 + \ell_2}{2}, \quad y_1 = \frac{m_1 + m_2}{2}, \quad x_2 = \frac{m_2 - m_1}{2}, \quad y_2 = \frac{\ell_1 - \ell_2}{2}. \tag{7.67}$$

Altogether we have an isomorphism of linear spaces $\phi : \mathbb{R}^4_{\vec{x}} \to \mathbb{R}^4_{\vec{\ell}}$ defined by

$$(x_1, y_1, x_2, y_2) \overset{\phi}{\longmapsto} (\ell_1, m_1, \ell_2, m_2) \tag{7.68}$$

with the inverse linear map $\phi^{-1} : \mathbb{R}^4_{\vec{\ell}} \to \mathbb{R}^4_{\vec{x}}$ given by (7.66). Because ϕ is a linear isomorphism, the matrix is the Jacobi matrix of ϕ, so we denote it by $J_{\vec{x}}[\phi]$. Note that considering the matrix transpose of (7.65) we can write:

$$\vec{\ell} = \vec{x} \cdot J_{\vec{x}}[\phi]^t \,.$$

Similarly, we denote the 4×4 matrix from the right-hand side of (7.66) by $J_{\vec{\ell}}[\phi^{-1}]$, which is nothing but the inverse matrix of $J_{\vec{x}}[\phi]$, having determinant equal to $\dfrac{1}{4}$. Hence, we can simply write (7.66) as

$$\vec{x} = \vec{\ell} \cdot J_{\vec{\ell}}[\phi^{-1}]^t \,.$$

Remark 7.7.1. *Because the determinant of $J_{\vec{x}}[\phi]$ is 4, then ϕ does not preserve Euclidean distances, i.e., it is not an isometry between $\mathbb{R}^4_{\vec{x}}$ and $\mathbb{R}^4_{\vec{\ell}}$: if the Euclidean norm of $\vec{x}$ is 1, then the Euclidean norm of $\vec{\ell}$ is*

$$|\vec{\ell}| = \sqrt{\ell_1^2 + m_1^2 + \ell_2^2 + m_2^2} = \sqrt{2}\sqrt{x_1^2 + y_1^2 + x_2^2 + y_2^2} = \sqrt{2}|\vec{x}| = \sqrt{2}\,.$$

Let us denote by $\mathcal{F}(V)$ any bicomplex module of $\mathbb{BC}$-valued functions defined on $V \subset \mathbb{R}^4_{\vec{\ell}}$, i.e., $\mathcal{F}(V)$ is a bicomplex module under the usual operations of addition and multiplication by bicomplex constant. We define the change of variables operator W_ϕ by

$$W_\phi : \mathcal{F}(U) \to \mathcal{F}(V), \qquad W_\phi[g](\vec{x}) := (g \circ \phi)(\vec{x}) = g(\phi(\vec{x}))\,, \qquad (7.69)$$

where $U := \phi^{-1}(V) \subset \mathbb{R}^4_{\vec{x}}$.

Since ϕ is a linear isomorphism, W_ϕ is a well defined linear operator, i.e., for all $\lambda, \mu \in \mathbb{BC}$ and for all $g_1, g_2 \in \mathcal{F}(U)$ we have:

$$W_\phi[\lambda g_1 + \mu g_2] = \lambda W_\phi[g_1] + \mu W_\phi[g_2]\,.$$

The operator W_ϕ has an inverse defined by

$$W_\phi^{-1} : \mathcal{F}(U) \to \mathcal{F}(V), \qquad W_\phi^{-1}[f](\vec{\ell}) := (f \circ \phi^{-1})(\vec{\ell}) = f(\phi^{-1}(\vec{\ell}))\,, \qquad (7.70)$$

since $(W_\phi \circ W_\phi^{-1})[f] = f$ and $(W_\phi^{-1} \circ W_\phi)[g] = g$. Therefore W_ϕ is a linear isomorphism between $\mathcal{F}(V)$ and $\mathcal{F}(U)$.

Having established an isomorphism between the two $\mathbb{BC}$-modules we can establish now an isomorphism between the corresponding $\mathbb{BC}$-algebras of linear operators acting on these $\mathbb{BC}$-modules. Indeed, given an operator B acting on $\mathcal{F}(V)$, we define the mapping

$$B \longmapsto A := W_\phi \circ B \circ W_\phi^{-1}\,,$$

where A is an operator acting on $\mathcal{F}(U)$. Then any linear combination or composition of operators is preserved under the mapping above: if B_1 and B_2 are two operators acting on $\mathcal{F}(V)$ and Λ_1 and Λ_2 are two bicomplex numbers, then

$$W_\phi \circ (\Lambda_1 B_1 + \Lambda_2 B_2) \circ W_\phi^{-1} = \Lambda_1 (W_\phi \circ B_1 \circ W_\phi^{-1}) + \Lambda_2 (W_\phi \circ B_2 \circ W_\phi^{-1})$$

$$=: \Lambda_1 A_1 + \Lambda_2 A_2$$

and

$$W_\phi \circ (B_1 \circ B_2) \circ W_\phi^{-1} = (W_\phi \circ B_1 \circ W_\phi^{-1}) \circ (W_\phi \circ B_2 \circ W_\phi^{-1})$$

$$=: A_1 \circ A_2 .$$

The above general reasoning is specified now for the case of our interest, namely, for $\mathcal{F}(V) = \mathcal{C}^\infty(V)$ and $\mathcal{F}(U) = \mathcal{C}^\infty(U)$ where the operators of partial derivatives act with respect to the corresponding variables.

Let g be a function in $\mathcal{C}^\infty(V)$, then $g(\phi(\vec{x}))$ is in $\mathcal{C}^\infty(U)$ and the chain rule gives:

$$\frac{\partial(W_\phi[g])}{\partial x_1}(\vec{x}) = \frac{\partial}{\partial x_1}[g(\phi(\vec{x}))]$$

$$= \frac{\partial g}{\partial \ell_1}(\phi(\vec{x})) \cdot \frac{\partial \ell_1}{\partial x_1} + \frac{\partial g}{\partial m_1}(\phi(\vec{x})) \cdot \frac{\partial m_1}{\partial x_1}$$

$$+ \frac{\partial g}{\partial \ell_2}(\phi(\vec{x})) \cdot \frac{\partial \ell_2}{\partial x_1} + \frac{\partial g}{\partial m_2}(\phi(\vec{x})) \cdot \frac{\partial m_2}{\partial x_1}$$

$$= W_\phi \left[\frac{\partial g}{\partial \ell_1} \right](\vec{x}) + W_\phi \left[\frac{\partial g}{\partial \ell_2} \right](\vec{x})$$

$$= W_\phi \left[\frac{\partial g}{\partial \ell_1} + \frac{\partial g}{\partial \ell_2} \right](\vec{x}) .$$

Since both the function g and the variable $\vec{x}$ are arbitrary, we obtain:

$$\frac{\partial}{\partial x_1} \circ W_\phi = W_\phi \circ \left(\frac{\partial}{\partial \ell_1} + \frac{\partial}{\partial \ell_2} \right) .$$

Similar computations are made for the other variables. Summarizing, we get:

$$\frac{\partial}{\partial x_1} = W_\phi \circ \left(\frac{\partial}{\partial \ell_1} + \frac{\partial}{\partial \ell_2} \right) \circ W_\phi^{-1} ,$$

$$\frac{\partial}{\partial y_1} = W_\phi \circ \left(\frac{\partial}{\partial m_1} + \frac{\partial}{\partial m_2} \right) \circ W_\phi^{-1} ,$$

$$\frac{\partial}{\partial x_2} = W_\phi \circ \left(\frac{\partial}{\partial m_2} - \frac{\partial}{\partial m_1} \right) \circ W_\phi^{-1} ,$$

$$\frac{\partial}{\partial y_2} = W_\phi \circ \left(\frac{\partial}{\partial \ell_1} - \frac{\partial}{\partial \ell_2} \right) \circ W_\phi^{-1} . \tag{7.71}$$

Briefly, we have established what the partial derivatives with respect to the canonical cartesian coordinates turn out to be when one makes the change of variables passing to the idempotent coordinates.

Considering the usual gradient operators in $\mathbb{R}^4_{\bar{x}}$ and $\mathbb{R}^4_{\bar{\ell}}$,

$$\nabla_{\bar{x}} = \left(\frac{\partial}{\partial x_1}, \frac{\partial}{\partial y_1}, \frac{\partial}{\partial x_2}, \frac{\partial}{\partial y_2}\right) \quad \text{and} \quad \nabla_{\bar{\ell}} = \left(\frac{\partial}{\partial \ell_1}, \frac{\partial}{\partial m_1}, \frac{\partial}{\partial \ell_2}, \frac{\partial}{\partial m_2}\right)$$

respectively, we rewrite the formulas above in the following compressed form:

$$\nabla_{\bar{x}} = W_\phi \circ \left(\nabla_{\bar{\ell}} \cdot J_{\bar{x}}[\phi]\right) \circ W_\phi^{-1}. \tag{7.72}$$

If F is a bicomplex function, seen as a mapping $F = (f_{11}, f_{12}, f_{21}, f_{22})$ from $\mathbb{R}^4_{\bar{x}}$ to $\mathbb{R}^4_{\bar{x}}$, then we get a relation between the Jacobi matrices of F in the two coordinates systems considered:

$$J_{\bar{x}}[F] = J_{\bar{\ell}}\left[W_\phi^{-1}[F]\right] \cdot J_{\bar{x}}[\phi]. \tag{7.73}$$

It is useful to have explicitly the reciprocal relations. Let us note that from (7.71) we obtain:

$$W_\phi^{-1} \circ \frac{\partial}{\partial x_1} \circ W_\phi = \frac{\partial}{\partial \ell_1} + \frac{\partial}{\partial \ell_2},$$

$$W_\phi^{-1} \circ \frac{\partial}{\partial y_2} \circ W_\phi = \frac{\partial}{\partial \ell_1} - \frac{\partial}{\partial \ell_2},$$

which leads to

$$\frac{\partial}{\partial \ell_1} = W_\phi^{-1} \circ \frac{1}{2}\left(\frac{\partial}{\partial x_1} + \frac{\partial}{\partial y_2}\right) \circ W_\phi, \tag{7.74}$$

$$\frac{\partial}{\partial \ell_2} = W_\phi^{-1} \circ \frac{1}{2}\left(\frac{\partial}{\partial x_1} - \frac{\partial}{\partial y_2}\right) \circ W_\phi. \tag{7.75}$$

Similarly, we have:

$$\frac{\partial}{\partial m_1} = W_\phi^{-1} \circ \frac{1}{2}\left(\frac{\partial}{\partial y_1} - \frac{\partial}{\partial x_2}\right) \circ W_\phi,$$

$$\frac{\partial}{\partial m_2} = W_\phi^{-1} \circ \frac{1}{2}\left(\frac{\partial}{\partial y_1} + \frac{\partial}{\partial x_2}\right) \circ W_\phi. \tag{7.76}$$

We rewrite the formulas above at the level of gradients:

$$\nabla_{\bar{\ell}} = W_\phi^{-1} \circ \left(\nabla_{\bar{x}} \cdot J_{\bar{\ell}}[\phi^{-1}]\right) \circ W_\phi. \tag{7.77}$$

Now for a bicomplex function G, seen again as a mapping from $\mathbb{R}^4_{\bar{\ell}}$ to $\mathbb{R}^4_{\bar{\ell}}$, we get:

$$J_{\bar{\ell}}[G] = J_{\bar{x}}\left[W_\phi[G]\right] \cdot J_{\bar{\ell}}[\phi^{-1}]. \tag{7.78}$$

The reader should immediately notice that the formula above is consistent with (7.73) for $F = W_\phi[G]$ and using the fact that $J_{\bar{x}}[\phi]$ and $J_{\vec{\ell}}[\phi^{-1}]$ are matrices inverse to each other.

In conclusion: if we want to translate a differential expression written in the cartesian coordinates (x_1, y_1, x_2, y_2) into an expression with the idempotent coordinates $(\ell_1, m_1, \ell_2, m_2)$, then we use formulas (7.71); if we start with an expression given in the idempotent coordinates, then (7.74), (7.75) and (7.76) give its equivalent in the cartesian coordinates. In other words, we have obtained direct relations between the differential operators acting on two copies of $\mathbb{R}^4$, one copy with the cartesian coordinates and another with the idempotent coordinates.

The next step is to extend the ideas above onto the $\mathbb{C}(\mathbf{i})$-complex differential operators (7.40) in the complex variables z_1 and z_2. Writing now the $\mathbb{C}(\mathbf{i})$-complex idempotent coordinates $\beta_1 := \ell_1 + \mathbf{i}m_1$ and $\beta_2 := \ell_2 + \mathbf{i}m_2$, the usual associated $\mathbb{C}(\mathbf{i})$-complex differential operators are:

$$\frac{\partial}{\partial \beta_1} := \frac{1}{2}\left(\frac{\partial}{\partial \ell_1} - \mathbf{i}\frac{\partial}{\partial m_1}\right), \qquad \frac{\partial}{\partial \beta_2} := \frac{1}{2}\left(\frac{\partial}{\partial \ell_2} - \mathbf{i}\frac{\partial}{\partial m_2}\right),$$
$$\frac{\partial}{\partial \bar{\beta}_1} := \frac{1}{2}\left(\frac{\partial}{\partial \ell_1} + \mathbf{i}\frac{\partial}{\partial m_1}\right), \qquad \frac{\partial}{\partial \bar{\beta}_2} := \frac{1}{2}\left(\frac{\partial}{\partial \ell_2} + \mathbf{i}\frac{\partial}{\partial m_2}\right). \tag{7.79}$$

Introducing the $\mathbb{C}(\mathbf{i})$-complex gradient operator

$$\nabla_{\vec{z}} := \left(\frac{\partial}{\partial z_1}, \frac{\partial}{\partial \bar{z}_1}, \frac{\partial}{\partial z_2}, \frac{\partial}{\partial \bar{z}_2}\right),$$

and similarly $\nabla_{\vec{\beta}}$, we summarize these formulas as follows:

$$\nabla_{\vec{x}} = \nabla_{\vec{z}} \cdot M, \qquad \nabla_{\vec{z}} = \nabla_{\vec{x}} \cdot M^{-1},$$
$$\nabla_{\vec{\ell}} = \nabla_{\vec{\beta}} \cdot M, \qquad \nabla_{\vec{\beta}} = \nabla_{\vec{\ell}} \cdot M^{-1}, \tag{7.80}$$

where

$$M := \begin{pmatrix} 1 & \mathbf{i} & 0 & 0 \\ 1 & -\mathbf{i} & 0 & 0 \\ 0 & 0 & 1 & \mathbf{i} \\ 0 & 0 & 1 & -\mathbf{i} \end{pmatrix}, \qquad M^{-1} := \frac{1}{2}\begin{pmatrix} 1 & 1 & 0 & 0 \\ -\mathbf{i} & \mathbf{i} & 0 & 0 \\ 0 & 0 & 1 & 1 \\ 0 & 0 & -\mathbf{i} & \mathbf{i} \end{pmatrix}.$$

Now, if we want to derive the formulas relating, say, the differential operators in $\vec{z}$ with the ones in $\vec{\ell}$, we just have to combine the relations (7.80) above with the formulas (7.72) and (7.77):

$$\nabla_{\vec{z}} = \nabla_{\vec{x}} \cdot M^{-1} = \left(W_\phi \circ (\nabla_{\vec{\ell}} \cdot J_{\bar{x}}[\phi]) \circ W_\phi^{-1}\right) \cdot M^{-1}$$
$$= W_\phi \circ \left(\nabla_{\vec{\ell}} \cdot J_{\bar{x}}[\phi] \cdot M^{-1}\right) \circ W_\phi^{-1}. \tag{7.81}$$

A simple matrix multiplication yields the matrix $J_{\bar{x}}[\phi]^t \cdot M^{-1}$ to be equal to

$$
\frac{1}{2}\begin{pmatrix}
1 & 1 & -\mathbf{i} & \mathbf{i} \\
-\mathbf{i} & \mathbf{i} & -1 & -1 \\
1 & 1 & \mathbf{i} & -\mathbf{i} \\
-\mathbf{i} & \mathbf{i} & 1 & 1
\end{pmatrix}.
$$

From this we can deduce, for example, the formula:

$$
\frac{\partial}{\partial z_1} = W_\phi \circ \frac{1}{2}\left(\frac{\partial}{\partial \ell_1} - \mathbf{i}\frac{\partial}{\partial m_1} + \frac{\partial}{\partial \ell_2} - \mathbf{i}\frac{\partial}{\partial m_2} \right) \circ W_\phi^{-1}.
$$

If we want now to express the operators in $\vec{z}$ in terms of the ones in $\vec{\beta}$, we incorporate in (7.81) the relation between the gradient in $\vec{\ell}$ and the one in $\vec{\beta}$:

$$
\nabla_{\vec{z}} = W_\phi \circ \left(\nabla_{\vec{\ell}} \cdot J_{\bar{x}}[\phi] \cdot M^{-1} \right) \circ W_\phi^{-1}
$$
$$
= W_\phi \circ \left(\nabla_{\vec{\beta}} \cdot M \cdot J_{\bar{x}}[\phi] \cdot M^{-1} \right) \circ W_\phi^{-1}.
$$

   To get a hands-on experience with the computations above, we could also write directly from (7.71):

$$
\frac{\partial}{\partial z_1} = \frac{1}{2}\left(\frac{\partial}{\partial x_1} - \mathbf{i}\frac{\partial}{\partial y_1} \right)
$$
$$
= W_\phi \circ \frac{1}{2}\left(\frac{\partial}{\partial \ell_1} + \frac{\partial}{\partial \ell_2} - \mathbf{i}\frac{\partial}{\partial m_1} - \mathbf{i}\frac{\partial}{\partial m_2} \right) \circ W_\phi^{-1}
$$
$$
= W_\phi \circ \left(\frac{\partial}{\partial \beta_1} + \frac{\partial}{\partial \beta_2} \right) \circ W_\phi^{-1}. \tag{7.82}
$$

Similarly, we obtain the following formulas:

$$
\frac{\partial}{\partial z_2} = W_\phi \circ (-\mathbf{i})\left(\frac{\partial}{\partial \beta_1} - \frac{\partial}{\partial \beta_2} \right) \circ W_\phi^{-1},
$$
$$
\frac{\partial}{\partial \bar{z}_1} = W_\phi \circ \left(\frac{\partial}{\partial \bar{\beta}_1} + \frac{\partial}{\partial \bar{\beta}_2} \right) \circ W_\phi^{-1},
$$
$$
\frac{\partial}{\partial \bar{z}_2} = W_\phi \circ \mathbf{i}\left(\frac{\partial}{\partial \bar{\beta}_1} - \frac{\partial}{\partial \bar{\beta}_2} \right) \circ W_\phi^{-1}. \tag{7.83}
$$

   A similar computation leads to the reciprocal transformations:

$$
\frac{\partial}{\partial \beta_1} = W_\phi^{-1} \circ \frac{1}{2}\left(\frac{\partial}{\partial z_1} + \mathbf{i}\frac{\partial}{\partial z_2} \right) \circ W_\phi,
$$
$$
\frac{\partial}{\partial \beta_2} = W_\phi^{-1} \circ \frac{1}{2}\left(\frac{\partial}{\partial z_1} - \mathbf{i}\frac{\partial}{\partial z_2} \right) \circ W_\phi,
$$

$$\frac{\partial}{\partial \overline{\beta}_1} = W_\phi^{-1} \circ \frac{1}{2} \left(\frac{\partial}{\partial \overline{z}_1} - \mathbf{i} \frac{\partial}{\partial \overline{z}_2} \right) \circ W_\phi \,,$$

$$\frac{\partial}{\partial \overline{\beta}_2} = W_\phi^{-1} \circ \frac{1}{2} \left(\frac{\partial}{\partial \overline{z}_1} + \mathbf{i} \frac{\partial}{\partial \overline{z}_2} \right) \circ W_\phi \,. \tag{7.84}$$

We can write the formulas above in terms of matrix multiplication:

$$\left(\frac{\partial}{\partial z_1}, \frac{\partial}{\partial \overline{z}_1}, \frac{\partial}{\partial z_2}, \frac{\partial}{\partial \overline{z}_2} \right) = \begin{pmatrix} 1 & 0 & -\mathbf{i} & 0 \\ 0 & 1 & 0 & \mathbf{i} \\ 1 & 0 & \mathbf{i} & 0 \\ 0 & 1 & 0 & -\mathbf{i} \end{pmatrix} \cdot \left(\frac{\partial}{\partial \beta_1}, \frac{\partial}{\partial \overline{\beta}_1}, \frac{\partial}{\partial \beta_2}, \frac{\partial}{\partial \overline{\beta}_2} \right),$$

where the 4×4 matrix above is $M \cdot J_{\overline{x}}[\phi] \cdot M^{-1}$, as expected. The reciprocal transformation is given by

$$\left(\frac{\partial}{\partial \beta_1}, \frac{\partial}{\partial \overline{\beta}_1}, \frac{\partial}{\partial \beta_2}, \frac{\partial}{\partial \overline{\beta}_2} \right) = \frac{1}{2} \begin{pmatrix} 1 & 0 & 1 & 0 \\ 0 & 1 & 0 & 1 \\ \mathbf{i} & 0 & -\mathbf{i} & 0 \\ 0 & -\mathbf{i} & 0 & \mathbf{i} \end{pmatrix} \cdot \left(\frac{\partial}{\partial z_1}, \frac{\partial}{\partial \overline{z}_1}, \frac{\partial}{\partial z_2}, \frac{\partial}{\partial \overline{z}_2} \right),$$

where the 4×4 matrix above is $M \cdot J_{\overline{\ell}}[\phi^{-1}] \cdot M^{-1}$.

In conclusion: we have obtained direct relations between the complex differential operators acting on two copies of $\mathbb{C}^2(\mathbf{i})$, one is with cartesian coordinates (z_1, z_2) and another is with idempotent coordinates (β_1, β_2).

Remark 7.1. *Note that we could have started these calculations from the* $\mathbb{C}(\mathbf{j})$*-idempotent representation of bicomplex numbers,* $Z = \gamma_1 \mathbf{e} + \gamma_2 \mathbf{e}^\dagger$*, where* $\gamma_1, \gamma_2 \in \mathbb{C}(\mathbf{j})$ *are given by*

$$\gamma_1 = \ell_1 + \mathbf{j}(-m_1), \qquad \gamma_2 = \ell_2 + \mathbf{j} m_2 \,.$$

Notice that the computations will not *yield the same formulas, since one has to start with a different linear isomorphism* $\widetilde{\phi}$*, for which the defining matrix is obtained from* (7.65) *by multiplying the second column by* (-1)*. For example, the formulas for the bicomplex differential operators in terms of the* $\mathbb{C}(\mathbf{j})$*-complex idempotent operators are:*

$$\frac{\partial}{\partial Z} = W_\phi \circ \left(\frac{\partial}{\partial \gamma_1} \mathbf{e} + \frac{\partial}{\partial \gamma_2} \mathbf{e}^\dagger \right) \circ W_\phi^{-1} \,,$$

$$\frac{\partial}{\partial Z^\dagger} = W_\phi \circ \left(\frac{\partial}{\partial \gamma_1^*} \mathbf{e} + \frac{\partial}{\partial \gamma_2^*} \mathbf{e}^\dagger \right) \circ W_\phi^{-1} \,,$$

$$\frac{\partial}{\partial \overline{Z}} = W_\phi \circ \left(\frac{\partial}{\partial \gamma_2} \mathbf{e} + \frac{\partial}{\partial \gamma_1} \mathbf{e}^\dagger \right) \circ W_\phi^{-1} \,,$$

$$\frac{\partial}{\partial Z^*} = W_\phi \circ \left(\frac{\partial}{\partial \gamma_2^*} \mathbf{e} + \frac{\partial}{\partial \gamma_1^*} \mathbf{e}^\dagger \right) \circ W_\phi^{-1} \,.$$

We leave it as an exercise to the reader to develop in full detail the above computations in the language of $\mathbb{C}(\mathbf{j})$-complex differential operators.

Let us turn now to the bicomplex differential operators (bicomplex analogues of the complex conjugate Cauchy–Riemann operators). They were defined by (7.47) and we have noted several times that they do not depend on the representation of the bicomplex variable Z. Let us comment on this in more detail.

Introduce the bicomplex gradient operator

$$\nabla_Z := \left(\frac{\partial}{\partial Z}, \frac{\partial}{\partial Z^\dagger}, \frac{\partial}{\partial \overline{Z}}, \frac{\partial}{\partial Z^*} \right) .$$

In standard real coordinates $\vec{x}$, we have: $\nabla_Z = \nabla_{\vec{x}} \cdot T$, where

$$T := \frac{1}{4} \begin{pmatrix} 1 & 1 & 1 & 1 \\ -\mathbf{i} & -\mathbf{i} & \mathbf{i} & \mathbf{i} \\ -\mathbf{j} & \mathbf{j} & -\mathbf{j} & \mathbf{j} \\ \mathbf{k} & -\mathbf{k} & -\mathbf{k} & \mathbf{k} \end{pmatrix} . \tag{7.85}$$

If we want to write ∇_Z in terms of $\vec{z}$ differential operators, we simply combine the formulas for the gradients in question:

$$\nabla_Z = \nabla_{\vec{x}} \cdot T = \nabla_{\vec{z}} \cdot M \cdot T ,$$

where

$$M \cdot T = \frac{1}{2} \begin{pmatrix} 1 & 1 & 0 & 0 \\ 0 & 0 & 1 & 1 \\ -\mathbf{j} & \mathbf{j} & 0 & 0 \\ 0 & 0 & -\mathbf{j} & \mathbf{j} \end{pmatrix} , \tag{7.86}$$

i.e., we get exactly formulas (7.47) which express the bicomplex differential operators in terms of the complex differential operators in z_1 and z_2.

If we want to obtain ∇_Z in the idempotent coordinates $\vec{\beta}$, we compute:

$$\nabla_Z = \nabla_{\vec{x}} \cdot T = \left(W_\phi \circ \left(\nabla_{\vec{\ell}} \cdot J_{\vec{x}}[\phi] \right) \circ W_\phi^{-1} \right) \cdot T$$

$$= W_\phi \circ \left(\nabla_{\vec{\ell}} \cdot J_{\vec{x}}[\phi] \cdot T \right) \circ W_\phi^{-1} ,$$

where the matrix $J_{\vec{x}}[\phi] \cdot T$ is given by

$$J_{\vec{x}}[\phi] \cdot T = \frac{1}{4} \begin{pmatrix} 1+\mathbf{k} & 1-\mathbf{k} & 1-\mathbf{k} & 1+\mathbf{k} \\ -\mathbf{i}+\mathbf{j} & -\mathbf{i}-\mathbf{j} & \mathbf{i}+\mathbf{j} & \mathbf{i}-\mathbf{j} \\ 1-\mathbf{k} & 1+\mathbf{k} & 1+\mathbf{k} & 1-\mathbf{k} \\ -\mathbf{i}-\mathbf{j} & -\mathbf{i}+\mathbf{j} & \mathbf{i}-\mathbf{j} & \mathbf{i}+\mathbf{j} \end{pmatrix} . \tag{7.87}$$

Since

$$\mathbf{e} = \frac{1+\mathbf{k}}{2}, \qquad \mathbf{e}^\dagger = \frac{1-\mathbf{k}}{2},$$

$$\mathbf{i}+\mathbf{j} = 2\mathbf{i}\mathbf{e}^\dagger, \qquad \mathbf{i}-\mathbf{j} = 2\mathbf{i}\mathbf{e},$$

we obtain:

$$J_{\vec{x}}[\phi] \cdot T = \frac{1}{2} \begin{pmatrix} \mathbf{e} & \mathbf{e}^\dagger & \mathbf{e}^\dagger & \mathbf{e} \\ -\mathbf{ie} & -\mathbf{ie}^\dagger & \mathbf{ie}^\dagger & \mathbf{ie} \\ \mathbf{e}^\dagger & \mathbf{e} & \mathbf{e} & \mathbf{e}^\dagger \\ -\mathbf{ie}^\dagger & -\mathbf{ie} & \mathbf{ie} & \mathbf{ie}^\dagger \end{pmatrix}. \tag{7.88}$$

For example, if we want to write $\dfrac{\partial}{\partial Z}$ in the real idempotent coordinates, we get from above:

$$\frac{\partial}{\partial Z} = W_\phi \circ \frac{1}{2} \left(\frac{\partial}{\partial \ell_1} \mathbf{e} - \mathbf{i} \frac{\partial}{\partial m_1} \mathbf{e} + \frac{\partial}{\partial \ell_2} \mathbf{e}^\dagger - \mathbf{i} \frac{\partial}{\partial m_2} \mathbf{e}^\dagger \right) \circ W_\phi^{-1}.$$

Furthermore, writing ∇_Z in $\vec{\beta}$ coordinates we obtain

$$\nabla_Z = W_\phi \circ \left(\nabla_{\vec{\beta}} \cdot M \cdot J_{\vec{x}}[\phi] \cdot T \right) \circ W_\phi^{-1},$$

where the matrix $M \cdot J_{\vec{x}}[\phi] \cdot T$ is given by

$$\begin{pmatrix} \mathbf{e} & \mathbf{e}^\dagger & 0 & 0 \\ 0 & 0 & \mathbf{e}^\dagger & \mathbf{e} \\ \mathbf{e}^\dagger & \mathbf{e} & 0 & 0 \\ 0 & 0 & \mathbf{e} & \mathbf{e}^\dagger \end{pmatrix}. \tag{7.89}$$

Consider a bicomplex holomorphic function F, then

$$\nabla_Z[F] = \left(\frac{\partial F}{\partial Z}, 0, 0, 0 \right).$$

Writing F in the idempotent form $F = G_1 \mathbf{e} + G_2 \mathbf{e}^\dagger$, we get:

$$\frac{\partial F}{\partial Z} = \frac{\partial F}{\partial \beta_1} \mathbf{e} + \frac{\partial F}{\partial \beta_2} \mathbf{e}^\dagger = \frac{\partial G_1}{\partial \beta_1} \mathbf{e} + \frac{\partial G_2}{\partial \beta_2} \mathbf{e}^\dagger,$$

$$0 = \frac{\partial F}{\partial \beta_1} \mathbf{e}^\dagger + \frac{\partial F}{\partial \beta_2} \mathbf{e} = \frac{\partial G_1}{\partial \beta_2} \mathbf{e} + \frac{\partial G_2}{\partial \beta_1} \mathbf{e}^\dagger,$$

$$0 = \frac{\partial F}{\partial \overline{\beta}_1} \mathbf{e}^\dagger + \frac{\partial F}{\partial \overline{\beta}_2} \mathbf{e} = \frac{\partial G_1}{\partial \overline{\beta}_2} \mathbf{e} + \frac{\partial G_2}{\partial \overline{\beta}_1} \mathbf{e}^\dagger,$$

$$0 = \frac{\partial F}{\partial \overline{\beta}_1} \mathbf{e} + \frac{\partial F}{\partial \overline{\beta}_2} \mathbf{e}^\dagger = \frac{\partial G_1}{\partial \overline{\beta}_1} \mathbf{e} + \frac{\partial G_2}{\partial \overline{\beta}_2} \mathbf{e}^\dagger,$$

which yields once more the conclusion of Theorem 7.6.3.

Remark 7.7.2. *Under the same hypothesis that* $F = f_{11} + \mathbf{i} f_{12} + \mathbf{j} f_{21} + \mathbf{k} f_{22}$ *is a bicomplex holomorphic function written in the standard coordinates* $\vec{x}$, *the derivative of* F *is given by*

$$\frac{\partial F}{\partial Z} = \frac{\partial F}{\partial x_1} = a + b\mathbf{i} + c\mathbf{j} + d\mathbf{k},$$

where a, b, c, d are real partial derivatives of the functions $f_{k\ell}$ related by (7.18). Written in the idempotent coordinates $\vec{l}$, the expression above becomes

$$W_\phi^{-1}\left[\frac{\partial F}{\partial Z}\right] = \frac{1}{2}\left(\frac{\partial \hat{F}}{\partial \ell_1}\mathbf{e} - \mathbf{i}\frac{\partial \hat{F}}{\partial m_1}\mathbf{e} + \frac{\partial \hat{F}}{\partial \ell_2}\mathbf{e}^\dagger - \mathbf{i}\frac{\partial \hat{F}}{\partial m_2}\mathbf{e}^\dagger\right),$$

where $\hat{F} = W_\phi^{-1}[F]$.

Remark 7.7.3. *Let us analyze in more detail the crucial difference between the complex differential operators in the $\vec{z}$ and $\vec{\beta}$ variables. For example, if we look side-by-side at the formulas:*

$$\frac{\partial}{\partial z_1} = \frac{\partial}{\partial Z} + \frac{\partial}{\partial Z^\dagger},$$

$$\frac{\partial}{\partial \beta_1} = W_\phi^{-1} \circ \left(\frac{\partial}{\partial Z} \cdot \mathbf{e} + \frac{\partial}{\partial Z^\dagger} \cdot \mathbf{e}^\dagger\right) \circ W_\phi,$$

we notice immediately a huge difference between them: the first one mixes the one-dimensional bicomplex operators in Z and $Z^\dagger$, while the second one keeps them separate (!). For example, consider the bicomplex function $F(Z) = (Z - Z^\dagger)^2$; then

$$\frac{\partial F}{\partial z_1}(Z) = 0, \qquad but \qquad \frac{\partial F}{\partial \beta_1}(Z) = 2(\beta_1 - \beta_2),$$

for all $Z \in \mathbb{BC}$, thus F is constant with respect to z_1, but not with respect to β_1.

    This is tightly related, of course, with the fact that a $\mathbb{BC}$-holomorphic function when seen as a mapping from $\mathbb{C}^2$ to $\mathbb{C}^2$ with the cartesian coordinates is a pair of holomorphic functions which depend on two complex variables and which have a Cauchy–Riemann type relation, meanwhile the same function written in the idempotent coordinates becomes a pair of holomorphic functions in one variable which moreover are independent of each other.

Remark 7.7.4. *Let us see now the relation between bicomplex differential operators and the idempotent representation of hyperbolic numbers. Recalling the formula for $\dfrac{\partial}{\partial Z}$, we regroup the terms in a different way:*

$$\frac{\partial}{\partial Z} = W_\phi \circ \frac{1}{2}\left(\left(\frac{\partial}{\partial \ell_1} - \mathbf{i}\frac{\partial}{\partial m_1}\right)\mathbf{e} + \left(\frac{\partial}{\partial \ell_2} - \mathbf{i}\frac{\partial}{\partial m_2}\right)\mathbf{e}^\dagger\right) \circ W_\phi^{-1}$$

$$= W_\phi \circ \frac{1}{2}\left(\left(\frac{\partial}{\partial \ell_1}\mathbf{e} + \frac{\partial}{\partial \ell_2}\mathbf{e}^\dagger\right) - \mathbf{i}\left(\frac{\partial}{\partial m_1}\mathbf{e} + \frac{\partial}{\partial m_2}\mathbf{e}^\dagger\right)\right) \circ W_\phi^{-1}. \qquad (7.90)$$

But if we write $Z = \mathfrak{z}_1 + \mathbf{i}\mathfrak{z}_2$, where

$$\mathfrak{z}_1 = x_1 + \mathbf{k}y_2, \qquad \mathfrak{z}_2 = y_1 - \mathbf{k}x_2,$$

and if we use the intrinsic hyperbolic idempotent representations

$$\mathfrak{z}_1 = (x_1 + y_2)\mathbf{e} + (x_1 - y_2)\mathbf{e}^\dagger = \ell_1\mathbf{e} + \ell_2\mathbf{e}^\dagger,$$
$$\mathfrak{z}_2 = (y_1 - x_2)\mathbf{e} + (y_1 + x_2)\mathbf{e}^\dagger = m_1\mathbf{e} + m_2\mathbf{e}^\dagger,$$

then the hyperbolic differential operators have the expressions:

$$\frac{\partial}{\partial \mathfrak{z}_1} = \frac{1}{2}\left(\frac{\partial}{\partial x_1} + \mathbf{k}\frac{\partial}{\partial y_2}\right) = W_\phi \circ \left(\frac{\partial}{\partial \ell_1}\mathbf{e} + \frac{\partial}{\partial \ell_2}\mathbf{e}^\dagger\right) \circ W_\phi^{-1},$$
$$\frac{\partial}{\partial \mathfrak{z}_2} = \frac{1}{2}\left(\frac{\partial}{\partial y_1} - \mathbf{k}\frac{\partial}{\partial x_2}\right) = W_\phi \circ \left(\frac{\partial}{\partial m_1}\mathbf{e} + \frac{\partial}{\partial m_2}\mathbf{e}^\dagger\right) \circ W_\phi^{-1}.$$

Going back to formula (7.90), we obtain the following formulation of the bicomplex differential operator with respect to Z, in terms of the hyperbolic derivatives:

$$\frac{\partial}{\partial Z} = W_\phi \circ \frac{1}{2}\left(\frac{\partial}{\partial \mathfrak{z}_1} - \mathbf{i}\frac{\partial}{\partial \mathfrak{z}_2}\right) \circ W_\phi^{-1},$$

and similarly for the other operators.

Remark 7.7.5. *Analogous computations can be made for all bicomplex differential operators, starting with either one of the other writings in terms of the $\mathbb{C}(\mathbf{j})$ or hyperbolic operators. We leave the details for the reader, with the note that all formulas are consistent from all points of view. The plethora of formulas in different writings, coordinates or change of bases of $\mathbb{BC}$, all of which agree perfectly with each other, is again a phenomenon very specific to the bicomplex setup. Think of having multiple pairs of glasses which allow you to see a picture (or a movie) from different points of view, in different dimensions, distinguishing different colorings, etc.*

Chapter 8

Some Properties of Bicomplex Holomorphic Functions

In what follows, we investigate several properties of bicomplex holomorphic functions defined on a set Ω in $\mathbb{BC}$. Also we assume that Ω can be written in the form

$$\Omega = \Omega_1 \mathbf{e} + \Omega_2 \mathbf{e}^\dagger := \left\{ \beta_1 \mathbf{e} + \beta_2 \mathbf{e}^\dagger \,\middle|\, \beta_1 \in \Omega_1,\ \beta_2 \in \Omega_2 \right\}$$

with Ω_1 and Ω_2 domains, that is, open and connected sets, in $\mathbb{C}(\mathbf{i})$. We will call such Ω "a product-type domain" in $\mathbb{BC}$. Note that such an Ω should be seen as the cartesian product $\Omega_1 \times \Omega_2$ in the "idempotent $\mathbb{C}^2(\mathbf{i})$" (which is $\left\{ (\beta_1, \beta_2) \,\middle|\, \beta_1, \beta_2 \in \mathbb{C}(\mathbf{i}) \right\}$) where the elements are written as bicomplex numbers in their idempotent form. The topological boundary $\partial\Omega$ is the union of the three sets: $\partial\Omega_1 \times \Omega_2$, $\Omega_1 \times \partial\Omega_2$ and $\partial\Omega_1 \times \partial\Omega_2$. The latter is called, sometimes, the distinguished boundary, or Shilov boundary, of Ω.

8.1 Zeros of bicomplex holomorphic functions

We will study the zeros of bicomplex holomorphic functions. For this, let F be such a function defined on a domain $\Omega = \Omega_1 \mathbf{e} + \Omega_2 \mathbf{e}^\dagger$, so in its idempotent form

$$F(Z) = G_1(\beta_1)\mathbf{e} + G_2(\beta_2)\mathbf{e}^\dagger,$$

for all $Z = \beta_1 \mathbf{e} + \beta_2 \mathbf{e}^\dagger \in \Omega$. Assume first that neither G_1 nor G_2 are identically zero. Because G_1 and G_2 are complex holomorphic functions in β_1 and β_2, each has isolated zeros. This means that for $\ell = 1, 2$, if $G_\ell(\beta_{\ell,0}) = 0$, then there exists a disk B_ℓ of center $\beta_{\ell,0}$ and radius r_ℓ such that $G_\ell(\beta_\ell) \neq 0$, for all $\beta_\ell \in B_\ell \setminus \{\beta_{\ell,0}\}$. Then on the open set $B := B_1 \mathbf{e} + B_2 \mathbf{e}^\dagger$, the only zero of F is at $Z_0 := \beta_{1,0}\mathbf{e} + \beta_{2,0}\mathbf{e}^\dagger$. Therefore F has isolated zeros in the set $\Omega := \Omega_1 \mathbf{e} + \Omega_2 \mathbf{e}^\dagger$.

Now if, say, $G_2 \equiv 0$ but G_1 is not identically zero, then $F(Z) = G_1(\beta_1)\mathbf{e}$, so $F(Z) = 0$ if and only if $Z = \beta_{1,0}\mathbf{e} + \lambda\mathbf{e}^\dagger$, where $\beta_{1,0}$ is a zero of G_1 and λ is an arbitrary $\mathbb{C}(\mathbf{i})$-complex number. Therefore the set of zeros of F is a countable (or finite) union of portions of complex lines parallel to the zero-divisor line $L_{\mathbf{e}^\dagger} = \mathbb{B}\mathbb{C}_{\mathbf{e}^\dagger}$; in this case the complex line that passes through $\beta_{1,0}\mathbf{e}$ and which is parallel to $\mathbb{B}\mathbb{C}_{\mathbf{e}^\dagger}$ is of the form $\beta_{1,0}\mathbf{e} + \mathbb{B}\mathbb{C}_{\mathbf{e}^\dagger}$. A similar statement is true for $G_1 \equiv 0$.

Summarizing, we have just proved the following

Theorem 8.1.1. *Consider a bicomplex holomorphic function $F : \Omega \to \mathbb{B}\mathbb{C}$ which is not identically zero. Then*

1. *if F takes at least one invertible value, then its zero set is either the empty set or a set of isolated points in Ω;*

2. *if $F(Z) \in \mathbb{B}\mathbb{C}_{\mathbf{e}}$ for all $Z \in \Omega$, then its zero set is a countable (or finite) union of portions of complex lines parallel to the zero-divisor line $L_{\mathbf{e}^\dagger}$;*

3. *if $F(Z) \in \mathbb{B}\mathbb{C}_{\mathbf{e}^\dagger}$ for all $Z \in \Omega$, then its zero set is a countable (or finite) union of portions of complex lines parallel to the zero-divisor line $L_{\mathbf{e}}$.*

Example 8.1.2. Consider the bicomplex holomorphic function $F(Z) = Z^2 = f_1(Z) + \mathbf{j}f_2(Z)$, with $f_1(z_1, z_2) = 2z_1 z_2$ and $f_2(z_1, z_2) = z_2^2 - z_1^2$, which are defined and holomorphic on $\mathbb{C}^2(\mathbf{i})$. Their zero sets are:

$$V_1 = \{(z_1, z_2) \in \mathbb{C}^2(\mathbf{i}) \,|\, z_1 = 0\} \cup \{(z_1, z_2) \in \mathbb{C}^2(\mathbf{i}) \,|\, z_2 = 0\},$$
$$V_2 = \{(z_1, z_2) \in \mathbb{C}^2(\mathbf{i}) \,|\, z_1 = z_2\} \cup \{(z_1, z_2) \in \mathbb{C}^2(\mathbf{i}) \,|\, z_1 = -z_2\},$$

respectively. Each of them is a union of two complex lines in $\mathbb{C}^2(\mathbf{i})$, respectively. Note that the intersection $V_1 \cap V_2 = \{(0,0)\}$, which is just a point and which is the only zero of F.

A direct computation shows that using the idempotent variables $\beta_1 = z_1 - \mathbf{i}z_2$ and $\beta_2 = z_1 + \mathbf{i}z_2$ we get:

$$F(Z) = \beta_1^2 \mathbf{e} + \beta_2^2 \mathbf{e}^\dagger =: G_1(\beta_1)\mathbf{e} + G_2(\beta_2)\mathbf{e}^\dagger,$$

thus the functions G_1 and G_2, which are holomorphic on $\mathbb{C}(\mathbf{i})$ with respect to their corresponding variable, have only one zero: $\beta_1 = 0$ and $\beta_2 = 0$, respectively.

Example 8.1.3. Consider now the bicomplex exponential function $F(Z) = e^Z$, which is bicomplex holomorphic on $\mathbb{B}\mathbb{C}$ and takes only invertible values for all $Z \in \mathbb{B}\mathbb{C}$. The functions $f_1(z_1, z_2) = e^{z_1} \cos z_2$ and $f_2(z_1, z_2) = e^{z_1} \sin z_2$ have their zero sets

$$V_1 = \{(z_1, z_2) \in \mathbb{C}^2(\mathbf{i}) \,|\, \cos z_2 = 0\} = \left\{\left(z_1, \frac{(2n+1)\pi}{2}\right) \in \mathbb{C}^2(\mathbf{i}) \,|\, n \in \mathbb{Z}\right\},$$
$$V_2 = \{(z_1, z_2) \in \mathbb{C}^2(\mathbf{i}) \,|\, \sin z_2 = 0\} = \{(z_1, n\pi) \in \mathbb{C}^2(\mathbf{i}) \,|\, n \in \mathbb{Z}\},$$

which are obviously non-compact sets in $\mathbb{C}^2(\mathbf{i})$. But $V_1 \cap V_2 = \emptyset$, so F has no zeros, as we already know.

8.2 When bicomplex holomorphic functions reduce to constants

In this section we investigate some conditions under which a bicomplex holomorphic function reduces to a constant function.

Theorem 8.2.1. *Let $F : \Omega \to \mathbb{BC}$ be a bicomplex holomorphic function on a product-type domain Ω. Then $F'(Z) = 0$ for all $Z \in \Omega$ if and only if F is a constant function.*

Proof. If F is a constant function, its derivative is zero on Ω. Conversely, using the notation and result of Corollary 7.6.6, we are assured that

$$0 = F'(Z) = G_1'(\beta_1)\mathbf{e} + G_2'(\beta_2)\mathbf{e}^\dagger ,$$

for all $Z = \beta_1\mathbf{e} + \beta_2\mathbf{e}^\dagger \in \Omega$. This is equivalent to $G_1'(\beta_1) = 0$ on Ω_1 and $G_2'(\beta_2) = 0$ on Ω_2. Because Ω_1 and Ω_2 are domains in the complex plane, this implies that G_1, G_2 are constant functions with respect to the variables β_1, β_2, respectively, i.e., $G_1(\beta_1) = a_1 \in \mathbb{C}(\mathbf{i})$ for all $\beta_1 \in \Omega_1$ and $G_2(\beta_2) = a_2 \in \mathbb{C}(\mathbf{i})$ for all $\beta_2 \in \Omega_2$, therefore $F(Z) = a_1\mathbf{e} + a_2\mathbf{e}^\dagger$, a bicomplex constant for all $Z \in \Omega$. $\square$

Remark 8.2.2. *The proof of the theorem above could have been obtained without using the idempotent representation of F. For example, writing $F = f_1 + \mathbf{j}f_2$ on Ω, the existence of F' on Ω implies the existence of the complex partial derivatives of f_1 and f_2 on Ω with respect to z_1 and z_2, related by the Cauchy-Riemann type conditions (7.55). Using the fact $F'(Z) = \dfrac{\partial F}{\partial Z} = 0$, we can derive easily that the partial derivatives of f_1 and f_2 with respect to both z_1 and z_2 are zero, which implies that f_1 and f_2 are constant on the domain Ω, so F is constant on Ω.*

Now we are interested in examining under what conditions a bicomplex holomorphic function F has zero derivative on a domain Ω.

Theorem 8.2.3. *Let F be a bicomplex holomorphic function on a product-type simply connected domain Ω in $\mathbb{BC}$. Then F is a constant function if and only if any of the following equivalent conditions holds:*

1. *Writing $F = f_{11} + \mathbf{i}f_{12} + \mathbf{j}f_{21} + \mathbf{k}f_{22}$, one of the real functions $f_{k\ell}$ is constant. In particular, F can be "missing" one (or more) of the real components $f_{k\ell}$, i.e., $f_{k\ell} = 0$.*

2. *Writing $F = f_1 + \mathbf{j}f_2$, either f_1 or f_2 is a constant $\mathbb{C}(\mathbf{i})$-function. Similarly for $F = g_1 + \mathbf{k}g_2$. In particular, F can be either $\mathbb{C}(\mathbf{i})$-valued (i.e., $f_2 = g_2 = 0$), or $\mathbf{j} \cdot \mathbb{C}(\mathbf{i})$-valued (i.e. $f_1 = 0$), or $\mathbf{k} \cdot \mathbb{C}(\mathbf{i})$-valued (i.e., $g_1 = 0$).*

3. *Writing $F = \rho_1 + \mathbf{i}\rho_2$, either ρ_1 or ρ_2 is constant. Similarly for $F = \gamma_1 + \mathbf{k}\gamma_2$. In particular, F can be either $\mathbb{C}(\mathbf{j})$-valued (i.e., $\rho_2 = \gamma_2 = 0$), or $\mathbf{i} \cdot \mathbb{C}(\mathbf{j})$-valued (i.e., $\rho_1 = 0$), or $\mathbf{k} \cdot \mathbb{C}(\mathbf{j})$-valued (i.e., $\gamma_1 = 0$).*

4. *Writing $F = \mathfrak{f}_1 + \mathbf{i}\mathfrak{f}_2$, either $\mathfrak{f}_1$ or $\mathfrak{f}_2$ is constant. Similarly, for $F = \mathfrak{g}_1 + \mathbf{j}\mathfrak{g}_2$. In particular, F can be either $\mathbb{D}$-valued (i.e., $\mathfrak{f}_2 = \mathfrak{g}_2 = 0$), or $\mathbf{i} \cdot \mathbb{D}$-valued (i.e., $\mathfrak{f}_1 = 0$), or $\mathbf{j} \cdot \mathbb{D}$-valued (i.e., $\mathfrak{g}_1 = 0$).*

5. *$F(Z) = A \cdot f(Z)$, where $A \in \mathbb{BC}$ is an invertible bicomplex number and f is either a $\mathbb{C}(\mathbf{i})$, $\mathbb{C}(\mathbf{j})$, $\mathbb{D}$, $\mathbf{j}\mathbb{C}(\mathbf{i})$, $\mathbf{k}\mathbb{C}(\mathbf{i})$, $\mathbf{i}\mathbb{C}(\mathbf{j})$, $\mathbf{k}\mathbb{C}(\mathbf{j})$, $\mathbf{i}\mathbb{D}$ or $\mathbf{j}\mathbb{D}$-valued function. Observe that if $A \in \mathfrak{S}$, then both $F(Z)$ and $F'(Z)$ are in $\mathfrak{S}$.*

6. *The function F is of the form $F = f + \mathbf{j}\lambda f$ with a $\mathbb{C}(\mathbf{i})$-valued f and $\lambda \in \mathbb{C}(\mathbf{i})\backslash\{\pm i\}$. Similarly F can be of the form $F = \rho + \mathbf{i}(\mu\rho)$ with $\mu \in \mathbb{C}(\mathbf{j})\backslash\{\pm j\}$ and a $\mathbb{C}(\mathbf{j})$-valued ρ; or $F = \mathfrak{f} + \mathbf{i}(\mathfrak{a}\mathfrak{f})$ with $\mathfrak{a} \in \mathbb{D}$. Observe that if $\lambda = \pm\mathbf{i}$, then the values of $F(Z)$ and of $F'(Z)$ are in $\mathfrak{S}_0$.*

7. *Either one of the complex moduli $|F(Z)|_{\mathbf{i}}$ or $|F(Z)|_{\mathbf{j}}$ equals a non-zero constant (in $\mathbb{C}(\mathbf{i})$ or $\mathbb{C}(\mathbf{j})$) for all $Z \in \Omega$. Observe that either one of the complex moduli is zero for all Z if and only if $F(Z)$ and so $F'(Z)$ are in $\mathfrak{S}_0$.*

8. *The hyperbolic modulus $|F(Z)|_{\mathbf{k}}$ equals an invertible hyperbolic constant for all $Z \in \Omega$. Observe that $|F(Z)|_{\mathbf{k}}$ is a hyperbolic zero-divisor if and only if $F(Z)$ and so $F'(Z)$ are in $\mathfrak{S}_0$.*

9. *$F(Z)$ is invertible for all $Z \in \Omega$ and either one of the complex arguments $\arg_{\mathbf{i}}(F(Z))$ or $\arg_{\mathbf{j}}(F(Z))$ are constant for all Z.*

Proof. Assume the function f_{11} is constant on Ω. Using relations (7.18) it follows that $a = b = c = d = 0$, i.e., all the other real functions $f_{k\ell}$ have real partial derivatives zero with respect to all four real variables x_1, y_1, x_2 and y_2. Since Ω is a domain, it follows that all $f_{k\ell}$ are constant functions, thus F is a constant bicomplex function which provides a proof of (1).

The proofs of the next items (2)–(4) are easily reduced to (1). For example if f_2 is constant, then f_{21} and f_{22} are constant, so F is constant, etc. But let us further investigate the case $f_2(z_1, z_2) = \lambda \in \mathbb{C}(\mathbf{i})$ in the idempotent representation $F = G_1\mathbf{e} + G_2\mathbf{e}^\dagger$. We obtain for any $Z \in \Omega$:

$$G_1(Z) = f_1(Z) - \mathbf{i}\lambda, \qquad G_2(Z) = f_1(Z) + \mathbf{i}\lambda.$$

Since $F'(Z)$ exists on all Ω, then G_1 depends only on $\beta_1 = Z\mathbf{e}$ and G_2 depends only on $\beta_2 = Z\mathbf{e}^\dagger$, i.e.,

$$0 = \frac{\partial G_1}{\partial \beta_2}(Z), \qquad 0 = \frac{\partial G_2}{\partial \beta_1}(Z).$$

Recalling the relation between the partial derivatives with respect to z_1 and z_2 and those with respect to β_1 and β_2 we have:

$$\frac{\partial}{\partial \beta_2} = W_\phi \circ \left[\frac{1}{2}\left(\frac{\partial}{\partial z_1} - \mathbf{i}\frac{\partial}{\partial z_2}\right)\right] \circ W_\phi^{-1},$$

$$\frac{\partial}{\partial \beta_1} = W_\phi \circ \left[\frac{1}{2}\left(\frac{\partial}{\partial z_1} + \mathbf{i}\frac{\partial}{\partial z_2}\right)\right] \circ W_\phi^{-1},$$

which implies that

$$\frac{1}{2}\left(\frac{\partial f_1}{\partial z_1} - \mathbf{i}\frac{\partial f_1}{\partial z_2}\right) = 0,$$

$$\frac{1}{2}\left(\frac{\partial f_1}{\partial z_1} + \mathbf{i}\frac{\partial f_1}{\partial z_2}\right) = 0,$$

hence the partial derivatives of f_1 with respect to z_1 and z_2 are zero, so f_1 is a constant. Similar arguments hold for the other cases (2)–(4).

We use the same type of calculations for the proof of the next assertion: let $F(Z) = A \cdot f(Z)$, where f is either a $\mathbb{C}(\mathbf{i})$-, $\mathbb{C}(\mathbf{j})$-, $\mathbb{D}$-, etc. valued function, and $A = a_1\mathbf{e} + a_2\mathbf{e}^\dagger$, say, with $a_1, a_2 \in \mathbb{C}(\mathbf{i})$, $a_1 a_2 \neq 0$. Then

$$F = (a_1 f)\mathbf{e} + (a_2 f)\mathbf{e}^\dagger =: G_1\mathbf{e} + G_2\mathbf{e}^\dagger.$$

Again, because $F'(Z)$ exists in Ω, then $\dfrac{\partial G_1}{\partial \beta_2}(Z) = \dfrac{\partial G_2}{\partial \beta_1}(Z) = 0$, which will force the partial derivatives of f with respect to the $\mathbb{C}(\mathbf{i})$-, $\mathbb{C}(\mathbf{j})$- or hyperbolic variables to be zero, i.e., f is a constant. Therefore F is constant.

Note that (5) can also be proved directly, using the fact that if F is derivable on Ω, then multiplying it by any bicomplex constant (in our case A^{-1}) will yield a bicomplex derivable function $G(Z)$. Therefore, if $F(Z) = A \cdot f(Z)$, then $G(Z) = f(Z)$ is a complex, hyperbolic, etc. valued $\mathbb{BC}$-holomorphic function, so it is constant.

The proofs of the case when, for example, $F = f + \mathbf{j}(\lambda f)$ is easily obtained from the previous one: we write $F = (1 + \mathbf{j}\lambda)f$, and note that $A := 1 + \mathbf{j}\lambda$ is invertible if and only if $\lambda \neq \pm\mathbf{i}$. We note also that a direct proof of this fact is obtained using the Cauchy-Riemann conditions for $f_1 = f$ and $f_2 = \lambda f$: at any Z we obtain

$$\frac{\partial f}{\partial z_1} = \frac{\partial f_1}{\partial z_1} = \frac{\partial f_2}{\partial z_2} = \lambda\frac{\partial f}{\partial z_2},$$

$$\frac{\partial f}{\partial z_2} = \frac{\partial f_1}{\partial z_2} = -\frac{\partial f_2}{\partial z_1} = -\lambda\frac{\partial f}{\partial z_1},$$

from which we get:

$$(1 + \lambda^2)\frac{\partial f}{\partial z_1} = (1 + \lambda^2)\frac{\partial f}{\partial z_2} = 0,$$

so if $\lambda \neq \pm\mathbf{i}$, then f is constant, and so on.

Assume now that $|F(Z)|_\mathbf{i} = a \in \mathbb{C}(\mathbf{i})$ for any $Z \in \Omega$. In the idempotent representation, we have:

$$a^2 = |F(Z)|_\mathbf{i}^2 = G_1(\beta_1) \cdot G_2(\beta_2),$$

for all $Z \in \Omega$. If $a = 0$, then we get that $G_1(\beta_1) = 0$ or $G_2(\beta_2) = 0$ for all $Z - \beta_1\mathbf{e} + \beta_2\mathbf{e}^\dagger$. Since G_1 and G_2 are complex holomorphic functions on Ω_1, respectively Ω_2, domains in $\mathbb{C}(\mathbf{i})$, their zeros are isolated, unless they are identically zero. Thus, it follows that either $G_1 \equiv 0$ on Ω_1 or $G_2 \equiv 0$ on Ω_2. In either case, the result is that $F(Z) \in \mathfrak{S}_0$ for all $Z \in \Omega$.

If $a \neq 0$, then the following argument applies: the partial derivatives with respect to both β_1 and β_2 of the product $G_1 \cdot G_2$ are zero:

$$0 = \frac{\partial(a^2)}{\partial \beta_1} = \frac{\partial(G_1(\beta_1) \cdot G_2(\beta_2))}{\partial \beta_1} = G_1'(\beta_1) \cdot G_2(\beta_2),$$

$$0 = \frac{\partial(a^2)}{\partial \beta_2} = \frac{\partial(G_1(\beta_1) \cdot G_2(\beta_2))}{\partial \beta_2} = G_1(\beta_1) \cdot G_2'(\beta_2),$$

for all $\beta_1 \in \Omega_1$ and $\beta_2 \in \Omega_2$, so in the case that neither G_1 nor G_2 are identically zero, then both G_1 and G_2 are constant functions on their domains. Therefore, F is constant. Similarly in the $\mathbb{C}(\mathbf{j})$ case.

A special computation is necessary when the hyperbolic modulus of F equals a hyperbolic constant $\mathfrak{a} = a_1 + \mathbf{k}a_2$, where now $a_1, a_2 \in \mathbb{R}$. Let $F = \mathfrak{f}_1 + \mathbf{i}\mathfrak{f}_2$, and write each hyperbolic function $\mathfrak{f}_1 = s_1\mathbf{e} + t_1\mathbf{e}^\dagger$ and $\mathfrak{f}_2 = s_2\mathbf{e} + t_2\mathbf{e}^\dagger$ in idempotent components (in the intrinsic hyperbolic writing), where s_1, s_2, t_1, t_2 are real-valued functions of the bicomplex variable Z. Then

$$F = (s_1\mathbf{e} + t_1\mathbf{e}^\dagger) + \mathbf{i}(s_2\mathbf{e} + t_2\mathbf{e}^\dagger) = (s_1 + \mathbf{i}s_2)\mathbf{e} + (t_1 + \mathbf{i}t_2)\mathbf{e}^\dagger =: G_1\mathbf{e} + G_2\mathbf{e}^\dagger.$$

Then the square of the hyperbolic modulus of F is given by

$$a_1^2\mathbf{e} + a_2^2\mathbf{e}^\dagger = \mathfrak{a}^2 = |F(Z)|_{\mathbf{k}}^2 = (s_1^2(Z) + s_2^2(Z))\mathbf{e} + (t_1^2(Z) + t_2^2(Z))\mathbf{e}^\dagger,$$

therefore

$$a_1^2 = s_1^2(Z) + s_2^2(Z) = G_1(Z) \cdot \overline{G_1(Z)}$$

and

$$a_2^2 = t_1^2(Z) + t_2^2(Z) = G_2(Z) \cdot \overline{G_2(Z)}$$

for all Z. Because s_1, s_2, t_1 and t_2 are real-valued functions, then $a_1 = 0$ or $a_2 = 0$ is equivalent to $s_1 = s_2 \equiv 0$ or $t_1 = t_2 \equiv 0$, respectively, which is equivalent to $F(Z)$ being a zero-divisor for all Z.

The function $G_1(Z)$ is a complex holomorphic function which depends on β_1 only: $G_1 = G_1(\beta_1)$, thus the function $\overline{G_1}$ is anti-holomorphic in β_1, so that in the identity $a_1^2 = G_1(\beta_1) \cdot \overline{G_1(\beta_1)}$ we cannot take the derivatives of both sides but we can apply the Cauchy–Riemann operator $\dfrac{\partial}{\partial \beta_1}$ which gives:

$$0 = \frac{\partial(G_1 \cdot \overline{G_1})}{\partial \beta_1}(\beta_1) = \frac{\partial G_1}{\partial \beta_1}(\beta_1) \cdot \overline{G_1(\beta_1)} = G_1'(\beta_1) \cdot \overline{G_1(\beta_1)},$$

where we used the fact that $\dfrac{\partial \overline{G_1}}{\partial \beta_1} = \dfrac{\overline{\partial G_1}}{\partial \overline{\beta_1}} = 0$. Since this happens for all Z, so for all $\beta_1 \in \Omega_1$, it follows that either $\overline{G_1} \equiv 0$ so $G_1 \equiv 0$, or that G_1 is constant. An analogous argument holds for G_2. Note that $G_1 \equiv 0$ if and only if $s_1 = s_2 = 0$, so $F(Z) \in \mathfrak{S}_0$ for all Z, and similarly for G_2. Otherwise, it follows that F is constant.

Consider now $F(Z)$ to be an invertible bicomplex number for all $Z \in \Omega$. Assume that, in its $\mathbb{C}(\mathbf{i})$-complex trigonometric form

$$F(Z) = |F(Z)|_{\mathbf{i}} e^{\mathbf{j}\Theta} = |F(Z)|_{\mathbf{i}}(\cos\Theta + \mathbf{j}\sin\Theta),$$

the principal argument $\Theta = \arg_{\mathbf{i}}(F(Z))$ is constant for all Z. Then the expression $e^{\mathbf{j}\Theta}$ is an invertible bicomplex number, say, A. Therefore $F(Z) = A \cdot f(Z)$, where $f(Z) = |F(Z)|_{\mathbf{i}}$ is a $\mathbb{C}(\mathbf{i})$-valued function. Because F is bicomplex holomorphic on Ω, it follows that F is constant. Similarly for the $\mathbb{C}(\mathbf{j})$ case. $\square$

Corollary 8.2.4. *Let F be a bicomplex holomorphic function on a product-type domain Ω. Each of the following conditions implies that $F'(Z) \in \mathfrak{S}$ for all $Z \in \Omega$:*

1. $F(Z) \in \mathfrak{S}$ *for all $Z \in \Omega$.*

2. *If we write $F = G_1\mathbf{e} + G_2\mathbf{e}^\dagger$, then either G_1 or G_2 is a constant function. This happens if and only if $F'(Z) \in \mathfrak{S}$ for all Z.*

3. *If $F(Z) = A \cdot f(Z)$, where $A \in \mathfrak{S}$ and f is either a $\mathbb{C}(\mathbf{i})$-, $\mathbb{C}(\mathbf{j})$-, $\mathbb{D}$-, $\mathbf{j}\mathbb{C}(\mathbf{i})$-, $\mathbf{k}\mathbb{C}(\mathbf{i})$-, $\mathbf{i}\mathbb{C}(\mathbf{j})$-, $\mathbf{k}\mathbb{C}(\mathbf{j})$-, $\mathbf{i}\mathbb{D}$- or $\mathbf{j}\mathbb{D}$-valued function.*

4. $|F(Z)|_{\mathbf{i}} = 0$ *for all $Z \in \Omega$.*

5. $|F(Z)|_{\mathbf{j}} = 0$ *for all $Z \in \Omega$.*

6. $|F(Z)|_{\mathbf{k}} \in \mathfrak{S}$ *for all $Z \in \Omega$.*

8.3 Relations among bicomplex, complex and hyperbolic holomorphies

We have proved that if F is a bicomplex holomorphic function on a bicomplex domain Ω, then its components in all bicomplex writings have very specific and rigid properties.

For example, if $F = f_1 + \mathbf{j}f_2$, then f_1 and f_2 are $\mathbb{C}(\mathbf{i})$-complex holomorphic functions in the two complex variables z_1 and z_2 and they are related by the Cauchy-Riemann type equation (7.21). Similarly, if $F = \rho_1 + \mathbf{i}\rho_2$, then ρ_1 and ρ_2 are $\mathbb{C}(\mathbf{j})$-complex holomorphic functions in the two complex variables ζ_1 and ζ_2 and they are related by the Cauchy-Riemann type equation (7.29). Next, if $F = \mathfrak{f}_1 + \mathbf{i}\mathfrak{f}_2$, then $\mathfrak{f}_1$ and $\mathfrak{f}_2$ are hyperbolic holomorphic functions in the two hyperbolic variables $\mathfrak{z}_1$ and $\mathfrak{z}_2$ and they are related by the Cauchy-Riemann type equation (7.34). Finally, in both idempotent representations the idempotent components of F are complex (in $\mathbb{C}(\mathbf{i})$ or $\mathbb{C}(\mathbf{j})$) holomorphic functions each of them of its corresponding complex variable.

Also we have seen that if we start with a $\mathcal{C}^1$-bicomplex function F, then all the holomorphy properties above are neatly obtained from the point of view of bicomplex, complex, and hyperbolic differential operators. In summary, if we use the different linear isomorphisms between $\mathbb{R}^4$, $\mathbb{C}^2$ (in $\mathbf{i}$ or $\mathbf{j}$), $\mathbb{D}^2$ and $\mathbb{BC}$, then all the differential operators whose annihilations lead to the notions of holomorphy in each case appear in the formulation of the bicomplex differential operators. So all four operators

$$\frac{\partial}{\partial Z}, \quad \frac{\partial}{\partial Z^\dagger}, \quad \frac{\partial}{\partial \overline{Z}} \quad \text{and} \quad \frac{\partial}{\partial Z^*}$$

encode multiple notions of complex and hyperbolic holomorphy when applied to the corresponding components of the function F.

In this respect, the differences between the theory of several complex variables and bicomplex analysis lie in the fact that in the latter case we work with pairs of complex holomorphic functions which are not independent: they are related by a Cauchy-Riemann type system. The same fact happens in the usual one complex variable theory, when the two real functions of two real variables are connected by the usual Cauchy-Riemann equations. By this reason and the numerous arguments we have already discovered so far, bicomplex analysis is in fact a one-variable theory inside the theory of mappings in two complex variables.

Therefore, having an example of a bicomplex holomorphic function F provides examples of two complex holomorphic functions and two hyperbolic holomorphic functions. If F is such a function, then writing $F = f_1 + \mathbf{j}f_2$, we have that f_1 and f_2 are complex holomorphic functions in two variables, which are related though, so f_2 is somehow almost determined by f_1. We will elaborate on this fact later on. Moreover, if we write $F = \mathfrak{f}_1 + \mathbf{i}\mathfrak{f}_2$, then $\mathfrak{f}_1$ and $\mathfrak{f}_2$ are hyperbolic holomorphic functions of two hyperbolic variables.

Example 8.3.1. *Consider the bicomplex exponential function $F(Z) = e^Z$. Then $f_1(z_1, z_2) = e^{z_1}\cos z_2$ and $f_2(z_1, z_2) = e^{z_1}\sin z_2$ are holomorphic in z_1 and z_2. Also $\mathfrak{f}_1(\mathfrak{z}_1, \mathfrak{z}_2) = e^{\mathfrak{z}_1}\cos\mathfrak{z}_2$ and $\mathfrak{f}_2(\mathfrak{z}_1, \mathfrak{z}_2) = e^{\mathfrak{z}_1}\sin\mathfrak{z}_2$ are hyperbolic holomorphic functions in two hyperbolic variables. Moreover, writing $Z = w_1 + \mathbf{k}w_2$, then also the functions $f_1(w_1, w_2) = e^{w_1}\cosh w_2$ and $f_2(w_1, w_2) = e^{w_1}\sinh w_2$ are holomorphic in w_1 and w_2.*

Of course, this example is simple but there is a fine point here: the pairs of functions that arise possess many additional properties which in no way are immediately seen from inside the theory of complex-valued holomorphic functions in $\mathbb{C}^2$. An illustration of this is the fact that the arising functions are null solutions to the complex Laplacian; we will comment on this later.

8.4   Bicomplex anti-holomorphies

We can define three notions of *anti-holomorphy* in the bicomplex context. Let us start with a bicomplex function F of class $\mathcal{C}^1$ on a domain Ω. We have proved that F is bicomplex holomorphic if and only if the three differential operators $\dfrac{\partial}{\partial Z^\dagger}, \dfrac{\partial}{\partial \overline{Z}}$

and $\dfrac{\partial}{\partial Z^*}$ anihilate the function F for all $Z \in \Omega$, i.e.,

$$\frac{\partial F}{\partial Z^\dagger}(Z) = \frac{\partial F}{\partial \overline{Z}}(Z) = \frac{\partial F}{\partial Z^*}(Z) = 0. \tag{8.1}$$

In this case the bicomplex derivative of F exists for all Z, and it is related to the fourth differential operator:

$$F'(Z) = \frac{\partial F}{\partial Z}(Z).$$

We have seen that the operators $\dfrac{\partial}{\partial Z}$ and $\dfrac{\partial}{\partial Z^\dagger}$ are intimately related, and the same is true for $\dfrac{\partial}{\partial \overline{Z}}$ and $\dfrac{\partial}{\partial Z^*}$.

One may wonder what happens if instead of having relation (8.1) we impose one of the following conditions for F for all Z in its domain:

$$\frac{\partial F}{\partial Z}(Z) = \frac{\partial F}{\partial \overline{Z}}(Z) = \frac{\partial F}{\partial Z^*}(Z) = 0, \tag{8.2}$$

$$\frac{\partial F}{\partial Z}(Z) = \frac{\partial F}{\partial Z^\dagger}(Z) = \frac{\partial F}{\partial Z^*}(Z) = 0, \tag{8.3}$$

$$\frac{\partial F}{\partial Z}(Z) = \frac{\partial F}{\partial Z^\dagger}(Z) = \frac{\partial F}{\partial \overline{Z}}(Z) = 0. \tag{8.4}$$

Equalities (8.2) define what can be called $\dagger$-anti-holomorphy while equalities (8.3) are about bar-anti-holomorphy and equalities (8.4) are about $*$-anti-holomorphy. It is clear how to introduce the corresponding derivatives.

We can interpret each of these conditions using all the formulas we developed so far, in all the possible bicomplex writings. Note that if we compose F with a change of bicomplex variable $W \mapsto Z^\dagger$, then (8.2) is equivalent to

$$\frac{\partial F^\dagger}{\partial W^\dagger}(W) = \frac{\partial F^\dagger}{\partial \overline{W}}(W) = \frac{\partial F^\dagger}{\partial W^*}(W) = 0,$$

where $F^\dagger(W) = (F(W))^\dagger$. This is equivalent to the fact that the function $F^\dagger$ is a bicomplex holomorphic function of a bicomplex variable $W = Z^\dagger$; more exactly, the function $F^\dagger(Z^\dagger)$ is simply $\mathbb{BC}$-holomorphic. In the same fashion, F is *bicomplex bar-anti-holomorphic* if and only if the function $\overline{F}$ is bicomplex holomorphic with respect to $\overline{Z}$; F is *bicomplex $*$-anti-holomorphic* if and only if the function F^* is bicomplex holomorphic with respect to Z^*. We leave the details to the reader.

Let us study in more detail the notion of bicomplex $\dagger$-anti-holomorphy. If we choose to write $F = \rho_1 + \mathbf{i}\rho_2$, then (8.2) leads to ρ_1 and ρ_2 being anti-holomorphic functions of the two $\mathbb{C}(\mathbf{j})$-complex variables ζ_1 and ζ_2, i.e., they are holomorphic with respect to the **j**-conjugate variables ζ_1^* and ζ_2^*, related by the Cauchy-Riemann type system:

$$\frac{\partial \rho_1}{\partial \zeta_1^*} = \frac{\partial \rho_2}{\partial \zeta_2^*}, \qquad \frac{\partial \rho_1}{\partial \zeta_2^*} = -\frac{\partial \rho_2}{\partial \zeta_1^*} \, .$$

For the same condition (8.2), using the idempotent representation $F = \Gamma_1 \mathbf{e} + \Gamma_2 \mathbf{e}^\dagger$, where $Z = \gamma_1 \mathbf{e} + \gamma_2 \mathbf{e}^\dagger$ with $\gamma_1, \gamma_2 \in \mathbb{C}(\mathbf{j})$, we obtain that Γ_1 and Γ_2 are complex anti-holomorphic functions depending on only one variable: γ_1^* for Γ_1 and γ_2^* for Γ_2, respectively. Similar statements are true for the other two types of bicomplex anti-holomorphy.

In the $\mathbb{C}(\mathbf{i})$-idempotent representation $F = G_1 \mathbf{e} + G_2 \mathbf{e}^\dagger$, where $Z = \beta_1 \mathbf{e} + \beta_2 \mathbf{e}^\dagger$ with $\beta_1, \beta_2 \in \mathbb{C}(\mathbf{i})$, equation (8.2) leads to the conclusion that G_1 and G_2 are complex holomorphic in variables β_2 and β_1, respectively: note here the reversed role of the complex variables β_1 and β_2.

8.5   Geometric interpretation of the derivative

We have described previously some properties of smooth hyperbolic curves which make them similar to the usual, real, curves in the complex plane. It turns out that this similarity extends quite far and it allows us to obtain a geometric interpretation of the derivative at a point.

Definition 8.5.1. *Let Ω be a domain in $\mathbb{BC}$, let Z_0 be a point in Ω and let $f : \Omega \subset \mathbb{BC} \longrightarrow \mathbb{BC}$ be a function. The function f is called a <u>hyperbolic angle preserving mapping</u> at Z_0, or a <u>$\mathbb{D}$-conformal</u> mapping at Z_0, whenever the following holds: if Γ_1 and Γ_2 are two hyperbolic smooth curves passing through Z_0 and if their images $f(\Gamma_1)$ and $f(\Gamma_2)$ are also hyperbolic smooth curves passing through $W_0 = f(Z_0)$, then the hyperbolic angle between Γ_1 and Γ_2 at Z_0 is the same as the hyperbolic angle between $f(\Gamma_1)$ and $f(\Gamma_2)$ at W_0.*

We begin with an auxiliary lemma which is of interest by itself.

Lemma 8.5.2. *If f is bicomplex holomorphic in Ω and Γ is a smooth hyperbolic curve in Ω, then $f(\Gamma)$ is also a smooth hyperbolic curve.*

Proof. Since f is bicomplex holomorphic, then it is of the form $f(Z) = F_1(\beta_1)\mathbf{e} + F_2(\beta_2)\mathbf{e}^\dagger$ for $Z = \beta_1 \mathbf{e} + \beta_2 \mathbf{e}^\dagger \in \Omega$. Since Γ is a smooth hyperbolic curve, then it has a parametrization of the form $\varphi(u, v) = \varphi_1(u)\mathbf{e} + \varphi_2(v)\mathbf{e}^\dagger$ where φ_1 and φ_2 are parametrizations of smooth curves in $\mathbb{C}(\mathbf{i})$; thus, $\varphi_1'(u) \neq 0$ and $\varphi_2'(v) \neq 0$ for any u and v.

It is clear that $f \circ \varphi$ is a parametrization of $f(\Gamma)$:

$$f \circ \varphi(u, v) = f\left(\varphi_1(u)\mathbf{e} + \varphi_2(v)\mathbf{e}^\dagger\right)$$
$$= F_1\left(\varphi_1(u)\right)\mathbf{e} + F_2\left(\varphi_2(v)\right)\mathbf{e}^\dagger.$$

Since both $F_1 \circ \varphi_1$ and $F_2 \circ \varphi_2$ are parametrizations of smooth curves in $\mathbb{C}(\mathbf{i})$, it follows that $f \circ \varphi$ is a parametrization of a smooth hyperbolic curve. $\qquad \square$

Theorem 8.5.3. *If f is bicomplex holomorphic at $Z_0 \in \Omega$ and $f'(Z_0)$ is neither zero nor a zero-divisor, then f is a $\mathbb{D}$-conformal mapping at Z_0.*

Proof. Since f is bicomplex holomorphic we can write $f(Z) = f(\beta_1\mathbf{e} + \beta_2\mathbf{e}^\dagger) = F_1(\beta_1)\mathbf{e} + F_2(\beta_2)\mathbf{e}^\dagger$; moreover

$$f'(Z_0) = F_1'(\beta_1^0)\mathbf{e} + F_2'(\beta_2^0)\mathbf{e}^\dagger$$

and

$$\arg_\mathbb{D} f'(Z_0) = \theta_1\mathbf{e} + \theta_2\mathbf{e}^\dagger \in \mathbb{D}.$$

Next, take a smooth hyperbolic curve Γ passing through Z_0 and let T be the hyperbolic tangent line to Γ at Z_0. By the previous Lemma, $f(\Gamma) = \Lambda = \lambda_1\mathbf{e} + \lambda_2\mathbf{e}^\dagger$ is also a smooth hyperbolic curve with hyperbolic tangent line P at $W_0 = f(Z_0)$.

Since $\Gamma = \gamma_1\mathbf{e} + \gamma_2\mathbf{e}^\dagger$ has a parametrization

$$\varphi(u, v) = \varphi_1(u)\mathbf{e} + \varphi_2(v)\mathbf{e}^\dagger,$$

where φ_1 and φ_2 are, respectively, the parametrizations of the smooth, real curves γ_1 and γ_2 in $\mathbb{C}(\mathbf{i})$, then one can take the projections of T and P on $\mathbb{BC}_\mathbf{e}$ and $\mathbb{BC}_{\mathbf{e}^\dagger}$ and write:

$$T = t_1\mathbf{e} + t_2\mathbf{e}^\dagger, \qquad P = p_1\mathbf{e} + p_2\mathbf{e}^\dagger.$$

Writing also $Z_0 = \beta_1^0\mathbf{e} + \beta_2\mathbf{e}^\dagger = \varphi_1(u_0)\mathbf{e} + \varphi_2(v_0)\mathbf{e}^\dagger$, we note that t_1 is the (real) tangent line to γ_1 at $\varphi_1(u_0)$ as well as t_2, p_1, p_2 are the (real) tangent lines to γ_2, λ_1 λ_2 at $\varphi_2(v_0)$, $F_1(\varphi_1(u_0))$, $F_2(\varphi_2(v_0))$ respectively.

Denote now by μ_1 and μ_2 the real numbers such that $\tan\mu_1$ and $\tan\mu_2$ are the (real) slopes of t_1 and t_2 respectively. Similarly, the real numbers κ_1 and κ_2 are such that $\tan\kappa_1$ and $\tan\kappa_2$ are the (real) slopes of p_1 and p_2. Hence, the hyperbolic slopes of the hyperbolic lines T and P are, respectively,

$$\tan\mu = \tan\mu_1\mathbf{e} + \tan\mu_2\mathbf{e}^\dagger,$$

$$\tan\kappa = \tan\kappa_1\mathbf{e} + \tan\kappa_2\mathbf{e}^\dagger.$$

But $\theta_1 = \arg F_1'(\beta_1^0)$ and $\theta_2 = \arg F_2'(\beta_2^0)$ hence

$$\kappa_1 = \mu_1 + \theta_1 \qquad \text{and} \qquad \kappa_2 = \mu_2 + \theta_2$$

which means that

$$\kappa = \mu + \arg_\mathbb{D} f'(Z_0). \tag{8.5}$$

In other words, the hyperbolic argument $\arg_\mathbb{D} f'(Z_0)$ can be seen as an "increment" which the hyperbolic angle μ, characterizing the tangent line T, gets when we pass to the tangent line P. As in the complex analysis case this increment depends on Z_0 only, and not on the curve Γ.

We are ready to complete the proof. Take now two curves Γ and $\widetilde{\Gamma}$ passing through Z_0, then for each of the (8.5) it holds that:

$$\kappa = \mu + \arg_{\mathbb{D}} f'(Z_0),$$
$$\widetilde{\kappa} = \widetilde{\mu} + \arg_{\mathbb{D}} f'(Z_0).$$

Substracting them gives:

$$\kappa - \widetilde{\kappa} = \mu - \widetilde{\mu}$$

which means that f is a $\mathbb{D}$-conformal mapping ar Z_0. $\square$

Remark 8.5.4. *Consider also an interpretation of the hyperbolic modulus of $f'(Z_0)$. By the properties of such moduli, one has:*

$$\left|f'(Z_0)\right|_{\mathbb{D}} = \lim_{Z \to Z_0} \frac{|f(Z) - f(Z_0)|_{\mathbb{D}}}{|Z - Z_0|_{\mathbb{D}}}$$

for $Z - Z_0 \notin \mathfrak{S}_0$. Hence, for any Z close to Z_0, $Z - Z_0 \notin \mathfrak{S}_0$, with respect to the hyperbolic modulus there holds:

$$\left|f(Z) - f(Z_0)\right|_{\mathbb{D}} \approx \left|f'(Z_0)\right|_{\mathbb{D}} \cdot \left|Z - Z_0\right|_{\mathbb{D}}. \tag{8.6}$$

In analogy with the case of complex holomorphic functions, we say that a bicomplex holomorphic function with the derivative at Z_0 different from zero and zero-divisors, realizes locally, at Z_0, a hyperbolic homothety, with the hyperbolic coefficient $\left|f'(Z_0)\right|_{\mathbb{D}}$, in all directions except those of the cone of zero-divisors generated by Z_0.

8.6   Bicomplex Riemann Mapping Theorem

In one complex variable the Riemann mapping theorem has different formulations. We will use the one from [47].

Theorem 8.6.1. *Let A be a simply connected domain such that $A \neq \mathbb{C}$. Then there exists a bijective conformal map $f : A \longrightarrow \mathbb{B}(0,1)$ where $\mathbb{B}(0,1)$ is the unit disk in the complex plane. Furthermore, for any fixed $z_0 \in A$, we can find a function f such that $f(z_0) = 0$ and $f'(z_0) > 0$. With such a specification f is unique.*

In the classical setting of holomorphic functions of two complex variables, there are no analogues of this theorem for domains in $\mathbb{C}^2$. It turns out that there exists an analogue of it in the context of bicomplex holomorphic functions.

Theorem 8.6.2 (Bicomplex Riemann mapping theorem). *Let $\Omega \subset \mathbb{BC}$ be a product-type domain $\Omega = \Omega_1 \mathbf{e} + \Omega_2 \mathbf{e}^{\dagger}$ such that Ω_1 and Ω_2 are simply connected domains in $\mathbb{C}(\mathbf{i})$ and $\Omega_1 \neq \mathbb{C}(\mathbf{i})$, $\Omega_2 \neq \mathbb{C}(\mathbf{i})$. Then there exists a bijective $\mathbb{D}$-conformal mapping $f : \Omega \to \mathbb{B}_1$, with $\mathbb{B}_1 := \{Z \in \mathbb{BC} \,|\, |Z|_{\mathbf{k}} \prec 1\}$ the bicomplex ball of hyperbolic radius 1 centered at the origin. Furthermore for any fixed $Z_0 \in \Omega$, we can find an f such that $f(Z_0) = 0$ and $f'(Z_0)$ is a strictly positive hyperbolic number. With such a specification f is unique.*

Proof. Since $\Omega = \Omega_1 \mathbf{e} + \Omega_2 \mathbf{e}^\dagger$ we can apply the complex Riemann mapping theorem to Ω_1 and Ω_2 obtaining the bijective complex conformal mappings F_1 and F_2 onto the unit disk $\mathbb{B}(0,1) \subset \mathbb{C}(\mathbf{i})$, which are complex holomorphic functions with derivatives different from zero. Set

$$F := F_1 \mathbf{e} + F_2 \mathbf{e}^\dagger.$$

Clearly F is a bicomplex holomorphic function on Ω such that the derivative

$$F'(Z) = F_1'(\beta_1)\mathbf{e} + F_2'(\beta_2)\mathbf{e}^\dagger$$

is not in $\mathfrak{S}_0$ for any $Z \in \Omega$. Moreover, F realizes a bijective correspondence between Ω and $\mathbb{B}_1$. Thus, the first part of the theorem is proved.

Take now $Z_0 \in \Omega$ hence $Z_0 = \beta_1^0 \mathbf{e} + \beta_2^0 \mathbf{e}^\dagger$, and let F_1 and F_2 be such that

$$F_1(\beta_1^0) = 0, \quad F_2(\beta_2^0) = 0 \quad \text{and} \quad F_1'(\beta_1^0) > 0, \ F_2'(\beta_2^0) > 0.$$

Hence $F(Z_0) = 0$ and $F'(Z_0)$ is a strictly positive hyperbolic number. $\square$

Definition 8.6.3. *Given two domains Ω and Ξ in $\mathbb{BC}$, they are called $\mathbb{D}$-conformally equivalent if there exists a bijective $\mathbb{D}$-conformal mapping from Ω onto Ξ.*

Corollary 8.6.4. *Any simply-connected product-type domain in $\mathbb{C}^2(\mathbf{i})$ such that none of its idempotent components is the whole $\mathbb{C}(\mathbf{i})$ is $\mathbb{D}$-conformally equivalent to the unitary bidisk.*

The analytic theory of bicomplex functions has also been carried out in Price [56]. Using computational algebra techniques, properties of bicomplex holomorphy such as removability of singularities, Cauchy-Kowalevsky extensions, and bicomplex hyperfunctions have been developed in [19, 20, 96].

There are numerous applications of bicomplex holomorphic functions in both mathematics and physics, from the study of bicomplex manifolds [7, 21, 22, 49], bicomplex pseudoanalytic functions [8], bicomplex fractal theory [12, 13, 29, 48, 59, 60, 61, 62, 102], bicomplex zeta functions [43, 42, 63], functional analysis with bicomplex scalars [2], all the way to applications in electromagnetism [3], homothetic motions [4, 5, 6, 38], and quantum mechanics [32, 33, 34, 64, 66, 67]. For example, the algebraic structure and the geometry of bicomplex numbers are a natural setup for the space of quantum mechanics projectors, as the bicomplex holomorphic functions are for the complex fields in relativistic quantum field theory.

Chapter 9

Second Order Complex and Hyperbolic Differential Operators

9.1 Holomorphic functions in $\mathbb{C}$ and harmonic functions in $\mathbb{R}^2$

It is well known that complex holomorphic functions are tightly related with harmonic functions of two real variables, a fact that proved to be of crucial importance for the theories of both classes of functions. On the general level, the same occurs with hyperholomorphic (synonymously - monogenic, regular) functions of (real) Clifford analysis and the harmonic functions of the respective number of (real) variables. By this reason, both one complex variable theory and Clifford analysis are considered as refinements of the corresponding harmonic function theories. This relation is due to the following factorizations of the respective Laplace operators. If

$$\Delta_{\mathbb{R}^2} := \frac{\partial^2}{\partial x^2} + \frac{\partial^2}{\partial y^2} ; \quad \frac{\partial}{\partial z} := \frac{1}{2} \left(\frac{\partial}{\partial x} - \mathbf{i} \frac{\partial}{\partial y} \right) , \quad \frac{\partial}{\partial \overline{z}} := \frac{1}{2} \left(\frac{\partial}{\partial x} + \mathbf{i} \frac{\partial}{\partial y} \right) ,$$

then

$$\frac{\partial}{\partial \overline{z}} \circ \frac{\partial}{\partial z} = \frac{\partial}{\partial z} \circ \frac{\partial}{\partial \overline{z}} = \frac{1}{4} \Delta_{\mathbb{R}^2} . \tag{9.1}$$

Similarly, if $\Delta_{\mathbb{R}^k}$ denotes the Laplace operator in $\mathbb{R}^k$, $\mathcal{D}_{CR}$ denotes the Cauchy-Riemann operator of Clifford analysis in $\mathbb{R}^{n+1}$, and $\mathcal{D}_{Dir}$ denotes the Dirac operator in $\mathbb{R}^n$ (see [23] and [9]), then

$$\mathcal{D}_{CR} \circ \overline{\mathcal{D}}_{CR} = \overline{\mathcal{D}}_{CR} \circ \mathcal{D}_{CR} = \Delta_{\mathbb{R}^{n+1}} ,$$

$$\mathcal{D}^2_{Dir} = -\Delta_{\mathbb{R}^n} .$$

Factorization (9.1) manifests the essence of the relation between complex holomorphic functions and harmonic functions. First of all, note that the Laplace operator $\Delta_{\mathbb{R}^2}$ acts initially on real-valued functions but its action extends onto complex valued functions component-wise: if $f = u + \mathbf{i}v$, then

$$\Delta_{\mathbb{R}^2}[f] := \Delta_{\mathbb{R}^2}[u] + \mathbf{i}\Delta_{\mathbb{R}^2}[v] .$$

Thus equality (9.1) holds on complex-valued functions (of class $\mathcal{C}^2$, not just $\mathcal{C}^1$).

Let $f = u + \mathbf{i}v$ be a holomorphic function, that is, $\dfrac{\partial f}{\partial \overline{z}} = 0$, hence by (9.1) $\Delta_{\mathbb{R}^2}[f] = 4\dfrac{\partial}{\partial z}\left(\dfrac{\partial f}{\partial \overline{z}}\right) = 0$ and thus f is a harmonic function. Reciprocally, taking a real-valued harmonic function u, consider $\dfrac{\partial u}{\partial z}$ (which is a formal derivative, that is, the result of the action of an operator, not the "honest" derivative of a holomorphic function), one has:

$$\frac{\partial}{\partial \overline{z}}\left[\frac{\partial u}{\partial z}\right] = \frac{1}{4}\Delta_{\mathbb{R}^2}[u] = 0 ,$$

which means that the complex-valued function $\dfrac{\partial u}{\partial z}$ generated by the harmonic function u is holomorphic.

9.2   Complex and hyperbolic Laplacians

Although we have noted already certain asymmetries in the properties of complex and hyperbolic numbers inside $\mathbb{BC}$, for the purposes of this chapter each of the sets $\mathbb{C}(\mathbf{i})$, $\mathbb{C}(\mathbf{j})$ and $\mathbb{D}$ plays for bicomplex numbers a role similar to that played by $\mathbb{R}$ for complex numbers. Thus, one can consider the three candidates for the corresponding Laplacians:

$$\Delta_{\mathbb{C}^2(\mathbf{i})} := \frac{\partial^2}{\partial z_1^2} + \frac{\partial^2}{\partial z_2^2} ; \tag{9.2}$$

$$\Delta_{\mathbb{C}^2(\mathbf{j})} := \frac{\partial^2}{\partial \zeta_1^2} + \frac{\partial^2}{\partial \zeta_2^2} ; \tag{9.3}$$

$$\Delta_{\mathbb{D}} := \frac{\partial^2}{\partial \mathfrak{z}_1^2} + \frac{\partial^2}{\partial \mathfrak{z}_2^2} . \tag{9.4}$$

Operators (9.2) and (9.3) are called complex ($\mathbb{C}(\mathbf{i})$ and $\mathbb{C}(\mathbf{j})$ respectively) Laplacians and (9.4) is the hyperbolic Laplacian. The first of them acts on $\mathbb{C}(\mathbf{i})$-valued holomorphic functions of two complex variables z_1 and z_2; the second acts on $\mathbb{C}(\mathbf{j})$-valued holomorphic functions of the complex variables ζ_1 and ζ_2; and the third acts on $\mathbb{D}$-valued holomorphic functions of the hyperbolic variables $\mathfrak{z}_1$ and $\mathfrak{z}_2$.

Although we wrote in (9.2)–(9.4) the formal partial derivatives $\dfrac{\partial}{\partial z_1}$, etc., in fact we mean authentic partial derivatives with respect to the corresponding variables $z_1, z_2, \zeta_1, \zeta_2, \mathfrak{z}_1, \mathfrak{z}_2$. Of course in this situation the formal and the authentic derivatives coincide but at the same time a confusion may arise, so that we rewrite (9.2)–(9.4) as

$$\Delta_{\mathbb{C}^2(\mathbf{i})} = \partial''_{z_1^2} + \partial''_{z_2^2}; \tag{9.5}$$

$$\Delta_{\mathbb{C}^2(\mathbf{j})} = \partial''_{\zeta_1^2} + \partial''_{\zeta_2^2}; \tag{9.6}$$

$$\Delta_{\mathbb{D}^2} = \partial''_{\mathfrak{z}_1^2} + \partial''_{\mathfrak{z}_2^2}. \tag{9.7}$$

This is in keeping with the notation we adopted where we wrote complex partial derivatives as f'_{z_1}, etc. Hence $\partial''_{z_1^2} f = f''_{z_1^2}$, etc.

The next step consists in extending the operators (9.5)–(9.7) onto bicomplex-valued functions. It turns out that how we write bicomplex numbers becomes important and it depends on the operators in (9.5)–(9.7). Indeed, for the operator $\Delta_{\mathbb{C}^2(\mathbf{i})}$ the bicomplex function F should be holomorphic in the sense of two complex variables and if we want the same for its components, then we are forced to consider F as $F = f_1 + \mathbf{j}f_2$. Now we set:

$$\Delta_{\mathbb{C}^2(\mathbf{i})}[F] := \Delta_{\mathbb{C}^2(\mathbf{i})}[f_1] + \mathbf{j}\Delta_{\mathbb{C}^2(\mathbf{i})}[f_2].$$

A similar reasoning holds for the other two operators: for $\Delta_{\mathbb{C}^2(\mathbf{j})}$ we take $F = \rho_1 + \mathbf{i}\rho_2$ and we set:

$$\Delta_{\mathbb{C}^2(\mathbf{j})}[F] := \Delta_{\mathbb{C}^2(\mathbf{j})}[\rho_1] + \mathbf{i}\Delta_{\mathbb{C}^2(\mathbf{j})}[\rho_2];$$

for $\Delta_{\mathbb{D}^2}$ we take $F = \mathfrak{f}_1 + \mathbf{i}\mathfrak{f}_2$ and we set:

$$\Delta_{\mathbb{D}^2}[F] := \Delta_{\mathbb{D}^2}[\mathfrak{f}_1] + \mathbf{i}\Delta_{\mathbb{D}^2}[\mathfrak{f}_2].$$

The analogues of the formula (9.1) arise if one uses the corresponding operators in formulas (7.49), (7.50), (7.51) and (7.52); more exactly the operators should be taken in an appropriate form. For instance, if we want to use the operators $\dfrac{\partial}{\partial Z^\dagger}$ and $\dfrac{\partial}{\partial Z}$ in the form

$$\frac{\partial}{\partial Z^\dagger} = \frac{1}{2}\left(\frac{\partial}{\partial z_1} + \mathbf{j}\frac{\partial}{\partial z_2}\right), \qquad \frac{\partial}{\partial Z} = \frac{1}{2}\left(\frac{\partial}{\partial z_1} - \mathbf{j}\frac{\partial}{\partial z_2}\right),$$

then we are forced to write the variable as $Z = z_1 + \mathbf{j}z_2$ and the function F as $F = f_1 + \mathbf{j}f_2$. With this understanding we have that

$$\Delta_{\mathbb{C}^2(\mathbf{i})} = \frac{\partial^2}{\partial z_1^2} + \frac{\partial^2}{\partial z_2^2} = 4\frac{\partial}{\partial Z}\frac{\partial}{\partial Z^\dagger}, \tag{9.8}$$

where the operators act on $\mathbb{BC}$-valued functions holomorphic in the sense of the complex variables z_1, z_2. Hence, the theory of $\mathbb{BC}$-holomorphic functions can (should?) be seen as a function theory for the $\mathbb{C}(\mathbf{i})$-complex Laplacian.

Analogously to what was done above, the factorization allows us to establish direct relations between $\mathbb{BC}$-holomorphic functions and complex harmonic functions, that is, null solutions to the operator $\Delta_{\mathbb{C}^2(\mathbf{i})}$. Indeed, let $F = f_1 + \mathbf{j} f_2$ be a $\mathbb{BC}$-holomorphic function, that is, it is holomorphic in the sense of two complex variables, but also $\dfrac{\partial F}{\partial Z^\dagger} = 0$. Then

$$\Delta_{\mathbb{C}^2(\mathbf{i})}[F] = 4\frac{\partial}{\partial Z}\left(\frac{\partial F}{\partial Z^\dagger}\right) = 0\,,$$

and thus F is a $\mathbb{BC}$-valued complex harmonic function. Reciprocally, taking a $\mathbb{C}(\mathbf{i})$-valued complex harmonic function f_1, consider the $\mathbb{BC}$-valued function $\dfrac{\partial}{\partial Z}[f_1] = \dfrac{\partial f_1}{\partial Z}$ (which is a formal operation on a holomorphic function of two complex variables, not the bicomplex derivative of a $\mathbb{BC}$-holomorphic function). One has:

$$\frac{\partial}{\partial Z^\dagger}\left[\frac{\partial f_1}{\partial Z}\right] = \frac{1}{4}\Delta_{\mathbb{C}^2(\mathbf{i})}[f_1] = 0\,,$$

which means that the $\mathbb{BC}$-valued function $\dfrac{\partial f_1}{\partial Z}$ generated by the complex harmonic function f_1 is $\mathbb{BC}$-holomorphic.

The operators in (9.2) and (9.3) are dealt with in exactly the same way, although now other first-order operators enter into the game. In the case of $Z = \zeta_1 + \mathbf{j}\zeta_2$ we take

$$\frac{\partial}{\partial Z} = \frac{1}{2}\left(\frac{\partial}{\partial \zeta_1} - \mathbf{i}\frac{\partial}{\partial \zeta_2}\right)\,,\qquad \frac{\partial}{\partial \overline{Z}} = \frac{1}{2}\left(\frac{\partial}{\partial \zeta_1} + \mathbf{i}\frac{\partial}{\partial \zeta_2}\right)\,,$$

arriving at

$$\Delta_{\mathbb{C}^2(\mathbf{j})} = \frac{\partial^2}{\partial \zeta_1^2} + \frac{\partial^2}{\partial \zeta_2^2} = 4\frac{\partial}{\partial Z}\frac{\partial}{\partial \overline{Z}}\,;\tag{9.9}$$

in the case of $Z = \mathfrak{z}_1 + \mathbf{i}\mathfrak{z}_2$ we take

$$\frac{\partial}{\partial Z} = \frac{1}{2}\left(\frac{\partial}{\partial \mathfrak{z}_1} - \mathbf{i}\frac{\partial}{\partial \mathfrak{z}_2}\right)\,,\qquad \frac{\partial}{\partial Z^*} = \frac{1}{2}\left(\frac{\partial}{\partial \mathfrak{z}_1} + \mathbf{i}\frac{\partial}{\partial \mathfrak{z}_2}\right)\,,$$

arriving at

$$\Delta_{\mathbb{D}} = \frac{\partial^2}{\partial \mathfrak{z}_1^2} + \frac{\partial^2}{\partial \mathfrak{z}_2^2} = 4\frac{\partial}{\partial Z}\frac{\partial}{\partial Z^*}\,.\tag{9.10}$$

So, we have provided each of the operators (9.2)-(9.4) with an adequate function theory. There is a fine point here: it is, in a sense, one and the same theory of $\mathbb{BC}$-holomorphic functions, but where different aspects of the latter are taken into account. This "common" function theory allows us to realize the similarities and the differences between the three operators (9.2)-(9.4). Indeed, (9.2) and (9.3) look identical from the viewpoint of classical complex analysis, but viewing them from the bicomplex perspective reveals subtle differences between them. At the same time, (9.4) seems to be quite different with any of (9.2) and (9.3) but bicomplex functions, again, show that there exists a deep underlying unity between all three of them.

The direct relations between the null solutions of any of the operators $\Delta_{\mathbb{C}^2(\mathbf{j})}$ and $\Delta_{\mathbb{D}^2}$ and the $\mathbb{BC}$-holomorphic function theory is established in the same way as we did for the operator $\Delta_{\mathbb{C}^2(\mathbf{i})}$.

9.3   Complex and hyperbolic wave operators

Another well-known second-order operator, in the real case, is the wave operator, and one can be tempted to look at some of its complex analogues:

$$\Box_{\mathbb{C}^2(\mathbf{i})} := \frac{\partial^2}{\partial w_1^2} - \frac{\partial^2}{\partial w_2^2}, \tag{9.11}$$

$$\Box_{\mathbb{C}^2(\mathbf{j})} := \frac{\partial^2}{\partial \omega_1^2} - \frac{\partial^2}{\partial \omega_2^2}. \tag{9.12}$$

They do not produce great novelties, the same $\mathbb{BC}$-holomorphic function theory provides all the necessary information. This can be obtained in two ways. First of all, the simple holomorphic changes of variables

$$(w_1, w_2) \mapsto (z_1, -\mathbf{i} z_2), \qquad (\omega_1, \omega_2) \mapsto (\zeta_1, -\mathbf{j} \zeta_2)$$

turn the operator $\Box_{\mathbb{C}^2(\mathbf{i})}$ into $\Delta_{\mathbb{C}^2(\mathbf{i})}$ and the operator $\Box_{\mathbb{C}^2(\mathbf{j})}$ into $\Delta_{\mathbb{C}^2(\mathbf{j})}$. But they can be factorized directly:

$$\Box_{\mathbb{C}^2(\mathbf{i})} = \left(\frac{\partial}{\partial w_1} + \mathbf{k} \frac{\partial}{\partial w_2} \right) \cdot \left(\frac{\partial}{\partial w_1} - \mathbf{k} \frac{\partial}{\partial w_2} \right)$$

$$= 4 \frac{\partial}{\partial Z} \frac{\partial}{\partial Z^\dagger};$$

compare with (9.8). Although formally they are the same operators, they are employed in other forms taken in the table in Section 4. In the same fashion,

$$\Box_{\mathbb{C}^2(\mathbf{j})} = \left(\frac{\partial}{\partial \omega_1} + \mathbf{k} \frac{\partial}{\partial \omega_2} \right) \cdot \left(\frac{\partial}{\partial \omega_1} - \mathbf{k} \frac{\partial}{\partial \omega_2} \right)$$

$$= 4 \frac{\partial}{\partial Z} \frac{\partial}{\partial \overline{Z}},$$

where we take, again, another form of writing the operators involved. Somewhat paradoxically, the same bicomplex "tricks" do not work for the "hyperbolic wave operator"

$$\frac{\partial^2}{\partial \mathfrak{w}_1^2} - \frac{\partial^2}{\partial \mathfrak{w}_2^2}$$

which can neither be factorized directly nor reduced to (9.4); this is because in the hyperbolic world there is only one hyperbolic-type imaginary unit and there are no complex-type imaginary units.

One can consider another hyperbolic Laplacian:

$$\frac{\partial^2}{\partial \mathfrak{w}_1^2} + \frac{\partial^2}{\partial \mathfrak{w}_2^2}$$

but the change of variables

$$(\mathfrak{w}_1, \mathfrak{w}_2) \mapsto (\mathfrak{z}_1, -\mathbf{k}\mathfrak{z}_2)$$

reduces it to (9.4).

9.4   Conjugate (complex and hyperbolic) harmonic functions

Let F be a bicomplex holomorphic function in a domain $\Omega \subset \mathbb{BC}$ which can be written as $F(Z) = f_1(z_1, z_2) + \mathbf{j}f_2(z_1, z_2) = \rho_1(\zeta_1, \zeta_2) + \mathbf{i}\rho_2(\zeta_1, \zeta_2) = \mathfrak{f}_1(\mathfrak{z}_1, \mathfrak{z}_2) + \mathbf{i}\mathfrak{f}_2(\mathfrak{z}_1, \mathfrak{z}_2)$. It has been shown above that:

- f_1 and f_2 are $\mathbb{C}(\mathbf{i})$-complex harmonic functions in $\Omega \subset \mathbb{C}^2(\mathbf{i})$, and they are related by the Cauchy–Riemann conditions

$$\frac{\partial f_1}{\partial z_1} = \frac{\partial f_2}{\partial z_2}; \quad \frac{\partial f_2}{\partial z_1} = -\frac{\partial f_1}{\partial z_2}.$$

  Such a pair of functions f_1 and f_2 will be called a pair of conjugate $\mathbb{C}(\mathbf{i})$-complex harmonic functions.

- ρ_1 and ρ_2 are $\mathbb{C}(\mathbf{j})$-complex harmonic functions in $\Omega \subset \mathbb{C}^2(\mathbf{j})$, and they are related by the Cauchy–Riemann conditions

$$\frac{\partial \rho_1}{\partial \zeta_1} = \frac{\partial \rho_2}{\partial \zeta_2}; \quad \frac{\partial \rho_2}{\partial \zeta_1} = -\frac{\partial \rho_1}{\partial \zeta_2}.$$

  Such a pair of functions ρ_1 and ρ_2 will be called a pair of conjugate $\mathbb{C}(\mathbf{j})$-complex harmonic functions.

- $\mathfrak{f}_1$ and $\mathfrak{f}_2$ are $\mathbb{D}$-harmonic functions in $\Omega \subset \mathbb{D}^2$, and they are related by the Cauchy–Riemann conditions

$$\frac{\partial \mathfrak{f}_1}{\partial \mathfrak{z}_1} = \frac{\partial \mathfrak{f}_2}{\partial \mathfrak{z}_2}; \quad \frac{\partial \mathfrak{f}_2}{\partial \mathfrak{z}_1} = -\frac{\partial \mathfrak{f}_1}{\partial \mathfrak{z}_2}.$$

Such a pair of functions $\mathfrak{f}_1$ and $\mathfrak{f}_2$ will be called a pair of conjugate $\mathbb{D}$-harmonic functions.

Bicomplex holomorphic functions play an important role in understanding constant coefficient second-order complex and hyperbolic differential operators, as emphasized also in e.g. [19, 20, 96], where computational algebra techniques are used to study such objects.

Chapter 10

Sequences and Series of Bicomplex Functions

10.1 Series of bicomplex numbers

We consider here briefly the series of the form

$$\sum_{n=1}^{\infty} Z_n \tag{10.1}$$

with bicomplex numbers Z_n as general terms. Writing Z_n as $Z_n = \beta_{1,n}\mathbf{e} + \beta_{2,n}\mathbf{e}^\dagger$ equation (10.1) becomes

$$\sum_{n=1}^{\infty} \beta_{1,n}\mathbf{e} + \sum_{n=1}^{\infty} \beta_{2,n}\mathbf{e}^\dagger, \tag{10.2}$$

thus the series of bicomplex numbers (10.1) is convergent if and only if both complex series in (10.2) are convergent.

A typical example is the bicomplex geometric series $\displaystyle\sum_{n=1}^{\infty} Z^n = 1 + Z + Z^2 +$ $\cdots + Z^n + \cdots$ which converges for Z in the bicomplex unitary ball, that is, the bicomplex ball of hyperbolic radius 1, and which diverges outside this ball; this is because the bicomplex geometric series is equivalent to the pair of complex geometric series:

$$\sum_{n=1}^{\infty} Z^n = \sum_{n=1}^{\infty} \beta_1^n \mathbf{e} + \sum_{n=1}^{\infty} \beta_2^n \mathbf{e}^\dagger.$$

Theorem 10.1.1 (Cauchy criteria). *The series* $\displaystyle\sum_{n=1}^{\infty} Z_n$ *converges if and only if the sequence of its partial sums is a* $\mathbb{D}$-*Cauchy sequence, i.e., for any positive hyperbolic* *number ε there exists a natural number $N = N(\varepsilon)$ such that*

$$\left| \sum_{k=1}^{m} Z_{N+k} \right|_{\mathbb{D}} \prec \varepsilon$$

for any natural m.

Proof. Follows by applying the Cauchy criteria to the complex series in (10.2).   □

This theorem gives a necessary condition of convergence of the series (10.1): the sequence $\{Z_n\}$ $\mathbb{D}$-converges to zero. Another consequence of the theorem is: if $\{A_n\}_{n=1}^{\infty}$ is a sequence of non-negative hyperbolic numbers: $A_n \in \mathbb{D}^+$, and if $|Z_n|_{\mathbb{D}} \prec A_n$ for $n \geq N$, then the convergence of $\sum_{n=1}^{\infty} A_n$ implies the convergence of $\sum_{n=1}^{\infty} Z_n$.

A bicomplex series $\sum_{n=1}^{\infty} Z_n$ is called $\mathbb{D}$-absolutely convergent if the series $\sum_{n=1}^{\infty} |Z_n|_{\mathbb{D}}$ of hyperbolic numbers is convergent. Of course, a $\mathbb{D}$-absolutely convergent series is convergent.

It is easy to prove that:

- If the bicomplex series (10.1) is $\mathbb{D}$-absolutely convergent, then any series composed of its members is also convergent and has the same sum as the initial series.

- If the series $\sum_{n=1}^{\infty} Z_n$ converges $\mathbb{D}$-absolutely and has the sum Z and if the series $\sum_{n=1}^{\infty} W_n$ converges to W, then the Cauchy product of the series, that is, the series $\sum_{n=1}^{\infty} U_n$ with $U_n = \sum_{k=0}^{n} Z_k W_{n-k}$, is convergent and its sum equals ZW.

10.2   General properties of sequences and series of functions

Consider now a sequence of functions $f_n : \Omega \subset \mathbb{BC} \longrightarrow \mathbb{BC}$. We say that the sequence converges on Ω to a function f (called the limit function) if for any $Z \in \Omega$ the sequence of bicomplex numbers $\{f_n(Z)\}_{n=1}^{\infty}$ converges to $f(Z)$, that is, for any $\varepsilon \succ 0$ there exists $N = N(\varepsilon, Z)$ such that $\left|f_n(Z) - f(Z)\right|_{\mathbb{D}} \prec \varepsilon$ for $n \geq N(\varepsilon, Z)$. Since the functions f_n are not necessarily bicomplex holomorphic their idempotent components $F_{1,n}$ and $F_{2,n}$ depend, in general, on β_1 and β_2, not on one of them. But the convergence of $\{f_n\}_{n=1}^{\infty}$ is equivalent to the convergence of the two complex sequences $\{F_{1,n}\}_{n=1}^{\infty}$ and $\{F_{2,n}\}_{n=1}^{\infty}$ on Ω.

A sequence $\{f_n\}_{n=1}^{\infty}$ is $\mathbb{D}$-uniformly convergent to f if for any $\varepsilon \succ 0$ the number N depends on ε only: $N = N(\varepsilon)$, and thus for $n \geq N$ the inequality $\left|f_n(Z) - f(Z)\right|_{\mathbb{D}} \prec \varepsilon$ holds for all Z. The $\mathbb{D}$-uniform convergence holds if and only if for any $\varepsilon \succ 0$ there exists $N = N(\varepsilon)$ such that for $m \geq N$, $n \geq N$ one has: $\left|f_n(Z) - f_m(Z)\right|_{\mathbb{D}} \prec \varepsilon$. Of course, a $\mathbb{D}$-uniformly convergent sequence is convergent.

Theorem 10.2.1. *If all the functions f_n are continuous on Ω and the sequence converges $\mathbb{D}$-uniformly on Ω, then the limit function f is continuous on Ω.*

Proof. Mimics the complex functions case using the properties of hyperbolic positive numbers instead of real positive numbers. □

Theorem 10.2.2. *Let Γ be a rectifiable curve in $\mathbb{BC}$ without self-intersections and let a sequence $\{f_n\}_{n=1}^{\infty}$ converge $\mathbb{D}$-uniformly on Γ, then*

$$\lim_{n \to \infty} \int_{\Gamma} f_n(t)\, dt = \int_{\Gamma} \lim_{n \to \infty} f_n(t)\, dt.$$

Proof. Follows the same line as the previous one. □

Theorem 10.2.3 (Bicomplex holomorphy of the limit functions; an analogue of Weierstrass' theorem). *If every f_n is bicomplex holomorphic in $\Omega = \Omega_1 \mathbf{e} + \Omega_2 \mathbf{e}^{\dagger}$ and if the sequence $\{f_n\}_{n=1}^{\infty}$ converges $\mathbb{D}$-uniformly on every compact set in Ω to a function f, then f is bicomplex holomorphic in Ω and for every $k \in \mathbb{N}$ the sequence $\left\{f_n^{(k)}\right\}_{n=1}^{\infty}$ of the k-th derivatives converges $\mathbb{D}$-uniformly on compacts to $f^{(k)}$.*

Proof. Since each f_n is bicomplex holomorphic, then $f_n(Z) = F_{1,n}(\beta_1)\mathbf{e} + F_{2,n}(\beta_2)\mathbf{e}^{\dagger}$ where for any n the functions $F_{1,n}$ are holomorphic in Ω_1 and all $F_{2,n}$ are holomorphic in Ω_2. If K is a compact subset in Ω and $K = K_1\mathbf{e} + K_2\mathbf{e}^{\dagger}$, then K_1 is a compact subset in Ω_1 and K_2 is a compact subset in Ω_2. Moreoover, $\{f_n\}_{n=1}^{\infty}$ converges $\mathbb{D}$-uniformly on K if and only if $\{F_{1,n}\}_{n=1}^{\infty}$ converges uniformly on K_1 and $\{F_{2,n}\}_{n=1}^{\infty}$ converges uniformly on K_2.

Note that a priori the limit function f does not need to be bicomplex holomorphic, in general, thus, in its idempotent representation

$$f(Z) = f(\beta_1\mathbf{e} + \beta_2\mathbf{e}^{\dagger}) = F_1(\beta_1\mathbf{e} + \beta_2\mathbf{e}^{\dagger})\mathbf{e} + F_2(\beta_1\mathbf{e} + \beta_2\mathbf{e}^{\dagger})\mathbf{e}^{\dagger}$$

the components F_1 and F_2 would depend on both variables β_1 and β_2. But the complex Weierstrass theorem implies that the sequence $\{F_{1,n}\}_{n=1}^{\infty}$ converges to F_1 uniformly on compacts and $\{F_{2,n}\}_{n-1}^{\infty}$ converges to F_2 uniformly on compacts, thus, F_1 is holomorphic in Ω_1 and F_2 is holomorphic in Ω_2. Because of the uniqueness of the limit function, F_1 depends on β_1 only and F_2 depends on β_2 only. Recalling that $f = F_1\mathbf{e} + F_2\mathbf{e}^\dagger$ we conclude that f is bicomplex holomorphic in Ω.

Moreover, for any $k \in \mathbb{N}$ the sequences $\left\{F_{1,n}^{(k)}\right\}_{n=1}^\infty$ and $\left\{F_{2,n}^{(k)}\right\}_{n=1}^\infty$ converge uniformly on compacts to $F_1^{(k)}$ and $F_2^{(k)}$ respectively. Hence, the sequence $f_n^{(k)} = F_{1,n}^{(k)}\mathbf{e} + F_{2,n}^{(k)}\mathbf{e}^\dagger$ converges $\mathbb{D}$-uniformly on compacts to the function $F_1^{(k)}\mathbf{e} + F_2^{(k)}\mathbf{e}^\dagger = f^{(k)}$ which completes the proof. □

10.3 Convergent series of bicomplex functions

We now consider series of the form

$$\sum_{n=1}^\infty f_n(Z) = f_1(Z) + f_2(Z) + \cdots + f_n(Z) + \cdots \qquad (10.3)$$

where all the functions f_n are defined on the set Ω in $\mathbb{BC}$. Their behavior is characterized by the sequence $S_n(Z) = \sum_{k=1}^n f_k(Z)$ of partial sums, hence the properties of sequences (convergence at a point, $\mathbb{D}$-uniform convergence on the set, etc.) apply to the sequence $\{S_n(Z)\}_{n=1}^\infty$ and become the respective properties of series of functions. For instance, in order that the series of functions (10.3) converges $\mathbb{D}$-uniformly on Ω it is necessary and sufficient that for any positive hyperbolic number ε there exists $N = N(\varepsilon)$ such that $\left|f_{N+1}(Z) + \cdots + f_{N+m}(Z)\right|_{\mathbb{D}} \prec \varepsilon$ for any $Z \in \Omega$ and any $m \in \mathbb{N}$ (Cauchy's criteria).

A series $\sum_{n=1}^\infty f_n(Z)$ is called $\mathbb{D}$-absolutely convergent if the series $\sum_{n=1}^\infty \left|f_n(Z)\right|_{\mathbb{D}}$ converges on Ω.

Theorem 10.3.1 (Weierstrass test for $\mathbb{D}$-uniform convergence). *Given a series of functions* $\sum_{n=1}^\infty f_n(Z)$, *if there exists a convergent series* $\sum_{n=1}^\infty A_n$ *of hyperbolic non-negative numbers such that the inequality* $\left|f_n(Z)\right| \preceq A_n$ *holds for* $n = n^*, n^* + 1, \ldots,$ *for some* $n^* \in \mathbb{N}$ *and for all* Z *in* Ω, *then the series* $\sum_{n=1}^\infty f_n(Z)$ *converges $\mathbb{D}$-uniformly and $\mathbb{D}$-absolutely on* Ω.

Proof. Consists in applying Cauchy's criteria twice. □

The properties of sequences of functions give rise to the following statements about series of functions.

- If the functions f_n are continuous on Ω and the series converges $\mathbb{D}$-uniformly on Ω, then the sum $S(Z)$ is a continuous function.

- Let Γ be a rectifiable curve without self-intersections, let every function f_n be continuous and assume that the series $\sum_{n=1}^{\infty} f_n$ converges $\mathbb{D}$-uniformly on Γ, then

$$\int_{\Gamma} \sum_{n=1}^{\infty} f_n(t)\, dt = \sum_{n=1}^{\infty} \int_{\Gamma} f_n(t)\, dt.$$

- (Weierstrass theorem.) If every f_n is bicomplex holomorphic in Ω and $\sum_{n=1}^{\infty} f_n(Z)$ converges $\mathbb{D}$-uniformly on compact subsets of Ω, then the sum S is bicomplex holomorphic in Ω and the series $\sum_{n=1}^{\infty} f_n^{(k)}(Z)$ converges $\mathbb{D}$-uniformly on compact subsets in Ω to $S^{(k)}(Z)$ for any $k \in \mathbb{N}$.

10.4 Bicomplex power series

Here we deal with the series of the form

$$\sum_{n=0}^{\infty} A_n (Z - Z_0)^n \tag{10.4}$$

with bicomplex coefficients A_n. Its members $A_n (Z - Z_0)^n$ are defined for any $Z \in \mathbb{BC}$ but, of course, it is a separate question to determine where such a series is convergent.

Equivalently, one can consider the series

$$\sum_{n=0}^{\infty} A_n\, Z^n \tag{10.5}$$

which is, obviously, convergent at the origin $Z = 0$.

Theorem 10.4.1 (Analogue of Abel's theorem). *Consider a series (10.5) and let $\mathbb{S} = \mathbb{S}_\rho = \{ Z \mid |Z|_{\mathbb{D}} = \rho \}$ be a bicomplex sphere of a hyperbolic radius $\rho \succ 0$ which is assumed to be an invertible hyperbolic number. If there is a point $\widetilde{Z}$ in $\mathbb{S}$ such that the series $\sum_{n=0}^{\infty} A_n\, Z^n$ converges at $\widetilde{Z}$, then it converges $\mathbb{D}$-absolutely in the open bicomplex ball $\mathbb{B} = \mathbb{B}_\rho$ of the same radius, and it converges $\mathbb{D}$-uniformly on compact subsets of this ball.*

Proof. First of all, note that if ρ is a zero-divisor, then any point of $\mathbb{S}$ is a zero-divisor, while if ρ is invertible, then all the points of $\mathbb{S}$ are invertible bicomplex numbers.

Let $\widetilde{Z} \in \mathbb{S}$ be a point of convergence. Take K to be a closed subset of $\mathbb{B}_\rho$ and consider the set $\left\{ \left| \dfrac{Z}{\widetilde{Z}} \right|_{\mathbb{D}} \mid Z \in K \right\}$. It is $\mathbb{D}$-bounded and, hence, it has a $\mathbb{D}$-supremum, $\sup_{\mathbb{D}}$; what is more, $\sup_{\mathbb{D}} \left\{ \left| \dfrac{Z}{\widetilde{Z}} \right|_{\mathbb{D}} \mid Z \in K \right\} =: q \prec 1$. Since the series $\displaystyle\sum_{n=0}^{\infty} A_n \widetilde{Z}^n$ converges, then $\lim_{n \to \infty} A_n \widetilde{Z}^n = 0$, thus the sequence $A_n \widetilde{Z}^n$ is $\mathbb{D}$-bounded: there exists a positive hyperbolic number M such that $\left| A_n \widetilde{Z}^n \right|_{\mathbb{D}} \preceq M$ for $n \in \mathbb{N}$. For $Z \in K$ we get now an estimate

$$\left| A_n Z^n \right|_{\mathbb{D}} = \left| A_n \widetilde{Z}^n \cdot \left(\frac{Z}{\widetilde{Z}} \right)^n \right|_{\mathbb{D}} \preceq M q^n.$$

The series with general term $M q^n$ converges and it may serve as a majorant for the series $\displaystyle\sum_{n=0}^{\infty} A_n Z^n$ on K; the Weierstrass theorem allows us to conclude that $\displaystyle\sum_{n=0}^{\infty} A_n Z^n$ converges $\mathbb{D}$-absolutely and $\mathbb{D}$-uniformly on K. Since any point of $\mathbb{B}_\rho$ belongs to some subset K, the theorem is proved. $\square$

As was noted above, if ρ is a zero-divisor, i.e., $\rho = \rho_1 \mathbf{e}$ or $\rho = \rho_2 \mathbf{e}^\dagger$, then all the points of $\mathbb{S}_\rho$ are zero-divisors of the same form. What can be said if $\widetilde{Z} \in \mathbb{S}_\rho$ and the series $\displaystyle\sum_{n=0}^{\infty} A_n \widetilde{Z}^n$ converges? Writing $A_n = a_n \mathbf{e} + b_n \mathbf{e}^\dagger$ we get:

$$\sum_{n=0}^{\infty} A_n \widetilde{Z}^n = \sum_{n=0}^{\infty} \left(a_n \mathbf{e} + b_n \mathbf{e}^\dagger \right) \left(\widetilde{\beta}_1 \mathbf{e} + \widetilde{\beta}_2 \mathbf{e}^\dagger \right) = \left(\sum_{n=0}^{\infty} a_n \widetilde{\beta}_1^n \right) \mathbf{e} + \left(\sum_{n=0}^{\infty} b_n \widetilde{\beta}_2^n \right) \mathbf{e}^\dagger.$$

Since $\widetilde{Z}$ is of one of the forms: $\widetilde{Z} = \widetilde{\beta}_1 \mathbf{e}$ or $\widetilde{Z} = \widetilde{\beta}_2 \mathbf{e}^\dagger$, then we can make a conclusion for one of the bicomplex components of the series $\displaystyle\sum_{n=0}^{\infty} A_n Z^n = \left(\sum_{n=0}^{\infty} a_n \beta_1^n \right) \mathbf{e} + \left(\sum_{n=0}^{\infty} b_n \beta_2^n \right) \mathbf{e}^\dagger$. Indeed, if $\rho = \rho_1 \mathbf{e}$, then the series $\displaystyle\sum_{n=0}^{\infty} a_n \beta_1^n \mathbf{e}$ converges for $|\beta_1| < \rho_1$ but we have no information for the series $\displaystyle\sum_{n=0}^{\infty} b_n \beta_2^n \mathbf{e}^\dagger$; similarly for $\rho = \rho_2 \mathbf{e}^\dagger$. Of course, this does not mean that there is no information in principle; we have just taken an "unlucky" point $\widetilde{Z}$.

Proposition 10.4.2 (Sets of convergence of bicomplex power series). *Given a series* $\displaystyle\sum_{n=0}^{\infty} A_n Z^n$, *one of the following options holds:*

(1) the series converges at the origin only;

(2) *the series converges on the whole* $\mathbb{BC}$;

(3) *the series converges on one of the idempotent complex lines* $\mathbb{BC}_{\mathbf{e}}$ *or* $\mathbb{BC}_{\mathbf{e}^\dagger}$ *and diverges on the other;*

(4) *the series converges on the disk centered at the origin and of (real) radius* r *situated on* $\mathbb{BC}_{\mathbf{e}}$ *or* $\mathbb{BC}_{\mathbf{e}^\dagger}$ *and diverges on the complement of its closure;*

(5) *the series converges on* $\mathbb{B}(0,r)\mathbf{e} + \mathbb{BC}_{\mathbf{e}^\dagger}$ *or on* $\mathbb{BC}_{\mathbf{e}} + \mathbb{B}(0,r)\mathbf{e}^\dagger$ *with* $r > 0$ *and diverges on the complement of its closure;*

(6) *the series converges on the bicomplex ball centered at the origin of hyperbolic radius* $R = r_1\mathbf{e} + r_2\mathbf{e}^\dagger \succ 0$ *and diverges on the complement of its closure.*

Proof. The first two options are illustrated with the series $\sum_{n=0}^{\infty} n!\, Z^n$ and $\sum_{n=0}^{\infty} \dfrac{Z^n}{n!}$.

Assume next that the first two cases are excluded and consider the idempotent representation

$$\sum_{n=0}^{\infty} A_n\, Z^n = \left(\sum_{n=0}^{\infty} a_n \beta_1^n\right) \mathbf{e} + \left(\sum_{n=0}^{\infty} b_n \beta_2^n\right) \mathbf{e}^\dagger.$$

Several possibilities arise now. Let, first, one of the idempotent series converge at the origin only, then the other may converge in a disk of finite radius which gives the case (4) or in the whole $\mathbb{C}(\mathbf{i})$ which gives the case (3); illustrations for (3) are

$$\sum_{n=0}^{\infty} \left(n!\, \beta_1^n \mathbf{e} + \frac{1}{n!}\, \beta_2^n \mathbf{e}^\dagger\right) \quad \text{or} \quad \sum_{n=0}^{\infty} \left(\frac{1}{n!}\, \beta_1^n \mathbf{e} + n!\, \beta_2^n \mathbf{e}^\dagger\right)$$

and illustrations for (4) are

$$\sum_{n=0}^{\infty} \left(\gamma_n\, \beta_1^n \mathbf{e} + n!\, \beta_2^n \mathbf{e}^\dagger\right) \quad \text{or} \quad \sum_{n=0}^{\infty} \left(n!\, \beta_1^n \mathbf{e} + \gamma_n\, \beta_2^n \mathbf{e}^\dagger\right)$$

where $\sum_{n=0}^{\infty} \gamma_n\, z^n$ is a complex series with a (real) radius of convergence $r > 0$. Second, both idempotent series converge on a disk which may be of a finite or infinite radius thus justifying the cases (5) and (6); for the same complex series $\sum_{n=0}^{\infty} \gamma_n\, z^n$

any of the bicomplex series $\sum_{n=0}^{\infty} \left(\gamma_n \beta_1^n \mathbf{e} + \frac{1}{n!}\, \beta_2^n \mathbf{e}^\dagger\right)$ or $\sum_{n=0}^{\infty} \left(\frac{1}{n!}\, \beta_1^n \mathbf{e} + \gamma_n\, \beta_2^n \mathbf{e}^\dagger\right)$

illustrates the former case. $\qquad\qquad\qquad\qquad\qquad\qquad\qquad\qquad\qquad\qquad\square$

Analogously to the case of complex analysis, we may speak in all the cases about bicomplex balls $\mathbb{B}_R$ of hyperbolic radii R where:

(1) $R = 0 = 0\mathbf{e} + 0\mathbf{e}^\dagger$, and the ball reduces to a point;

(2) $R = \infty = \infty\mathbf{e} + \infty\mathbf{e}^\dagger$, and the ball means the whole $\mathbb{BC}$;

(3) $R = 0\mathbf{e} + \infty\mathbf{e}^\dagger$ or $R = \infty\mathbf{e} + 0\mathbf{e}^\dagger$, and the ball means one of the complex lines $\mathbb{BC}_\mathbf{e}$ or $\mathbb{BC}_{\mathbf{e}^\dagger}$;

(4) $R = r\mathbf{e}$ or $R = r\mathbf{e}^\dagger$, and the ball means one of the "labeled" disks $\mathbb{B}(0, r)\mathbf{e}$ or $\mathbb{B}(0, r)\mathbf{e}^\dagger$;

(5) $R = r\mathbf{e} + \infty\mathbf{e}^\dagger$ or $R = \infty\mathbf{e} + r\mathbf{e}^\dagger$;

(6) $R = r_1\mathbf{e} + r_2\mathbf{e}^\dagger$, and this is an "authentic" bicomplex ball of a hyperbolic radius.

This R will be called the hyperbolic radius of convergence of a bicomplex power series.

Proposition 10.4.3. *If R is strictly positive (i.e., $R = r_1\mathbf{e} + r_2\mathbf{e}^\dagger$, or $R = \infty$, or $R = r\mathbf{e} + \infty\mathbf{e}^\dagger$ or $R = \infty\mathbf{e} + r\mathbf{e}^\dagger$), then the sum $S(Z)$ of a power series is a bicomplex holomorphic function and*

$$S'(Z) = \sum_{n=1}^\infty n\, A_n\, Z^{n-1}.$$

Proof. Follows directly from the theorems of Abel and Weierstrass.   □

Proposition 10.4.4. *If R is strictly positive, then $A_n = \dfrac{S^{(n)}(0)}{n!}$.*

Proof. It follows from the previous proposition that

$$S^{(k)}(Z) = \sum_{n=k}^\infty n(n-1)\cdots(n-k+1)\, A_n\, Z^{(n-k)}$$

for $k \in \mathbb{N}$. In particular, $S^{(k)}(0) = k(k-1)\cdots 1\, A_k = k!\, A_k$.   □

10.5   Bicomplex Taylor Series

Definition 10.5.1. *Let Ω be a domain in $\mathbb{BC}$, and assume that a function $f : \Omega \longrightarrow \mathbb{C}$ is such that it has derivatives of any order at $Z_0 \in \Omega$. Then the power series $\sum_{n=0}^\infty \dfrac{f^{(n)}(Z_0)}{n!}\,(Z - Z_0)^n$ is called the Taylor series for f at Z_0.*

Theorem 10.5.2. *If a power series $\sum_{n=0}^\infty A_n\,(Z - Z_0)^n$ has a strictly positive radius of convergence R, then it is the Taylor series for its sum at Z_0.*

Proof. By Proposition 10.4.4, $A_n = \dfrac{S^{(n)}(Z_0)}{n!}$, thus

$$S(Z) = \sum_{n=0}^{\infty} A_n \, (Z - Z_0)^n = \sum_{n=0}^{\infty} \frac{S^{(n)}(Z_0)}{n!} \cdot (Z - Z_0)^n \qquad \qquad \square$$

Theorem 10.5.3. *Let Ω be a product-type domain in $\mathbb{BC}$, $\Omega = \Omega_1 \mathbf{e} + \Omega_2 \mathbf{e}^\dagger$, let f be bicomplex holomorphic in Ω. For a fixed Z_0 in Ω consider the Taylor series of f at Z_0. Denote by d the hyperbolic distance between Z_0 and the distinguished boundary of Ω, that is,*

$$d := \inf{}_{\mathbb{D}} \left\{ \left| Z - Z_0 \right|_{\mathbb{D}} \mid Z \in \partial\Omega_1 \mathbf{e} + \partial\Omega_2 \mathbf{e}^\dagger \right\} \in \mathbb{D}.$$

Then the Taylor series converges in the bicomplex ball $\mathbb{B}_d(Z_0)$ where its sum coincides with $f(Z)$:

$$f(Z) = \sum_{n=0}^{\infty} \frac{f^{(n)}(Z_0)}{n!} \, (Z - Z_0)^n$$

for any $Z \in \mathbb{B}_d(Z_0)$.

Proof. As f is bicomplex holomorphic we can write

$$f(Z) = f(\beta_1 \mathbf{e} + \beta_2 \mathbf{e}^\dagger) = F_1(\beta_1)\mathbf{e} + F_2(\beta_2)\mathbf{e}^\dagger,$$

where F_1 and F_2 are (complex) holomorphic, respectively, in Ω_1 and Ω_2, and

$$Z_0 = \beta_1^0 \mathbf{e} + \beta_2^0 \mathbf{e}^\dagger,$$

where $\beta_1^0 \in \Omega_1$, $\beta_2^0 \in \Omega_2$. Denote by d_1 the Euclidean distance between β_1^0 and $\partial\Omega_1$, and by d_2 the Euclidean distance between β_2^0 and $\partial\Omega_2$, then by the complex Taylor theorem one has:

$$F_1(\beta_1) = \sum_{n=0}^{\infty} \frac{F_1^{(n)}(\beta_1^0)}{n!} \left(\beta_1 - \beta_1^0\right)^n$$

for $\beta_1 \in \mathbb{B}\left(\beta_1^0, d_1\right)$, and

$$F_2(\beta_2) = \sum_{n=0}^{\infty} \frac{F_2^{(n)}(\beta_2^0)}{n!} \left(\beta_2 - \beta_2^0\right)^n$$

for $\beta_2 \in \mathbb{B}\left(\beta_2^0, d_2\right)$. Hence

$$f(Z) = F_1(\beta_1)\mathbf{e} + F_2(\beta_2)\mathbf{e}^\dagger$$

$$= \sum_{n=0}^{\infty} \frac{F_1^{(n)}(\beta_1^0)}{n!} \left(\beta_1 - \beta_1^0\right)^n \mathbf{e} + \sum_{n=0}^{\infty} \frac{F_2^{(n)}(\beta_2^0)}{n!} \left(\beta_2 - \beta_2^0\right)^n \mathbf{e}^\dagger$$

$$= \sum_{n=0}^{\infty} \frac{1}{n!} \left(F_1^{(n)}(\beta_1^0)\,\mathbf{e} + F_2^{(n)}(\beta_2^0)\,\mathbf{e}^\dagger\right) \left(\left(\beta_1 - \beta_1^0\right)^n \mathbf{e} + \left(\beta_2 - \beta_2^0\right)^n \mathbf{e}^\dagger\right)$$

$$= \sum_{n=0}^{\infty} \frac{f^{(n)}(Z_0)}{n!} \left(Z - Z_0\right)^n$$

for any Z in $\mathbb{B}\left(\beta_1^0, d_1\right)\mathbf{e} + \mathbb{B}\left(\beta_2^0, d_2\right)\mathbf{e}^\dagger = \mathbb{B}_d(Z_0)$. The proof is completed.   □

Example 10.5.4. *Recalling the formulas for the derivatives of elementary functions we obtain the following expansions into bicomplex Taylor series:*

$$e^Z = \sum_{n=0}^{\infty} \frac{Z^n}{n!}; \quad \sin Z = \sum_{n=0}^{\infty} (-1)^n \frac{Z^{2n+1}}{(2n+1)!}; \quad \cos Z = \sum_{n=0}^{\infty} (-1)^n \frac{Z^{2n}}{(2n)!}.$$

In addition to the book of Price [56], we refer the reader also to publications such as [40, 41, 84], where certain properties of bicomplex sequences and series are studied.

Chapter 11

Integral Formulas and Theorems

In this chapter we establish bicomplex analogues of the main integral theorems and formulas of one-dimensional complex analysis. We are not going to reach the highest level of generality for curves and surfaces involved since our aim is to present some basic ideas and structures for those formulas; the more general setting will be presented elsewhere.

11.1 Stokes' formula compatible with the bicomplex Cauchy–Riemann operators

Let Ω be a domain in $\mathbb{BC}$ and consider a function F of class $\mathcal{C}^1(\Omega, \mathbb{BC})$. It is thus real differentiable and, recalling (7.48), we have that its real differential dF is

$$dF = \frac{\partial F}{\partial Z} dZ + \frac{\partial F}{\partial \overline{Z}} d\overline{Z} + \frac{\partial F}{\partial Z^\dagger} dZ^\dagger + \frac{\partial F}{\partial Z^*} dZ^*.$$

Note that this is just a "bicomplex combination" of the real differentials of the components of F; hence the function itself and the variable Z can be written in any form as well as all the differentials and the differential operators; for instance, $dZ = dx_1 + \mathbf{i}dy_1 + \mathbf{j}dx_2 + \mathbf{k}dy_2 = dz_1 + \mathbf{j}dz_2 = d\zeta_1 + \mathbf{i}d\zeta_2 = \ldots,\quad d\overline{Z} = dx_1 - \mathbf{i}dy_1 + \mathbf{j}dx_2 - \mathbf{k}dy_2 = d\overline{z}_1 + \mathbf{j}d\overline{z}_2 = \ldots.$.

Next, consider the action of the exterior differentiation operator on the differential form FdZ; we have that

$$d(FdZ) = dF \wedge dZ$$

$$= \left(\frac{\partial F}{\partial Z} dZ + \frac{\partial F}{\partial \overline{Z}} d\overline{Z} + \frac{\partial F}{\partial Z^\dagger} dZ^\dagger + \frac{\partial F}{\partial Z^*} dZ^* \right) \wedge dZ \qquad (11.1)$$

$$= \frac{\partial F}{\partial \overline{Z}} d\overline{Z} \wedge dZ + \frac{\partial F}{\partial Z^\dagger} dZ^\dagger \wedge dZ + \frac{\partial F}{\partial Z^*} dZ^* \wedge dZ.$$

The same can be done with the three other differential forms, that is, $Fd\overline{Z}$, $FdZ^\dagger$, FdZ^*, arriving at

$$d(Fd\overline{Z}) = \frac{\partial F}{\partial Z} dZ \wedge d\overline{Z} + \frac{\partial F}{\partial Z^\dagger} dZ^\dagger \wedge d\overline{Z} + \frac{\partial F}{\partial Z^*} dZ^* \wedge d\overline{Z}; \qquad (11.2)$$

$$d(FdZ^\dagger) = \frac{\partial F}{\partial Z} dZ \wedge dZ^\dagger + \frac{\partial F}{\partial \overline{Z}} d\overline{Z} \wedge dZ^\dagger + \frac{\partial F}{\partial Z^*} dZ^* \wedge dZ^\dagger; \qquad (11.3)$$

$$d(FdZ^*) = \frac{\partial F}{\partial Z} dZ \wedge dZ^* + \frac{\partial F}{\partial \overline{Z}} d\overline{Z} \wedge dZ^* + \frac{\partial F}{\partial Z^\dagger} dZ^\dagger \wedge dZ^*. \qquad (11.4)$$

Of course, each of the formulas (11.2), (11.3) and (11.4) can be obtained also by making the corresponding change of variable.

Next, take Γ to be a two-dimensional, piecewise smooth, oriented surface in Ω whose boundary $\gamma = \partial\Gamma$ is a piecewise smooth curve.

Then integrating both parts of (11.1) and applying Stokes theorem we get:

$$\int_\gamma FdZ = \int_\Gamma d(FdZ)$$
$$= \int_\Gamma \left(\frac{\partial F}{\partial \overline{Z}} d\overline{Z} \wedge dZ + \frac{\partial F}{\partial Z^\dagger} dZ^\dagger \wedge dZ + \frac{\partial F}{\partial Z^*} dZ^* \wedge dZ \right). \qquad (11.5)$$

Similar formulas arise if one uses (11.2), (11.3) and (11.4).

Formula (11.5) contains several special cases which are of interest by themselves. In particular, we can take F as

$$F(z_1, z_2) = f_1(z_1, z_2) + \mathbf{j}f_2(z_1, z_2)$$

where $f_2(z_1, z_2) = 0$ in Ω and f_1 is holomorphic in the sense of two complex variables z_1 and z_2 which leads to

$$\int_\gamma f_1(z_1, z_2)(dz_1 + \mathbf{j}dz_2) = \int_\Gamma \left(\frac{\partial f_1}{\partial z_1} + \mathbf{j}\frac{\partial f_1}{\partial z_2} \right) \mathbf{j}\, dz_1 \wedge dz_2$$

since $\dfrac{\partial f_1}{\partial \overline{Z}} = \dfrac{\partial f_1}{\partial Z^*} = 0$ in Ω and

$$dZ^\dagger \wedge dZ = 2\mathbf{j}\, dz_1 \wedge dz_2.$$

Separation of the complex components results in the formulas

$$\int_\gamma f_1(z_1, z_2)dz_1 = -\int_\Gamma \frac{\partial f_1}{\partial z_2} dz_1 \wedge dz_2;$$

$$\int_\gamma f_1(z_1, z_2)dz_2 = \int_\Gamma \frac{\partial f_1}{\partial z_1} dz_1 \wedge dz_2.$$

The formulas remain true if, in addition, f does not depend on z_2:

$$\int_\gamma f_1(z_1)dz_1 = 0,$$

$$\int_\gamma f_1(z_1)dz_2 = \int_\Gamma f_1'(z_1)dz_1 \wedge dz_2.$$

Analogously, if f_1 does not depend on z_1, then

$$\int_\gamma f_1(z_2)dz_1 = -\int_\Gamma f_1'(z_2)dz_1 \wedge dz_2,$$

$$\int_\gamma f_1(z_2)dz_2 = 0.$$

But F can be taken also as

$$F(\zeta_1, \zeta_2) = \rho_1(\zeta_1, \zeta_2) + i\rho_2(\zeta_1, \zeta_2)$$

where $\rho_2(\zeta_1, \zeta_2) = 0$ in Ω and ρ_1 is holomorphic in the sense of two complex variables ζ_1 and ζ_2; now we can repeat the reasoning and arrive at very similar formulas and conclusions.

Finally, F can be taken as

$$F(\mathfrak{z}_1, \mathfrak{z}_2) = \mathfrak{f}_1(\mathfrak{z}_1, \mathfrak{z}_2) + i\mathfrak{f}_2(\mathfrak{z}_1, \mathfrak{z}_2)$$

where $\mathfrak{z}_1$ and $\mathfrak{z}_2$ are hyperbolic variables, $\mathfrak{f}_1$ and $\mathfrak{f}_2$ are $\mathbb{D}$-valued functions; what is more, we assume that $\mathfrak{f}_2(\mathfrak{z}_1, \mathfrak{z}_2) = 0$ in Ω and that $\mathfrak{f}_1$ is holomorphic in the sense of two hyperbolic variables $\mathfrak{z}_1$ and $\mathfrak{z}_2$. Then $\dfrac{\partial \mathfrak{f}_1}{\partial \overline{Z}} = \dfrac{\partial \mathfrak{f}_1}{\partial Z^\dagger} = 0$ and $dZ^* \wedge dZ = 2\,i d\mathfrak{z}_1 \wedge d\mathfrak{z}_2$; thus (11.5) gives the equality

$$\int_\gamma \mathfrak{f}_1(\mathfrak{z}_1, \mathfrak{z}_2)(d\mathfrak{z}_1 + i d\mathfrak{z}_2) = \int_\Gamma \left(\frac{\partial \mathfrak{f}_1}{\partial \mathfrak{z}_1} + i\frac{\partial \mathfrak{f}_1}{\partial \mathfrak{z}_2} \right) i\, d\mathfrak{z}_1 \wedge d\mathfrak{z}_2.$$

Separation of the hyperbolic components gives:

$$\int_\gamma \mathfrak{f}_1(\mathfrak{z}_1, \mathfrak{z}_2)d\mathfrak{z}_1 = -\int_\Gamma \frac{\partial \mathfrak{f}_1}{\partial \mathfrak{z}_2} d\mathfrak{z}_1 \wedge d\mathfrak{z}_2,$$

$$\int_\gamma \mathfrak{f}_1(\mathfrak{z}_1, \mathfrak{z}_2)d\mathfrak{z}_2 = \int_\Gamma \frac{\partial \mathfrak{f}_1}{\partial \mathfrak{z}_1} d\mathfrak{z}_1 \wedge d\mathfrak{z}_2.$$

The same comments as above can be made although the situation is rather different because holomorphic functions of two hyperbolic variables do not have yet their theory.

But, of course, for us the most important consequence of (11.5) is the bicomplex Cauchy integral theorem.

Theorem 11.1.1 (Bicomplex Cauchy integral theorem). *Let F be a bicomplex holo-morphic function in a product-type domain $\Omega \subset \mathbb{BC}$. If γ is any piecewise smooth curve which is the boundary of a two-dimensional, piecewise smooth surface $\Gamma \subset \Omega$, then*

$$\int_{\gamma} f(Z)\, dZ = 0.$$

Proof. Since F is bicomplex holomorphic, then $\dfrac{\partial F}{\partial \overline{Z}} = \dfrac{\partial F}{\partial Z^{\dagger}} = \dfrac{\partial F}{\partial Z^{*}} = 0$ in Ω, and the result follows from (11.5). $\qquad\qquad\qquad\qquad\qquad\qquad\qquad\qquad\qquad\qquad\quad\square$

11.2   Bicomplex Borel–Pompeiu formula

Having in mind further developments and applications, we are not looking for a high level of generality; our aim is to obtain a bicomplex analogue of the com-plex Borel–Pompeiu formula which is based on the idempotent representation of bicomplex numbers. This requires us to work with curves and surfaces in $\mathbb{BC}$ of particular shape and with $\mathbb{BC}$-valued $\mathcal{C}^{1}$-functions of a particular structure. It is explained more precisely immediately after this brief introduction.

Let Ω be a domain in $\mathbb{BC}$, and consider a two-dimensional, simply connected, piecewise smooth surface $\Gamma \subset \Omega$ with boundary $\gamma = \partial\Gamma \subset \Omega$ which has the following properties: Γ has a parametrization $\psi = \psi(u,v)$ such that $\psi = \psi_{1}\mathbf{e}+\psi_{2}\mathbf{e}^{\dagger}$ where ψ_{1} and ψ_{2} are the parametrizations, respectively, of domains Γ_{1} and Γ_{2} in $\mathbb{C}(\mathbf{i})$ which are simply connected; γ has a parametrization $\varphi = \varphi(t)$ which is the restriction of ψ onto $\partial\Gamma$ and is such that $\varphi = \varphi_{1}\mathbf{e} + \varphi_{2}\mathbf{e}^{\dagger}$ where φ_{1} and φ_{2} are the parametrizations, respectively, of $\gamma_{1} := \partial\Gamma_{1}$ and of $\gamma_{2} := \partial\Gamma_{2}$, γ_{1} and γ_{2} being piecewise smooth, closed, Jordan curves in $\mathbb{C}(\mathbf{i})$.

We illustrate now with an example that the restrictions on Γ and γ are not contradictory. Let Γ_{1} be a domain in $\mathbb{C}(\mathbf{i})$ with the above described properties and let Γ_{2} be another domain in $\mathbb{C}(\mathbf{i})$ which is conformally equivalent to Γ_{1} via the mapping $\psi_{2} : \Gamma_{1} \to \Gamma_{2}$; if we denote as $\psi_{1} : u + \mathbf{i}v \in \Gamma_{1} \mapsto u + \mathbf{i}v \in \Gamma_{1}$ the identity mapping, then the parametrization $\psi := \psi_{1}(u,v)\mathbf{e} + \psi_{2}(u,v)\mathbf{e}^{\dagger}$ determines the surface Γ in $\mathbb{BC}$ with the necessary properties.

As the next step we consider the integrals over γ and Γ. If g is a continuous function on γ, then we assume additionally that g is of the form $g(Z) = g_{1}(\beta_{1})\mathbf{e}+ g_{2}(\beta_{2})\mathbf{e}^{\dagger}$ for $Z = \beta_{1}\mathbf{e} + \beta_{2}\mathbf{e}^{\dagger}$. Note that this property is true automatically for bicomplex holomorphic functions, but for continuous functions their idempotent coefficients depend, in general, on both variables β_{1} and β_{2}. The requirement above is caused by the method which we use below. Now, using the definition and the properties of the integral of the differential form $g(Z)\, dZ$ we have:

$$\int_\gamma g(Z)\,dZ = \int_a^b g(\varphi(t))\,\varphi'(t)\,dt$$

$$= \int_a^b \left(g_1(\varphi_1(t))\mathbf{e} + g_2(\varphi_2(t))\mathbf{e}^\dagger\right)\left(\varphi_1'(t)\mathbf{e} + \varphi_2'(t)\mathbf{e}^\dagger\right)\,dt$$

$$= \int_a^b \left(g_1(\varphi_1(t))\,\varphi_1'(t)\mathbf{e} + g_2(\varphi_2(t))\,\varphi_2'(t)\mathbf{e}^\dagger\right)\,dt$$

$$= \mathbf{e}\int_{\gamma_1} g_1(\beta_1)\,d\beta_1 + \mathbf{e}^\dagger \int_{\gamma_2} g_2(\beta_2)\,d\beta_2,$$

that is,

$$\int_\gamma g(Z)\,dZ = \mathbf{e}\int_{\gamma_1} g_1(\beta_1)\,d\beta_1 + \mathbf{e}^\dagger \int_{\gamma_2} g_2(\beta_2)\,d\beta_2. \tag{11.6}$$

As a matter of fact, the first equality is the definition of the integral of the bicomplex differential form $g(Z)\,dZ$ along γ, and it is easy to see that this definition is well posed.

A similar reasoning applies to the surface Γ and the differential form $g(Z)\,dZ \wedge dZ^*$. Indeed,

$$\int_\Gamma g(Z)\,dZ \wedge dZ^* = \int_\Gamma g(Z)\left(d\beta_1 \mathbf{e} + d\beta_2 \mathbf{e}^\dagger\right) \wedge \left(d\overline{\beta}_1 \mathbf{e} + d\overline{\beta}_2 \mathbf{e}^\dagger\right)$$

$$= \int_\Gamma \left(g_1(\beta_1)\mathbf{e} + g_2(\beta_2)\mathbf{e}^\dagger\right)\left(d\beta_1 \wedge d\overline{\beta}_1 \mathbf{e} + d\beta_2 \wedge d\overline{\beta}_2 \mathbf{e}^\dagger\right)$$

$$= \int_\Gamma \left(g_1(\beta_1)\,d\beta_1 \wedge d\overline{\beta}_1 \mathbf{e} + g_2(\beta_2)\,d\beta_2 \wedge d\overline{\beta}_2 \mathbf{e}^\dagger\right)$$

$$= \int_{(u,v)} \left(g_1\left(\psi_1(u,v)\right)\left(\frac{\partial\psi_1}{\partial u}\cdot\frac{\partial\overline{\psi}_1}{\partial v} - \frac{\partial\psi_1}{\partial v}\cdot\frac{\partial\overline{\psi}_1}{\partial u}\right)\mathbf{e}\right.$$

$$\left. +\; g_2\left(\psi_2(u,v)\right)\left(\frac{\partial\psi_2}{\partial u}\cdot\frac{\partial\overline{\psi}_2}{\partial v} - \frac{\partial\psi_2}{\partial v}\cdot\frac{\partial\overline{\psi}_2}{\partial u}\right)\mathbf{e}^\dagger\right)\,du\,dv$$

$$= \int_{\Gamma_1} g_1(\beta_1)\,d\beta_1 \wedge d\overline{\beta}_1 \mathbf{e} + \int_{\Gamma_2} g_2(\beta_2)\,d\beta_2 \wedge d\overline{\beta}_2 \mathbf{e}^\dagger,$$

that is,

$$\int_\Gamma g(Z)\,dZ \wedge dZ^* = \left(\int_{\Gamma_1} g_1(\beta_1)\,d\beta_1 \wedge d\overline{\beta}_1\right)\mathbf{e} + \left(\int_{\Gamma_2} g_2(\beta_2)\,d\beta_2 \wedge d\overline{\beta}_2\right)\mathbf{e}^\dagger. \tag{11.7}$$

Remark 11.2.1. *Under the same hypotheses, we say that the function g is integrable in the improper sense along Γ if the function g_ℓ is integrable in the improper sense along the domain Γ_ℓ for $\ell = 1$ and $\ell = 2$.*

Theorem 11.2.2 (Bicomplex Borel–Pompeiu formula). *Let $g \in \mathcal{C}^1(\Omega)$ be such that $g(Z) = g_1(\beta_1)\mathbf{e} + g_2(\beta_2)\mathbf{e}^\dagger$, $Z = \beta_1\mathbf{e} + \beta_2\mathbf{e}^\dagger$, and let γ and Γ be as described above.*

Then for any $Z \in \Gamma \setminus \gamma$,

$$g(Z) = \frac{1}{2\pi i} \int_\gamma \frac{g(t)\, dt}{t - Z} + \frac{1}{2\pi i} \int_\Gamma \frac{\frac{\partial g}{\partial t^*}}{t - Z}\, dt \wedge dt^*, \qquad (11.8)$$

where $t = t_1 \mathbf{e} + t_2 \mathbf{e}^\dagger$ and $\dfrac{\partial}{\partial t^*} = \mathbf{e}\dfrac{\partial}{\partial \bar{t}_1} + \mathbf{e}^\dagger \dfrac{\partial}{\partial \bar{t}_2}$.

Proof. One has:

$$g(Z) = g_1(\beta_1)\mathbf{e} + g_2(\beta_2)\mathbf{e}^\dagger.$$

Using the complex Borel–Pompeiu formula gives:

$$g(Z) = \frac{1}{2\pi i} \int_{\gamma_1} \frac{g_1(t_1)\, dt_1}{t_1 - \beta_1}\, \mathbf{e} + \frac{1}{2\pi i} \int_{\Gamma_1} \frac{\frac{\partial g_1}{\partial \bar{t}_1}}{t_1 - \beta_1}\, dt_1 \wedge d\bar{t}_1\, \mathbf{e}$$

$$+ \frac{1}{2\pi i} \int_{\gamma_2} \frac{g_2(t_2)\, dt_2}{t_2 - \beta_2}\, \mathbf{e}^\dagger + \frac{1}{2\pi i} \int_{\Gamma_2} \frac{\frac{\partial g_2}{\partial \bar{t}_2}}{t_2 - \beta_2}\, dt_2 \wedge d\bar{t}_2\, \mathbf{e}^\dagger$$

which, by regrouping the terms, equals

$$\frac{1}{2\pi i} \int_\gamma \frac{\left(g_1(t_1)\,\mathbf{e} + g_2(t_2)\,\mathbf{e}^\dagger\right)\cdot\left(dt_1\,\mathbf{e} + dt_2\,\mathbf{e}^\dagger\right)}{(t_1\mathbf{e} + t_2\mathbf{e}^\dagger) - (\beta_1\mathbf{e} + \beta_2\mathbf{e}^\dagger)}$$

$$+ \frac{1}{2\pi i} \int_\Gamma \frac{\left(\mathbf{e}\dfrac{\partial}{\partial \bar{t}_1} + \mathbf{e}^\dagger\dfrac{\partial}{\partial \bar{t}_2}\right)[g_1\mathbf{e} + g_2\mathbf{e}^\dagger]}{(t_1\mathbf{e} + t_2\mathbf{e}^\dagger) - (\beta_1\mathbf{e} + \beta_2\mathbf{e}^\dagger)}\left(dt_1 \wedge d\bar{t}_1\mathbf{e} + dt_2 \wedge d\bar{t}_2\,\mathbf{e}^\dagger\right)$$

$$= \frac{1}{2\pi i} \int_\gamma \frac{g(t)\, dt}{t - Z} + \frac{1}{2\pi i} \int_\Gamma \frac{\frac{\partial g}{\partial t^*}(t)}{t - Z}\, dt \wedge dt^*.$$

This completes the proof.   □

Note that we used in the proof formulas (11.6) and (11.7) and Remark 11.2.1.

Theorem 11.2.3 (The bicomplex Cauchy integral representation). *Let Ω be a product-type domain in $\mathbb{BC}$, let f be a bicomplex holomorphic function in Ω, and let Z be an arbitrary point in Ω. Then, for any surface $\Gamma \subset \Omega$ passing through Z and with the above described properties the Cauchy representation formula holds:*

$$F(Z) = \int_\gamma \frac{f(t)}{t - Z}\, dt,$$

where $\gamma = \partial\Gamma$.

Proof. Since a bicomplex holomorphic function is of the form $f(Z) = f_1(\beta_1)\mathbf{e} + f_2(\beta_2)\mathbf{e}^\dagger$, then we can apply (11.8). Moreover, as we know, $\dfrac{\partial f}{\partial \overline{Z}} = \dfrac{\partial f}{\partial Z^\dagger} = \dfrac{\partial f}{\partial Z^*} = 0$ and thus the surface integral in (11.8) vanishes. $\qquad\square$

Note that if a $\mathcal{C}^1$-function g is of the form $g(Z) = g_1(\beta_1)\mathbf{e} + g_2(\beta_2)\mathbf{e}^\dagger$, then the conditions $\dfrac{\partial g}{\partial \overline{Z}} = \dfrac{\partial g}{\partial Z^\dagger} = 0$ are valid. Hence the bicomplex holomorphic functions are singled out among the $\mathcal{C}^1$-functions of this form by the unique condition $\dfrac{\partial g}{\partial Z^*} = 0$, which explains why in the Borel–Pompeiu formula just one operator $\dfrac{\partial}{\partial Z^*}$ remains.

We have shown how Stokes' formula can be used to obtain the bicomplex Cauchy Integral Theorem, the bicomplex Borel-Pompeiu formula, and the bicomplex Cauchy Integral Representation formula.

Integration theory in the context of complexified Clifford Analysis, in which bicomplex analysis constitutes a first step, have been extensively studied by Ryan [68]–[81].

Bibliography

[1] L.V. AHLFORS. *Complex Analysis.* McGraw–Hill Book Co. (1966).

[2] D. ALPAY, M.E. LUNA, M. SHAPIRO, D.C. STRUPPA. *Basics of functional analysis with bicomplex scalars, and bicomplex Schur analysis.* Series SpringerBriefs in Mathematics (2014).

[3] H.T. ANASTASSIU, P.E. ATLAMAZOGLOU, D.I. KAKLAMANI. *Application of bicomplex (quaternion) algebra to fundamental electromagnetics: a lower order alternative to the Helmholtz equation.* IEEE Trans. Antennas and Propagation, v. 51, No. 8 (2003), 2130–2136.

[4] F. BABADAG, Y. YAYLI, N. EKMEKCI. *Homothetic motions at (E^8) with bicomplex numbers (C_3).* Int. J. Contemp. Math. Sci. v. 4, No. 33–36 (2009), 1619–1626.

[5] F. BABADAG, Y. YAYLI, N. EKMEKCI. *Homothetic motions and bicomplex numbers.* Commun. Fac. Sci. Univ. Ank. Ser A1 Math. Stat. v. 58, No. 1 (2009), 23–28.

[6] F. BABADAG. *Homothetic motions and bicomplex numbers.* Algebras Groups Geom. v. 26, No. 2 (2009), 193–201.

[7] P. BAIRD, J.C. WOOD. *Harmonic morphisms and bicomplex manifolds.* J. Geom. Phys. v. 61 (2011), 46–61.

[8] P. BERGLEZ. *On some classes of bicomplex pseudoanalytic functions.* Progress in analysis and its applications, Hackensack, NJ, World Sci. Publ. (2010), 81–88.

[9] J. BORY-REYES, M. SHAPIRO. *Clifford analysis versus its quaternionic counterparts.* Math. Methods Appl. Sciences, v. 33, issue 9, (2010), 1089-1101.

[10] F. BRACKX, H. DE SCHEPPER, V. SOUCEK. *On the Structure of Complex Clifford Algebra.* Adv. Appl. Clifford Algebras v. 21 (2011), 477–492.

[11] F. CATONI, D. BOCCALETTI, R. CANNATA, V. CATONI, E. NICHELATTI, P. ZAMPETTI. *The Mathematics of Minkowski Space-Time.* Birkhäuser-Basel, 2008.

[12] K.S. CHARAK, D. ROCHON, N. SHARMA. *Normal families of bicomplex holomorphic functions.* Fractals v. 17 No. 3 (2009), 257–268.

[13] K.S. CHARAK, D. ROCHON, N. SHARMA. *Normal families of bicomplex meromorphic functions.* Ann. Polon. Math. v. 103 No. 3 (2012), 303–317.

[14] K.S. CHARAK, D. ROCHON. *On factorization of bicomplex meromorphic functions.* Hypercomplex Analysis, Series Trends Math., Birkhäuser, Basel (2009), 55–68.

[15] J. COCKLE. *On Certain Functions Resembling Quaternions and on a New Imaginary in Algebra.* London-Dublin-Edinburgh Philosophical Magazine, series 3, v. 33 (1848), 43–59.

[16] J. COCKLE. *On a New Imaginary in Algebra.* London-Dublin-Edinburgh Philosophical Magazine, series 3, v. 34 (1849), 37–47.

[17] J. COCKLE. *On the Symbols of Algebra and on the Theory of Tessarines.* London-Dublin-Edinburgh Philosophical Magazine, series 3, v. 34 (1849), 40610.

[18] J. COCKLE. *On Impossible Equations, on Impossible Quantities and on Tessarines.* London-Dublin-Edinburgh Philosophical Magazine, series 3, v. 37 (1850), 2813.

[19] F. COLOMBO, I. SABADINI, D.C. STRUPPA, A. VAJIAC, M.B. VAJIAC. *Bicomplex hyperfunctions.* Ann. Mat. Pura Appl. (4) v. 190, No. 2 (2011), 247–261.

[20] F. COLOMBO, I. SABADINI, D.C. STRUPPA, A. VAJIAC, M.B. VAJIAC. *Singularities of functions of one and several bicomplex variables.* Ark. Mat. v. 49, No. 2 (2011), 277–294.

[21] V. CRUCEANU. *Almost product bicomplex structures on manifolds.* An. Stiint. Univ. Al. I. Cuza Iasi. Mat. (N.S.) v. 51, No. 1 (2005), 99–118.

[22] V. CRUCEANU. *A product bicomplex structure on the total space of a vector bundle.* An. Stiint. Univ. Al. I. Cuza Iasi. Mat. (N.S.) v. 53, No. 2 (2007), 315–324.

[23] R. DELANGHE, F. SOMMEN, V. SOUČEK. *Clifford Algebra and Spinor–Valued Functions.* Kluwer Academic Publishers, (1992).

[24] S. DIMIEV, R. LAZOV, S. SLAVOVA. *Remarks on bicomplex variables and other similar variables.* Topics in contemporary differential geometry, complex analysis and mathematical physics, World Sci. Publ. Hackensack, NJ. (2007), 50–56.

[25] M.A. DZAVADOV, N.T. ABBASOV. *Bicomplex and biquaternionic hyperbolic spaces.* Azerbaidzan. Gos. Univ. Ucen. Zap. Ser. Fiz.-Mat. i Him. Nauk, No. 1 (1964), 9–15.

[26] L.D. EGOROVA, L.I. KRJUCKOVA, L.B. LOBANOVA. *Bicomplex and bidual spaces.* Moskov. Oblast. Ped. Inst. Ucen. Zap. v. 262 (1969), 76–103.

[27] R. FUETER. *Analytische Funktionen einer Quaternionen Variablen.* Comm. Math. Helv. **4** (1932), 9-20.

[28] S.G. GAL. *Introduction to geometric function theory of hypercomplex variables.* Nova Science Publishers, Inc. Chapter 4 (2004), xvi + 319.

[29] V. GARANT-PELLETIER, D. ROCHON. *On a generalized Fatou-Julia theorem in multicomplex spaces.* Fractals v. 17, No. 3 (2009), 241–255.

[30] G. GENTILI, D.C. STRUPPA. *A new theory of regular functions of a quaternionic variable.* Adv. Math. v. 216, No. 1 (2007), 279-301.

[31] G. GENTILI, C. STOPPATO, D.C. STRUPPA. *Regular Functions of a Quaternionic Variable.* Springer Verlag (2013).

[32] R. GERVAIS-LAVOIE, L. MARCHILDON, D. ROCHON. *Infinite-dimensional bicomplex Hilbert spaces.* Ann. Funct. Anal. v. 1 No. 2 (2010), 75–91.

[33] R. GERVAIS-LAVOIE, L. MARCHILDON, D. ROCHON. *The bicomplex quantum harmonic oscillator.* Nuovo Cimento Soc. Ital. Fis. B, v. 125, No. 10 (2010), 1173–1192.

[34] R. GERVAIS-LAVOIE, L. MARCHILDON, D. ROCHON. *Finite-dimensional bicomplex Hilbert spaces.* Adv. Appl. Clifford Algebr. v. 21, No. 3 (2011), 561–581.

[35] K. GÜRLEBECK, F. KIPPIG. *Complex Clifford-Analysis and Elliptic Boundary Problems.* Adv. Appl. Clifford Algebras, v. 5, No. 1 (1995), 51–62.

[36] W.R. HAMILTON. *On quaternions, or on a new system of imaginaries in algebra.* Philosophical Magazine. Vol. 25, no. 3 (1844), 489–495.

[37] W.R. HAMILTON. *Lectures on Quaternions: Containing a Systematic Statement of a New Mathematical Method.* Dublin: Hodges and Smith (1853).

[38] H. KABADAYI, Y. YAYLI. *Homothetic Motions at $\mathbb{E}^4$ with Bicomplex Numbers.* Adv. Appl. Clifford Algebras, v. 21, No. 2 (2002), 541–546.

[39] KRANTZ. *Several Complex Variables.* Second Edition, AMS Chelsea Publishing (2001).

[40] R.S. KRAUSSHAR. *Eisenstein Series in Complexified Clifford Analysis.* Comp. Methods and Func. Theory, v. 2, No. 1 (2002), 29–65.

[41] J. KUMAR, R.K. SRIVASTAVA. *On a class of entire bicomplex sequences.* South East Asian J. Math. Math. Sci. v. 5, No. 3 (2007), 47–67.

[42] J. KUMAR, R.K. SRIVASTAVA. *A note on poles of the bicomplex Riemann zeta function.* South East Asian J. Math. Math. Sci. v. 9, No. 1 (2010), 65–75.

[43] J. KUMAR, R.K. SRIVASTAVA. *On entireness of bicomplex Dirichlet series.* Int. J. Math. Sci. Eng. Appl. v. 5, No. 2 (2011), 221–228.

[44] D.A. LAKEW, J. RYAN. *The Intrinsic π-Operator on Domain Manifolds in $\mathbb{C}^{n+1}$.* Complex Anal. Oper. Theory, (2010), 271–280.

[45] M.E. LUNA–ELIZARRARÁS, M. SHAPIRO, D.C. STRUPPA, A. VAJIAC. *Bicomplex numbers and their elementary functions.* Cubo A Mathematical Journal v. 14. No. 2 (2012), 61-80.

[46] M.E. LUNA–ELIZARRARÁS, M. SHAPIRO, D.C. STRUPPA, A. VAJIAC. *Complex Laplacian and derivatives of bicomplex functions.* Complex Analysis and Operator Theory, v. 7, No. 5, (2013), 1675-1711.

[47] J.E. MARSDEN, M.J. HOFFMAN. *Basic Complex Analysis.* Third edition, 1999, W. H. Freeman and Company.

[48] E. MARTINEAU, D. ROCHON. *On a bicomplex distance estimation for the Tetrabrot.* Internat. J. Bifur. Chaos Appl. Sci. Engrg. v. 15, No. 9 (2005), 3039–3050.

[49] A.I. MELENT'EV. *A normalized double sphere and a real model for a connection over the algebra of bicomplex numbers.* Proceedings of the Seminar of the Department of Geometry, Izdat. Kazan. Univ., Kazan, No. VI (1971), 57–69.

[50] G. MOISIL. *Sur les quaternions monogenes.* Bull. Sci. Math. Paris v. 55 (2) (1931), 169-194.

[51] G. MOISIL, N. TEODORESCU. *Fonctions holomorphes dans l'espace.* Mathematica (Cluj) **5** (1931), 142-159.

[52] I.A. MOREV. *A generalisation of the Cauchy-Riemann equations and the harmonicity of monogenic hypercomplex functions.* Izv. Vyss. Ucebn. Zaved. Matematika, v. 3, No. 4 (1958), 176–182

[53] I.A. MOREV. *A class of monogenic functions.* Mat. Sb. (N.S.), v. 50, No. 92 (1960), 233–240.

[54] S. OLARIU. *Complex numbers in n Dimensions.* North-Holland, Series Mathematics Studies First edition, v. 290 (2002), chapter 3.

[55] A.A. POGORUI, R.M. RODRIGUEZ-DAGNINO. *On the set of zeros of bicomplex polynomials.* Complex Variables and Elliptic Equations, v. 51, No. 7 (2006), 725–730.

[56] G.B. PRICE. *An Introduction to Multicomplex Spaces and Functions.* Monographs and Textbooks in Pure and Applied Mathematics, **140**, Marcel Dekker, Inc., New York, 1991.

[57] J.D. RILEY. *Contributions to the theory of functions of bicomplex variable.* Tohoku Math. J. v. 2 (1953), 132–165.

[58] D. ROCHON. *Sur une généralisation des nombres complexes: les tétranombres.* Master Thesis, Université de Montréal (1997), 66pp.

[59] D. ROCHON. *A generalized Mandelbrot set for bicomplex numbers.* Fractals, v. 8, No. 4 (2000), 355–368.

[60] D. ROCHON. *Dynamique bicomplexe et théorème de Bloch pour fonctions hyperholomorphes.* PhD Thesis, Université de Montréal (Canada), (2001), 77pp.

[61] D. ROCHON. *A Bloch constant for hyperholomorphic functions.* Complex Variables Theory Appl. v. 44, No. 2 (2001), 85–101.

[62] D. ROCHON. *On a generalized Fatou-Julia theorem.* Fractals, v. 11, No. 3 (2003), 213–219.

[63] D. ROCHON. *A bicomplex Riemann zeta function.* Tokyo J. Math. v. 27, No. 2 (2004), 357-369.

[64] D. ROCHON. *On a relation of bicomplex pseudoanalytic function theory to the complexified stationary Schrödinger equation.* Complex Var. Elliptic Equ. v. 53, No. 6 (2008), 501–521

[65] D. ROCHON, M. SHAPIRO. *On algebraic properties of bicomplex and hyperbolic numbers.* An. Univ. Oradea Fasc. Mat. v. 11 (2004), 71–110.

[66] D. ROCHON, S. TREMBLAY. *Bicomplex quantum mechanics. I. The generalized Schrödinger equation.* Adv. Appl. Clifford Algebr. v. 14, No. 2 (2004), 231–248.

[67] D. ROCHON, S. TREMBLAY. *Bicomplex quantum mechanics. II. The Hilbert space.* Adv. Appl. Clifford Algebr. v. 16, No. 2 (2006), 135–157.

[68] J. RYAN. *Topics in Hypercomplex Analysis.* Doctoral Thesis, University of York, Britain (1982).

[69] J. RYAN. *Complexified Clifford Analysis.* Complex Variables, v. 1 (1982), 119–149.

[70] J. RYAN. *Singularities and Laurent expansions in complex Clifford Analysis.* Appl. Anal. v. 16 (1983), 33–49.

[71] J. RYAN. *Special functions and relations within complex Clifford Analysis, I.* Complex Variables Theory Applications, v. 2 (1983), 177–198.

[72] J. RYAN. *Hilbert modules with reproducing kernels within complex Clifford analysis.* Mathematical Structures - Computational Mathematics- Mathematical Modelling, v. 2 (1984) 273–278.

[73] J. RYAN. *Conformal Clifford Manifolds arising in Clifford Analysis.* Proceedings of the Royal Irish Academy, v. 85A, No. 1 (1985), 1–23.

[74] J. RYAN. *Duality in Complex Clifford Analysis.* Journal of Functional Analysis, v. 61, No. 2 (1985), 117–135.

[75] J. RYAN. *Cells of harmonicity and generalized Cauchy integral formulae.* Proc. Lond. Math. Soc. III Ser. v. 60, No. 2 (1990), 295–318.

[76] J. RYAN. *Complex Clifford analysis and domains of holomorphy.* J. Aust. Math. Soc. Ser. A, v. 48, No. 3 (1990), 413–433.

[77] J. RYAN. *Plemelj formula and Transformations Associated to Plane Wave Decompositions in Complex Clifford Analysis.* Proc. London Math. Soc. V. 64, No. 3 (1992), 70–94.

[78] J. RYAN. *Intertwining operators for iterated Dirac operators over Minkowski-type spaces.* J. Math. Anal. Appl. v. 177 No. 1 (1993), 1–23.

[79] J. RYAN. *Intrinsic Dirac Operators in $\mathbb{C}^n$.* Advances in Mathematics, v. 118 (1996), 99–133.

[80] J. RYAN. *Basic Clifford Analysis.* Cubo Matemática Educacional, v. 2 (2000), 226–256.

[81] J. RYAN. $\mathbb{C}^2$ *Extensions of Analytic Functions Defined in the Complex Plane.* Adv. Appl. Clifford Algebras, v. 11 S1 (2001), 137–145.

[82] C. SEGRE. *Le rappresentazioni reali delle forme complesse e gli enti iperalgebrici.* Math. Ann. v. 40 (1892), 413–467.

[83] G. SCORZA DRAGONI. *Sulle funzioni olomorfe di una variabile bicomplessa.* Reale Accad. d'Italia, Mem. Classe Sci. Nat. Fis. Mat. v. 5 (1934), 597–665.

[84] K.D. SHOINBEKOV, B.I. MAUKEEV. *Sequences of bicomplex numbers and the notion of a function of a bicomplex variable.* Investigations in the theory of functions and differential equations, v. 202, Kazakh. Gos. Univ., Alma Ata (1985), 173–180.

[85] K.D. SHOINBEKOV, B.I. MAUKEEV. *Algebra of bicomplex numbers.* Investigations in the theory of functions and differential equations, v. 203, Kazakh. Gos. Univ., Alma Ata (1985), 180–190.

[86] G.L. SHPILKER. *Some differential properties of a commutative hypercomplex potential. (Russian).* Dokl. Akad. Nauk SSSR, v. 293, No. 3 (1987), 578–583.

[87] G. SOBCZYK. *The Hyperbolic Number Plane.* The College Mathematics Journal, v. 26, No. 4 (1995), 268–280.

[88] N. SPAMPINATO. *Estensione nel campo bicomplesso di due teoremi, del Levi-Civita e del Severi, per le funzioni olomorfe di due variabili bicomplesse I, II.* Reale Accad. Naz. Lincei, v. 22 No. 38–43 (1935), 96–102.

[89] N. SPAMPINATO. *Sulla rappresentazione di funzioni di variabile bicomplessa totalmente derivabili.* Ann. Mat. Pura Appli. v. 14 (1936), 305–325.

[90] R.K. SRIVASTAVA. *Certain points in the theory of bicomplex numbers.* Math. Student, v. 70, No. 1–4 (2001), 153–160.

[91] R.K. SRIVASTAVA. *Bicomplex numbers: analysis and applications.* Math. Student, v. 72, No. 1–4 (2003), 69–87.

[92] R.K. SRIVASTAVA. *Certain topological aspects of bicomplex space.* Bull. Pure Appl. Math.v. 2, No. 2 (2008), 222–234.

[93] R.K SRIVASTAVA, S. SINGH. *Certain bicomplex dictionary order topologies.* Int. J. Math. Sci. Eng. Appl. v. 4, No. 3 (2010), 245–258.

[94] N.T. STELMASUK. *On certain linear partial differential equations in dual and bicomplex algebras.* An. Sti. Univ. "Al. I. Cuza" Iasi Sect. I (N.S.), v. 9 (1963), 63–72.

[95] N.T. STELMASUK. *Some linear partial differential equations in dual and bicomplex algebras.* Izv. Vyss. Ucebn. Zaved. Matematika, v. 40, No. 2 (1964), 136–142.

[96] D.C. STRUPPA, A. VAJIAC, M.B. VAJIAC. *Remarks on Holomorphicity in Three Settings: Complex, Quaternionic, and Bicomplex.* Hypercomplex Analysis and Applications, Trends in Mathematics, Springer, I. Sabadini, F. Sommen editors (2011), 261–274.

[97] A. SUDBERY *Quaternionic Analysis.* Mathematical Proceedings of the Cambridge Philosophical Society v. 85 (1979), 199?225.

[98] V.A. TRET'JAKOV. *Some properties of mappings in the space accompaning the algebra of bicomplex numbers in modulus (Russian).* Application of functional analysis in the approximation theory (Russian), Kalinin. Gos. Univ., Kalinin (1979), 129–136.

[99] V.A. TRET'JAKOV. *On the properties of some elementary functions that are defined on the algebra of bicomplex numbers (Russian).* Mathematical analysis and the theory of functions (Russian), Moskov. Oblast. Ped. Inst., Moscow (1980), 99–106.

[100] I.M. YAGLOM. *Complex Numbers in Geometry.* Academic Press, New York-London (1968), 243 pp.

[101] K. YONEDA. *An integration theory in the general bicomplex function theory.* Yokohama Math. J. v. 1 (1953), 225–262.

[102] A. ZIREH. *A generalized Mandelbrot set of polynomials of type E_d for bicomplex numbers.* Georgian Math. J. v. 15, No. 1 (2008), 189–194.

[103] V.A. ZORICH. *Mathematical Analysis, Volumes I and II.* Springer–Verlag Berlin Heidelberg (2004).

Index